Mathematical Quantization

Studies in Advanced Mathematics

Titles Included in the Series

Steven R. Bell, The Cauchy Transform, Potential Theory, and Conformal Mapping
John J. Benedetto, Harmonic Analysis and Applications
John J. Benedetto and Michael W. Frazier, Wavelets: Mathematics and Applications
Albert Boggess, CR Manifolds and the Tangential Cauchy–Riemann Complex
Goong Chen and Jianxin Zhou, Vibration and Damping in Distributed Systems
Vol. 1: Analysis, Estimation, Attenuation, and Design
Vol. 2: WKB and Wave Methods, Visualization, and Experimentation
Carl C. Cowen and Barbara D. MacCluer, Composition Operators on Spaces of Analytic Functions
John P. D'Angelo, Several Complex Variables and the Geometry of Real Hypersurfaces
Lawrence C. Evans and Ronald F. Gariepy, Measure Theory and Fine Properties of Functions
Gerald B. Folland, A Course in Abstract Harmonic Analysis
José García-Cuerva, Eugenio Hernández, Fernando Soria, and José-Luis Torrea, Fourier Analysis and Partial Differential Equations
Peter B. Gilkey, Invariance Theory, the Heat Equation, and the Atiyah-Singer Index Theorem, 2nd Edition
Alfred Gray, Modern Differential Geometry of Curves and Surfaces with Mathematica, 2nd Edition
Eugenio Hernández and Guido Weiss, A First Course on Wavelets
Steven G. Krantz, Partial Differential Equations and Complex Analysis
Steven G. Krantz, Real Analysis and Foundations
Kenneth L. Kuttler, Modern Analysis
Michael Pedersen, Functional Analysis in Applied Mathematics and Engineering
Clark Robinson, Dynamical Systems: Stability, Symbolic Dynamics, and Chaos, 2nd Edition
John Ryan, Clifford Algebras in Analysis and Related Topics
Xavier Saint Raymond, Elementary Introduction to the Theory of Pseudodifferential Operators
John Scherk, Algebra: A Computational Introduction
Robert Strichartz, A Guide to Distribution Theory and Fourier Transforms
André Unterberger and Harald Upmeier, Pseudodifferential Analysis on Symmetric Cones
James S. Walker, Fast Fourier Transforms, 2nd Edition
James S. Walker, Primer on Wavelets and Their Scientific Applications
Gilbert G. Walter and Xiaoping Shen, Wavelets and Other Orthogonal Systems, Second Edition
Kehe Zhu, An Introduction to Operator Algebras
Dean G. Duffy, Green's Functions with Applications
Nik Weaver, Mathematical Quantization

Mathematical Quantization

NIK WEAVER

CHAPMAN & HALL/CRC

Boca Raton London New York Washington, D.C.

Library of Congress Cataloging-in-Publication Data

Weaver, Nik.
Mathematical quantization / Nik Weaver.
p. cm.— (Studies in advanced mathematics)
Includes bibliographical references and index.
ISBN 1-58488-001-5 (alk. paper)
1. Geometric quantization. I. Title. II. Series.
QC174.17 .G46 W43 2001
530.14′3—dc21 2001028679

Direct all inquiries to CRC Press LLC, 2000 N.W. Corporate Blvd., Boca Raton, Florida 33431.

International Standard Book Number 1-58488-001-5
Library of Congress Card Number 2001028679
Printed in the United States of America 1 2 3 4 5 6 7 8 9 0
Printed on acid-free paper

Preface

It has been roughly one hundred years since physicists began to realize that classical mechanics is fundamentally wrong in the atomic realm and hence cannot be a correct description of nature. The subsequent transition to quantum mechanics was rather rapid; despite some false starts, and the truly alien quality of the new theory, by the late 1920s its basic framework was complete.

Since that time there has been a parallel development in mathematics which has played out much more gradually.[1] It began around 1930. Not long after John von Neumann published the definitive mathematical treatment of quantum mechanics [52], he and Garrett Birkhoff pointed out that the logical structure of quantum systems was different from that of classical systems [7]. Their description of the former is now known as quantum logic. This was the first example of a quantum version of a classical mathematical subject.

Over the next several years von Neumann, together with Francis Murray, initiated the study of what are now called von Neumann algebras [49, 50, 53, 51], and quantum measure theory was born. But although Murray and von Neumann knew that their algebras were a noncommutative generalization of classical L^∞ spaces, it was only several decades later that this point of view was openly embraced in the popular expression "noncommutative measure theory."

As other types of quantum structures arose — quantum topological spaces, quantum groups, quantum Banach spaces, etc. — it became increasingly clear that they were all instances of a single basic phenomenon. Also, examples such as quantum computation and quantum logic showed that the central property they all shared was not noncommutativity, which only appears explicitly in algebraic structures, but

[1] See [22] for a similar historical take.

rather their common relation to Hilbert space. We have now reached a point where it is possible to give a simple, unified approach to the general concept of quantization in mathematics. That is the aim of this book.

The fundamental idea of mathematical quantization is that *sets are replaced by Hilbert spaces.* Thus, we regard lattice operations (join, meet, orthocomplement) on subspaces of a Hilbert space as corresponding to set-theoretic operations (union, intersection, complement) on subsets of a set. This already allows one to determine quantum analogs of some simple structures. But the real breakthrough is the fact that the quantum version of a complex-valued function on a set is an operator on a Hilbert space. The reason for this is not obvious, but it has dramatic consequences: since topologies and measure classes on a set can be defined in terms of scalar-valued functions, we are then able to transfer these constructions into the quantum realm.

With more work the analogy can be pushed even further. At each step one must formulate the given classical notion in just the right way to obtain a viable quantum version. This sometimes requires significant creativity. However, as it is done in case after case, general quantization principles emerge.

In this book I discuss the following correspondences.

classical theory	**quantum analog**
computation	quantum computation
propositional logic	quantum logic
entire functions	Bargmann-Segal multipliers
topological spaces	C*-algebras
measure spaces	von Neumann algebras
Banach spaces	operator spaces
Hilbert bundles	Hilbert modules
metric spaces	Lipschitz algebras
Riemannian geometry	noncommutative geometry
topological groups	quantum groups

I use the quantum plane and tori, which are arguably the most fundamental noncommutative examples, to illustrate several of these topics.

My identification of the set/Hilbert space analogy as a basic principle may surprise some C*-algebraists who view noncommutativity as the defining property of their subject. I hope they will appreciate the economy of my approach and the strength of its unifying power. Having said that, I should point out that no part of this approach is original. In fact, the basic analogy between sets and Hilbert spaces was brought out quite clearly in Birkhoff and von Neumann's original paper [7]. However, that paper gave rise to a line of research which developed in the direction of general lattice-theoretic issues which have little relevance to the main quantization program, and perhaps for this reason it has not received the attention it deserves. The next major observation, that real-valued functions correspond to self-adjoint operators, was made by George Mackey [45], but in this case the axiomatic formalism in which he cast his idea seems to have obscured its significance. Once the function/operator correspondence is granted, I think my interpretation of C*-algebras and their relatives is fairly standard.

My goal has been to write a sort of broad introductory survey, including some deep results but keeping the whole account as nontechnical as possible. I did not discuss several prominent related topics (nonselfadjoint operator algebras, subfactors, K-theory) because I could not fit them directly into my thematic framework, and I omitted quantum probability because there was not enough space. Likewise, with the exception of Chapter 7 on quantum field theory, I also avoided presenting much in the way of applications. But I want to emphasize that there are many other applications to physics; indeed, many of the ideas in this book were originally developed in connection with mathematical physics. Some references on this aspect are [2], [6], [10], [17], [24], [33], and [69]. The most significant applications within mathematics but outside analysis proper are probably in knot theory [38] and index theory for foliations [12].

The essential prerequisite is a good first-year graduate course in analysis along the lines of [28]. Readers need to be familiar with measure theory and functional analysis on at least the level of the Hahn-Banach, Stone-Weierstrass, and Riesz representation theorems. Graduate level topics outside functional analysis are touched on here and there, but I generally try to give enough background so that an unfamiliar reader can follow the development at some level. There are also occasional instances where a complete proof of some fact would have required a prohibitively long excursion. In these cases the reader will find references to full treatments in the notes given at the end of the chapter.

This work was partially supported by NSF grant DMS-0070634.

Contents

Chapter 1

Quantum Mechanics

1.1 Classical physics

No background in physics is really needed to understand the concept of mathematical quantization, which may be taken as nothing more than a formal analogy between sets and Hilbert spaces. However, knowledge of some physics adds a layer of meaning to the mathematics which can be quite illuminating. It can also help when one is unsure of the best way to translate some set theoretic construction into the Hilbert space setting.

The basic ideas of quantum mechanics are really quite simple, and in the setting of finite state systems the mathematical apparatus is elementary. It involves finite dimensional Hilbert spaces, which are concretely just the vector spaces $\mathbf{C}^n$ equipped with the euclidean inner product

$$\langle v, w\rangle = \sum_{i=1}^{n} a_i \bar{b}_i,$$

where $v = (a_1, \ldots, a_n)$ and $w = (b_1, \ldots, b_n)$. The key properties of the inner product are (1) $\langle v, v\rangle \geq 0$ for all $v \in \mathbf{C}^n$, and $\langle v, v\rangle = 0$ only if $v = 0$; (2) $\langle av_1 + bv_2, w\rangle = a\langle v_1, w\rangle + b\langle v_2, w\rangle$ for $a, b \in \mathbf{C}$ and $v, w \in \mathbf{C}^n$; and (3) $\langle v, w\rangle = \overline{\langle w, v\rangle}$ for all $v, w \in \mathbf{C}^n$.

Infinite dimensional Hilbert spaces are a bit deeper, but we can ignore them for now because the most fundamental principles of quantum mechanics are illustrated perfectly well in finite dimensions. Our goal in this chapter is just to give an informal introduction to the physical concepts that underlie the math which follows. We do this here at a very basic mathematical level, and in the following two chapters we will develop the essential mathematical tools used in the infinite dimensional case. Thus, all of the math in this chapter will appear again in a more general form later. When this is done the reader should easily be able to infer its physical interpretation by analogy with the finite dimensional case.

The following concepts appear in classical physics:

- phase space
- states
- events
- observables
- transformations

These are best explained in the context of a specific example. Consider a one-dimensional particle, that is, a particle constrained to move in one dimension. In this example the state of the particle is characterized by the pair of real numbers (q, p) where q is the position of the particle and p is its momentum. The particle's position alone is not enough, because a complete description of its state has to include how fast and in which direction it is moving.

The phase space of a physical system is the set of all possible states of the system. In our example the phase space can be identified with $\mathbf{R}^2$, as q and p can take on any real values independently of each other.

An event is a subset of phase space, and this corresponds to a single bit of information about the system. For instance, in the above example the right half-plane $\{(q,p) : q > 0\}$ describes the event that the particle lies to the right of the origin. Events are always true-or-false propositions. Any state either belongs to ("satisfies") a given event or it does not.

An observable is a real-valued function on phase space. This kind of structure has the effect of isolating some particular quality of a given state. For example, in the case of a one-dimensional particle, the coordinate functions $(q,p) \mapsto q$ and $(q,p) \mapsto p$ tell us, for a given state of the system, the particle's position and momentum in that state. If the particle has mass m and is free (meaning that there are no forces present, and hence no potential energy) then the function $(q,p) \mapsto p^2/2m$ describes its energy.

Finally, transformations are permutations of phase space. In general, a transformation describes the result of some action which can be taken on the system. This action could be simply letting the system evolve undisturbed for a certain length of time, or it could involve interaction with some external influence. In any case, it must be reversible if the corresponding map on phase space is to be a bijection.

1.2 States and events

Now we describe the corresponding concepts in quantum mechanics. The characteristic property of quantum systems is the possibility of superposition: distinct states can be superimposed. This is because the phase space of a quantum system is modelled by a vector space, in fact a Hilbert space. States are described by unit vectors.

In this chapter we will deal with finite state systems, systems whose corresponding Hilbert spaces are finite dimensional and hence of the form $\mathbf{C}^n$. The term "finite state" refers to the fact that every state is a linear combination of a finite set of states (i.e., a basis). Note that we take the scalar field to be complex; except in a handful of cases, we will continue to do so throughout the book, usually without comment.

For example, the polarization of a photon is represented by a two-dimensional complex Hilbert space $\mathbf{C}^2$. The vector $(1, 0)$ represents a horizontally polarized photon and the vector $(0, 1)$ represents a vertically polarized photon.

Events in finite state systems are modelled by subspaces of $\mathbf{C}^n$. In the photon example, the subspaces

$$\begin{aligned} E_1 &= \{(a, a) : a \in \mathbf{C}\} \\ E_2 &= \{(a, 0) : a \in \mathbf{C}\} \\ E_3 &= \{(0, a) : a \in \mathbf{C}\} \end{aligned}$$

indicate the events that the photon is diagonally, horizontally, or vertically polarized, respectively.

Before we explain how events are physically interpreted in relation to states, we need to introduce the operations that can be performed on them and establish the basic properties of these operations.

DEFINITION 1.2.1 *Let E, E_1, and E_2 be subspaces of $\mathbf{C}^n$. Then the orthocomplement of E is the subspace*

$$E^\perp = \{v \in \mathbf{C}^n : \langle v, w\rangle = 0 \text{ for all } w \in E\};$$

the join of E_1 and E_2 is the subspace

$$E_1 \vee E_2 = E_1 + E_2 = \{v + w : v \in E_1 \text{ and } w \in E_2\};$$

and the meet of E_1 and E_2 is the subspace

$$E_1 \wedge E_2 = E_1 \cap E_2.$$

We say that E_1 and E_2 are orthogonal if $\langle v, w\rangle = 0$ for all $v \in E_1$ and $w \in E_2$.

LEMMA 1.2.2
Let E be a subspace of $\mathbf{C}^n$ and let $v \in \mathbf{C}^n$. Then $v = v_1 + v_2$ for some $v_1 \in E$ and $v_2 \in E^\perp$.

PROOF Consider the function $f : E \to \mathbf{R}$ defined by $f(w) = ||v - w||$. This function is continuous (by the triangle inequality), and it satisfies $f(0) = ||v||$ and $f(w) > ||v||$ whenever $||w|| > 2||v||$. Therefore it attains a global minimum at some (perhaps not unique) vector w_0 in the compact set $\{w \in E : ||w|| \leq 2||v||\}$.

We claim that $v_0 = v - w_0$ belongs to $E^\perp$. To see this, choose a nonzero vector $w \in E$. Then $||v_0 - aw|| \geq ||v_0||$ for all $a \in \mathbf{C}$, by minimality of f at w_0. In particular, taking $a = \langle v_0, w\rangle/||w||^2$ and performing a short computation, we get

$$||v_0||^2 - \frac{|\langle v_0, w\rangle|^2}{||w||^2} = \left\| v_0 - \frac{\langle v_0, w\rangle}{||w||^2} w \right\|^2 \geq ||v_0||^2.$$

Thus $\langle v_0, w\rangle = 0$, as desired. We conclude that $v_0 \in E^\perp$, and so $v_1 = w_0$ and $v_2 = v_0$ verify the conclusion of the lemma. ∎

PROPOSITION 1.2.3

Let E be a subspace of $\mathbf{C}^n$. Then $E \vee E^\perp = \mathbf{C}^n$ and $E \wedge E^\perp = \{0\}$. Every $v \in \mathbf{C}^n$ can be expressed uniquely in the form $v = v_1 + v_2$ with $v_1 \in E$ and $v_2 \in E^\perp$. We have $||v||^2 = ||v_1||^2 + ||v_2||^2$.

PROOF The first assertion follows immediately from the lemma. Also, if $v \in E \wedge E^\perp$ then $\langle v, v\rangle = 0$, so $v = 0$, and thus $E \wedge E^\perp = \{0\}$.

To prove uniqueness, let $v \in \mathbf{C}^n$ and suppose $v = v_1 + v_2 = v_1' + v_2'$ with $v_1, v_1' \in E$ and $v_2, v_2' \in E^\perp$. Then $v_1 - v_1' = v_2' - v_2$. But $v_1 - v_1' \in E$ and $v_2' - v_2 \in E^\perp$, and we know that $E \wedge E^\perp = \{0\}$, so both sides must be zero. Thus $v_1 = v_1'$ and $v_2 = v_2'$. This proves that the decomposition $v = v_1 + v_2$ is unique. Finally, since $\langle v_1, v_2\rangle = 0$ it follows that

$$\langle v, v\rangle = \langle v_1 + v_2, v_1 + v_2\rangle = \langle v_1, v_1\rangle + \langle v_2, v_2\rangle,$$

so that $||v||^2 = ||v_1||^2 + ||v_2||^2$. ∎

We can now explain the naive probabilistic interpretation of states and events. Let $v \in \mathbf{C}^n$ be a unit vector and $E \subset \mathbf{C}^n$ a subspace. If $v \in E$ then we regard the state as *definitely* satisfying this event, but in general it satisfies the event only with probability $||v_1||^2$, where $v = v_1 + v_2$ with $v_1 \in E$, $v_2 \in E^\perp$. Since $||v_1||^2 + ||v_2||^2 = ||v||^2 = 1$, the probability of satisfying any event and the probability of satisfying its orthocomplement add up to one. For instance, a photon in the state (a, b) (with $|a|^2 + |b|^2 = 1$) will pass through a horizontally polarized filter with probability $|a|^2$, and through a vertically polarized filter with probability $|b|^2$.

In this naive interpretation, after one measures whether a state satisfies an event, the state actually *becomes* either $v_1/||v_1||$ or $v_2/||v_2||$, with probability $||v_1||^2$ or $||v_2||^2$. One says that the state has "collapsed" onto

E or $E^{\perp}$ as a result of the measurement. The problem with this idea is that it is difficult to identify which kinds of physical interactions should be classified as measurements and to justify why they should have such an effect. We will discuss this issue further in Section 1.5.

The sense of this interpretation is seen in the classroom demonstration that a horizontal filter followed immediately by a vertical filter passes no light, while a sequence of horizontal, diagonal, then vertical filters does allow some light through. If a photon in the state (a, b) passes through a horizontally polarized filter (which it does with probability $|a|^2$), then afterwards it is in the state $(1, 0)$ and cannot pass through a vertically polarized filter. But it still has a 50% chance of passing through a diagonally polarized filter, after which it would be in the state $(1/\sqrt{2}, 1/\sqrt{2})$ and have a 50% chance of passing through a vertically polarized filter.

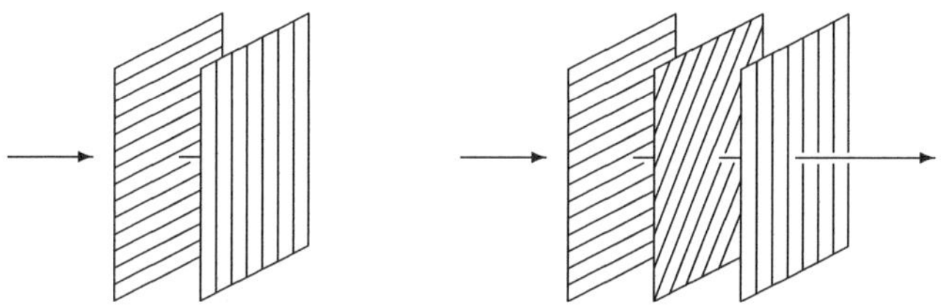

Figure 1.1 Passage of light through polarized filters

Orthocomplementation, join, and meet of subspaces correspond to complemention, union, and intersection of classical events (i.e., subsets of a set), and they can be interpreted physically in much the same way: the states in $E^{\perp}$ are precisely those states which have zero probability of satisfying the event E, and the states in $E_1 \wedge E_2$ are precisely those which satisfy both E_1 and E_2 with full probability. A physical interpretation of $E_1 \vee E_2$ is not immediate but can be obtained by reducing to the previous two operations via the following proposition.

PROPOSITION 1.2.4
Let E, E_1, and E_2 be subspaces of $\mathbf{C}^n$. Then

(a) $E^{\perp\perp} = E$;
(b) $(E_1 \vee E_2)^{\perp} = E_1^{\perp} \wedge E_2^{\perp}$; *and*
(c) $(E_1 \wedge E_2)^{\perp} = E_1^{\perp} \vee E_2^{\perp}$.

PROOF

(a) Let $v \in E$. Then $\langle v, w \rangle = 0$ for all $w \in E^\perp$, so $v \in E^{\perp\perp}$. This shows that $E \subset E^{\perp\perp}$. Conversely, any $v \in E^{\perp\perp}$ can be written $v = v_1 + v_2$ with $v_1 \in E$ and $v_2 \in E^\perp$; since $\langle v_2, v_2 \rangle = \langle v, v_2 \rangle = 0$ we have $v_2 = 0$, and hence $v = v_1 \in E$. So $E^{\perp\perp} \subset E$.

(b) Suppose $v \in (E_1 \vee E_2)^\perp$. Then $v \in E_1^\perp$ and $v \in E_2^\perp$, so that $v \in E_1^\perp \wedge E_2^\perp$. This shows one containment.

Conversely, suppose $v \in E_1^\perp \wedge E_2^\perp$. Then for any $w_1 \in E_1$ and $w_2 \in E_2$ we have

$$\langle v, w_1 + w_2 \rangle = \langle v, w_1 \rangle + \langle v, w_2 \rangle = 0,$$

so $v \in (E_1 \vee E_2)^\perp$. This verifies the reverse containment.

(c) By (a) and (b),

$$E_1^\perp \vee E_2^\perp = (E_1^\perp \vee E_2^\perp)^{\perp\perp} = (E_1^{\perp\perp} \wedge E_2^{\perp\perp})^\perp = (E_1 \wedge E_2)^\perp,$$

as desired. ∎

It follows from this proposition that $E_1 \vee E_2 = (E_1^\perp \wedge E_2^\perp)^\perp$ for any subspaces E_1 and E_2.

1.3 Observables

The quantum mechanical version of an observable is a self-adjoint matrix.

DEFINITION 1.3.1 *The* Hermitian transpose *or* adjoint *of an $n \times n$ complex matrix $A = [a_{ij}]$ is the matrix $A^* = [\bar{a}_{ji}]$. A is* self-adjoint *if $A = A^*$.*

A straightforward computation shows that $\langle Av, w \rangle = \langle v, A^* w \rangle$ for all $v, w \in \mathbf{C}^n$. Conversely, applying this equality to the canonical basis vectors of $\mathbf{C}^n$ recovers the fact that $A^* = [\bar{a}_{ji}]$. In particular, it follows that A is self-adjoint if and only if $\langle Av, w \rangle = \langle v, Aw \rangle$ for all v and w.

Now one might expect that since classical observables are real-valued functions on classical phase space, a quantum mechanical observable should be some kind of real-valued function on $\mathbf{C}^n$, perhaps a linear functional. Understanding why it is instead self-adjoint matrices which play this role requires the spectral theorem (Theorem 1.3.3). First we need the following lemma.

Let $I = I_n$ be the $n \times n$ matrix whose diagonal entries are all 1 and whose non-diagonal entries are all 0. We take it as known that the determinant of a matrix is a polynomial in the entries of the matrix and that $\det(AB) = \det(A)\det(B)$.

LEMMA 1.3.2
Every $n \times n$ complex matrix has at least one nonzero eigenvector.

PROOF Let A be an $n \times n$ matrix. The expression $\det(A - \lambda I)$ is a polynomial in λ, so it has at least one complex root. For such a root λ the matrix $A - \lambda I$ has null determinant. Fix such a λ and set $B = A - \lambda I$. We claim that $\ker(B)$ is nonzero. Suppose not. Then B is invertible and we have

$$\det(B)\det(B^{-1}) = \det(BB^{-1}) = \det(I) = 1.$$

Thus $\det(B) \neq 0$, a contradiction. Therefore there is a nonzero vector v such that $(A - \lambda I)v = Bv = 0$, i.e., $Av = \lambda v$. ∎

THEOREM 1.3.3
Let A be a self-adjoint $n \times n$ complex matrix. Then

(a) every eigenvalue of A is real;
(b) distinct eigenspaces are orthogonal; and
(c) the eigenspaces of A span $\mathbf{C}^n$.

Conversely, if $\lambda_1, \ldots, \lambda_k$ are distinct real numbers and $E_1, \ldots, E_k$ are orthogonal subspaces which collectively span $\mathbf{C}^n$, then there is a unique self-adjoint $n \times n$ complex matrix with these eigenvalues and eigenspaces.

PROOF
(a) Let $v \in \mathbf{C}^n$ be an eigenvector with eigenvalue λ. Then

$$\lambda\langle v, v\rangle = \langle Av, v\rangle = \langle v, Av\rangle = \bar{\lambda}\langle v, v\rangle,$$

so $\lambda = \bar{\lambda}$ and hence λ must be real.

(b) Let $v, w \in \mathbf{C}^n$ be eigenvectors belonging to distinct eigenvalues λ and μ. Then

$$\lambda\langle v, w\rangle = \langle Av, w\rangle = \langle v, Aw\rangle = \bar{\mu}\langle v, w\rangle.$$

Since λ and μ are real and distinct, this implies $\langle v, w\rangle = 0$.

(c) Suppose the eigenspaces do not span $\mathbf{C}^n$, and let $E \subset \mathbf{C}^n$ be the orthocomplement of their span. Observe that $A(E) \subset E$ because if $v \in E$ and w is an eigenvector with eigenvalue λ, we have $\langle v, w\rangle = 0$, and hence

$$\langle Av, w\rangle = \langle v, Aw\rangle = \lambda\langle v, w\rangle = 0;$$

so $\langle Av, w\rangle = 0$ also, and this shows that $Av \in E$. Thus $A|_E$ is a linear operator on the complex vector space E, and so $A|_E$ must have a nonzero eigenvector by the lemma, which means that A has a nonzero eigenvector in E. This is a contradiction and we conclude that the eigenspaces of A must span $\mathbf{C}^n$.

To prove the converse assertion, let $\lambda_1, \ldots, \lambda_k$ be distinct real numbers and let $E_1, \ldots, E_k$ be orthogonal subspaces which together span $\mathbf{C}^n$. Define a linear operator A by setting $Av = \lambda_i v$ for $v \in E_i$ and extending linearly. Then for any $v, w \in E_i$ we have $\langle Av, w\rangle = \lambda_i\langle v, w\rangle = \langle v, Aw\rangle$, and for any $v \in E_i$ and $w \in E_j$, $i \neq j$, we have $\langle Av, w\rangle = 0 = \langle v, Aw\rangle$. Since the E_i span $\mathbf{C}^n$, this implies that $\langle Av, w\rangle = \langle v, Aw\rangle$ for all $v, w \in \mathbf{C}^n$, so A is self-adjoint. ∎

We claim that the characterization of self-adjoint matrices given in the preceding theorem is the exact quantum mechanical analog of a real-valued function on a finite set. Why is this?

Let X be a finite set and $f : X \to \mathbf{R}$ a real-valued function. Define an equivalence relation on X by setting $x \sim y$ if $f(x) = f(y)$. Then the blocks of this equivalence relation partition X into disjoint subsets, and f labels each subset with a distinct real number λ_i. This is the structure that a real-valued function imparts to X.

The quantum mechanical analog of a partition into disjoint subsets is a decomposition into orthogonal subspaces. Therefore, the analog of a real-valued function should decompose $\mathbf{C}^n$ into orthogonal subspaces and label each subspace with a real number. By Theorem 1.3.3, this is precisely what self-adjoint matrices do.

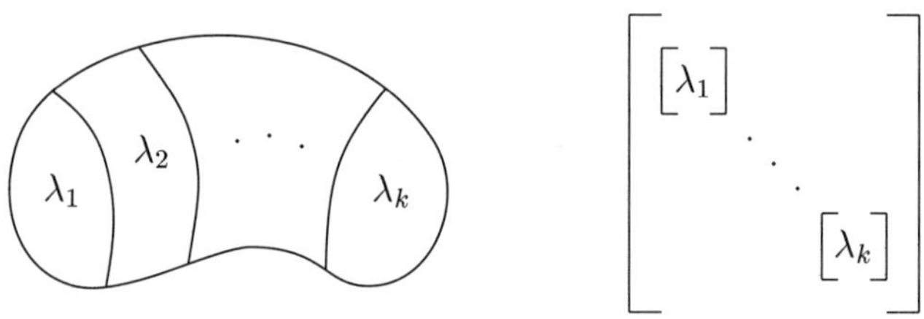

Figure 1.2 Functions versus matrices

Of course, there are other ways of looking at real-valued functions, and it may be possible to find another structure on $\mathbf{C}^n$ which is analogous to them in some different way. But the above analogy is the one which is physically correct.

In detail, here is the physical interpretation of a self-adjoint matrix as an observable. Let A be a self-adjoint matrix and let $E_1, \ldots, E_k$ be its eigenspaces with corresponding eigenvalues $\lambda_1, \ldots, \lambda_k$. Then the observable corresponding to A will take the value λ_i on any state $v \in E_i$, with certainty. In general, for any $v \in \mathbf{C}^n$ we can write $v = v_1 + \cdots + v_k$ with $v_i \in E_i$, and the observable will then take the value λ_i on v_i with probability $||v_i||^2$.

We define the expectation value of A for v to be $\sum_{i=1}^k \lambda_i ||v_i||^2$. That is, it is the average of all possible observations weighted by their likelihood. The following gives an alternative formula.

PROPOSITION 1.3.4
Let v be a state and A an observable. Then the expectation value of A for v equals $\langle Av, v\rangle$.

PROOF Let $E_1, \ldots, E_k$ be the eigenspaces of A, with corresponding eigenvectors λ_i, and write $v = v_1 + \cdots + v_k$ with $v_i \in E_i$. Then

$$\begin{aligned}\langle Av, v\rangle &= \big\langle A(v_1 + \cdots + v_k), (v_1 + \cdots + v_k)\big\rangle \\ &= \langle \lambda_1 v_1 + \cdots + \lambda_k v_k, v_1 + \cdots + v_k\rangle \\ &= \lambda_1 ||v_1||^2 + \cdots + \lambda_k ||v_k||^2,\end{aligned}$$

as desired. ∎

1.4 Dynamics

The dynamics of a quantum system are described using unitary matrices.

DEFINITION 1.4.1 *An $n\times n$ complex matrix U is unitary if $U^*U = UU^* = I$.*

In other words, U is invertible and $U^* = U^{-1}$.

Unitary matrices are isometries because

$$||Uv||^2 = \langle Uv, Uv\rangle = \langle U^*Uv, v\rangle = \langle v, v\rangle = ||v||^2$$

for all $v \in \mathbf{C}^n$. Thus, they are the natural analog for $\mathbf{C}^n$ of permutations of a set, and for this reason they are used to model transformations in quantum mechanics. In particular, they are used to model time evolution.

A complete description of the dynamics of a quantum system involves a family of unitaries $\{U_t : t \in \mathbf{R}\}$, where the operator U_t takes a state vector v to the state $U_t v$ that it will evolve into after t units of time. Obviously, we must have $U_0 = I$, and it is also natural to require that $U_{s+t} = U_s U_t$ for all $s, t \in \mathbf{R}$, on the grounds that allowing a system to evolve for successively t and then s units of time is equivalent to allowing it to evolve for $s+t$ units of time all at once. There is also a continuity requirement.

DEFINITION 1.4.2 *A <u>continuous one-parameter unitary group</u> on $\mathbf{C}^n$ is a family of unitary $n \times n$ matrices $\{U_t : t \in \mathbf{R}\}$ such that $U_0 = I$, $U_s U_t = U_{s+t}$ for all $s, t \in \mathbf{R}$, and $U_s \to U_t$ as $s \to t$.*

Here convergence of matrices is taken in the only reasonable sense, namely, pointwise convergence of each entry. Note that the above conditions imply $U_t^* = U_t^{-1} = U_{-t}$ for all $t \in \mathbf{R}$.

Example 1.4.3

Let A be a self-adjoint matrix and let $E_1, \ldots, E_k$ and $\lambda_1, \ldots, \lambda_k$ be its eigenspaces and eigenvalues. For $t \in \mathbf{R}$ define U_t by setting $U_t v = e^{it\lambda_j} v$ for all $v \in E_j$ $(1 \le j \le k)$ and extending linearly. Then $\{U_t\}$ is a continuous one-parameter unitary group. We write $U_t = e^{itA}$.

The notation $U_t = e^{itA}$ in the preceding example is well-taken in light of the following fact.

PROPOSITION 1.4.4

Let A be a self-adjoint $n \times n$ complex matrix. Then the infinite sum $\sum_{k=0}^{\infty} (itA)^k/k!$ converges to the unitary operator e^{itA}.

PROOF Let v be an eigenvector for A with eigenvalue λ. Then $A^k v = \lambda^k v$, and so

$$\left(\sum_{k=1}^{m} \frac{(itA)^k}{k!}\right) v = \left(\sum_{k=1}^{m} \frac{(it\lambda)^k}{k!}\right) v \to e^{it\lambda} v = e^{itA} v$$

as $m \to \infty$. Since the eigenvectors of A span $\mathbf{C}^n$, we can conclude that $(\sum (itA)^k/k!)v \to e^{itA} v$ for all $v \in \mathbf{C}^n$, which is enough. ∎

Remarkably, Example 1.4.3 is completely general. This result requires the following generalization of the spectral theorem.

LEMMA 1.4.5

Let $A_1, \ldots, A_m$ be self-adjoint $n \times n$ complex matrices which commute in pairs. Then there is a sequence of orthogonal subspaces $E_1, \ldots, E_k$ which span $\mathbf{C}^n$, such that each E_i is an intersection of eigenspaces of the A_j.

PROOF This is true for $m = 1$ by Theorem 1.3.3. To pass from m to $m+1$, suppose $E \subset \mathbf{C}^n$ is an intersection of eigenspaces of the A_j for $1 \leq j \leq m$. That is, there exist $\lambda_1, \ldots, \lambda_m \in \mathbf{R}$ such that $v \in E$ if and only if $A_j v = \lambda_j v$ for $1 \leq j \leq m$.

It follows that for any $v \in E$ and any $1 \leq j \leq m$ we have

$$A_j A_{m+1} v = A_{m+1} A_j v = \lambda_j A_{m+1} v,$$

so that $A_{m+1} v \in E$ as well. Applying Theorem 1.3.3 to $A_{m+1}|_E$, we may therefore decompose E into orthogonal eigenspaces of $A_{m+1}|_E$. Each of these is then the intersection of E with an eigenspace of A_{m+1}. Since the subspaces at the mth step spanned $\mathbf{C}^n$, it follows that the subspaces at the $(m+1)$st step also span $\mathbf{C}^n$. ∎

LEMMA 1.4.6

Let $\{U_t\}$ be a continuous one-parameter unitary group on $\mathbf{C}^n$. Then there is a sequence of orthogonal subspaces $E_1, \ldots, E_k$ which span $\mathbf{C}^n$, such that for every t each E_i is contained in an eigenspace of U_t.

PROOF Note first that $U_s U_t = U_{s+t} = U_t U_s$ for all $s, t \in \mathbf{R}$, so all of the unitaries commute. Writing $\operatorname{Re} U_t = \frac{1}{2}(U_t + U_{-t})$ and $\operatorname{Im} U_t = \frac{1}{2i}(U_t - U_{-t})$, we have $U_t = \operatorname{Re} U_t + i \operatorname{Im} U_t$, and $\operatorname{Re} U_t$ and $\operatorname{Im} U_t$ are self-adjoint. The matrices $\operatorname{Re} U_t$ and $\operatorname{Im} U_t$ commute pairwise.

By Lemma 1.4.5, for any $t_1, \ldots, t_m \in \mathbf{R}$ there is a sequence of orthogonal subspaces $E_1, \ldots, E_k$ which span $\mathbf{C}^n$, such that each E_i is an intersection of eigenspaces of the $\operatorname{Re} U_{t_j}$ and $\operatorname{Im} U_{t_j}$. Since k must be less than or equal to n, there is a maximum value of k which can be achieved in this way. For this k the subspaces $E_1, \ldots, E_k$ must each be contained in an eigenspace of $\operatorname{Re} U_t$ and $\operatorname{Im} U_t$, for all t. It follows that each E_i is contained in an eigenspace of every U_t. ∎

THEOREM 1.4.7

Let $\{U_t\}$ be a continuous one-parameter unitary group on $\mathbf{C}^n$. Then there is a self-adjoint $n \times n$ complex matrix A such that $U_t = e^{itA}$.

PROOF Let $E_1, \ldots, E_k$ be as in Lemma 1.4.6, and for each $1 \leq j \leq k$ let $\lambda_j(t)$ be the eigenvalue of U_t to which E_j belongs. Observe that $|\lambda_j(t)| = 1$ for all t (since U_t is unitary) and $\lambda_j(0) = 1$. By continuity, there exists $\delta > 0$ such that $\lambda_j(t) \neq -1$ for $t \in [0, \delta]$. Then for $0 \leq t \leq \delta$, define $f_j(t) = -i \ln(\lambda_j(t))$, taking the branch of the logarithm which yields $-\pi < f_j(t) < \pi$. Since $\lambda_j(s+t) = \lambda_j(s)\lambda_j(t)$ for all s and t, it follows that $f_j(s+t) = f_j(s) + f_j(t)$, provided $s, t \geq 0$ and $s + t \leq \delta$.

This weak form of linearity implies that if $t = r\delta$ where r is a rational number between 0 and 1, then $f_j(t) = rf_j(\delta) = a_jt$ where $a_j = f_j(\delta)/\delta$; by continuity $f_j(t) = a_jt$ for all t between 0 and δ. Thus, for $0 \le t \le \delta$ and $v \in E_j$ we have $\lambda_j(t) = e^{ia_jt}$ and $U_tv = e^{ia_jt}v$. Doing this for each j, it follows that for sufficiently small δ' we have $U_t = e^{itA}$ for all $t \in [0, \delta']$, where A is the self-adjoint operator such that $Av = a_jv$ for $v \in E_j$. By the group property of $\{U_t\}$ we conclude that the same formula holds for all t. ∎

The matrix iA is called the generator of the unitary group $\{U_t\}$. Since any initial state $v = v(0)$ evolves in time according to $v(t) = U_tv = e^{itA}v$, we have

$$\frac{d}{dt}v(t) = iAv(t).$$

This is an abstract form of Schrödinger's equation. It is a differential version of the fact that the dynamics of a quantum system are described by a continuous one-parameter unitary group.

There is an alternative approach to dynamics, path integration, which is popular in the physics literature. The basic idea is this. Let (e_i) be the standard basis of $\mathbf{C}^n$; we want to determine the inner product $\langle U_se_i, e_j\rangle$ for given i and j. These "transition amplitudes" are the matrix entries of the operator U_s of evolution by s units of time.

For any $v, w \in \mathbf{C}^n$, we have $\langle v, w\rangle = \sum_k \langle v, e_k\rangle\langle e_k, w\rangle$. Therefore

$$\begin{aligned}\langle U_se_i, e_j\rangle &= \langle U_{s/2}e_i, U_{s/2}^*e_j\rangle \\ &= \sum_k \langle U_{s/2}e_i, e_k\rangle\langle e_k, U_{s/2}^*e_j\rangle \\ &= \sum_k \langle U_{s/2}e_i, e_k\rangle\langle U_{s/2}e_k, e_j\rangle.\end{aligned}$$

A similar computation shows that

$$\langle U_se_i, e_j\rangle = \sum_{k,k'} \langle U_{s/3}e_i, e_k\rangle\langle U_{s/3}e_k, e_{k'}\rangle\langle U_{s/3}e_{k'}, e_j\rangle,$$

and so on. In general we may break up the total time interval $[0, s]$ into m short subintervals and sum over all "paths" from e_i to e_j, i.e. all sequences of basis vectors of length $m + 1$ whose first entry is e_i and whose last entry is e_j. The corresponding term in the sum is called the amplitude of that path.

The computational value in doing this lies in the fact that over short times $U_t = e^{iAt}$ is approximated by $I + iAt$. Thus one could hope to take a limit as $m \to \infty$ and work with the matrix A in place of U_t. One might also identify $\mathbf{C}^n$ with the complex-valued functions on a set of n points in $\mathbf{R}^3$, say, and then take $n \to \infty$, allowing the points to become dense in $\mathbf{R}^3$. Then the "paths" might really be thought of as physical paths in space.

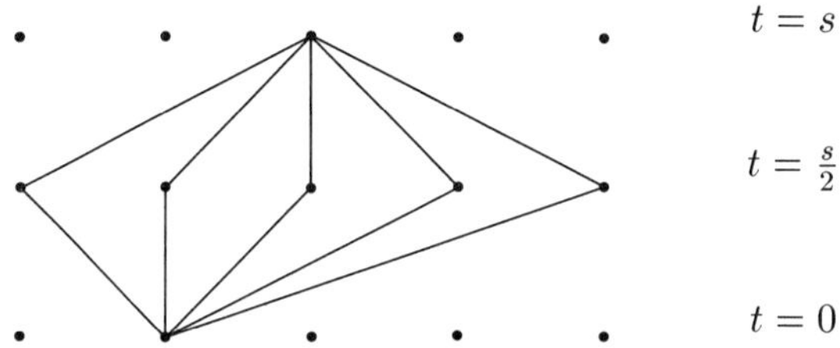

Figure 1.3 Summing over paths

The problem with this idea is that the sum which is used to evaluate $\langle U_s e_i, e_j \rangle$ tends to become a badly divergent integral in the limit. It is said to be "oscillatory" because the product

$$\langle (I + iAt)e_i, e_k \rangle \langle (I + iAt)e_k, e_{k'} \rangle \cdots \langle (I + iAt)e_{k^{(m)}}, e_j \rangle$$

is interpreted in the limit as an exponential factor, and the oscillation of this factor is responsible for the divergence of the path integral.

It is still possible to develop workable versions of the path integral approach in various rather circumscribed situations. However, it seems unlikely that there will ever be a truly general theory, in spite of the evident heuristic value of the idea. With this pessimistic observation we abandon path integrals.

1.5 Composite systems

Composite systems which are made up of interacting subsystems can be modelled using tensor products: if the Hilbert spaces $\mathbf{C}^m$ and $\mathbf{C}^n$ model two separate quantum systems then their tensor product $\mathbf{C}^m \otimes \mathbf{C}^n \cong \mathbf{C}^{mn}$ models the two systems together.

The basic properties of the tensor product are these. Corresponding to any states $v \in \mathbf{C}^m$ and $w \in \mathbf{C}^n$ there is an elementary tensor product state $v \otimes w \in \mathbf{C}^{mn}$ whose coordinate entries are $a_i b_j$, if $v = (a_1, \ldots, a_m)$ and $w = (b_1, \ldots, b_n)$. This state is regarded as modelling the situation where the first subsystem is in the state v and the second is in the state w. For any unit vector $w \in \mathbf{C}^n$, the map $v \mapsto v \otimes w$ is an isometric embedding of $\mathbf{C}^m$ in $\mathbf{C}^{mn}$, so the states of the composite system for which the second system is in the state w are in one-to-one correspondence with the states of the first system.

A general element of a tensor product can be expressed as a linear combination of elementary tensor product states. Thus, one usually cannot say what the state of one subsystem is independently of the state of the other subsystem.

Composite systems of identical particles require a minor variation on the preceding construction. For a system of k identical particles, each modelled on $\mathbf{C}^n$, rather than the full tensor power $\mathbf{C}^n \otimes \cdots \otimes \mathbf{C}^n \cong \mathbf{C}^{n^k}$ we use either the symmetric tensor power $\mathbf{C}^n \otimes_s \cdots \otimes_s \mathbf{C}^n$ or the antisymmetric tensor power $\mathbf{C}^n \otimes_a \cdots \otimes_a \mathbf{C}^n$. These consist of those vectors in the full tensor product whose components satisfy, respectively, $a_{i_1 \ldots i_k} = a_{\sigma(i_1) \ldots \sigma(i_k)}$ or $a_{i_1 \ldots i_k} = (-1)^{|\sigma|} a_{\sigma(i_1) \ldots \sigma(i_k)}$ for any permutation σ of the index set $\{1, \ldots, k\}$, where $|\sigma|$ is the parity of σ. The appropriate choice varies; particles are called Bosons or Fermions depending on which tensor product is used to model their multiple-particle systems.

Using the notion of a composite system, we can now clarify the naive interpretation of the quantum mechanical formalism given in Section 1.2. What makes a diagonally polarized photon abruptly become horizontally or vertically polarized when measured, and how does it choose which to become? The answer to these questions involves the fact that we, the observers of the system, are ourselves quantum systems. Thus, when measurement issues such as these arise, we must consider the composite Hilbert space $\mathbf{C}^m \otimes \mathbf{C}^n$ where $\mathbf{C}^m$ is the Hilbert space of the observed system and $\mathbf{C}^n$ is the Hilbert space of the observer (assumed here, with apologies, to be finite dimensional). For example, say that $\mathbf{C}^2$ models the polarization of a photon, and let $\{e_1, e_2\}$ be the canonical basis of $\mathbf{C}^2$. Then any state of the composite Hilbert space $\mathbf{C}^2 \otimes \mathbf{C}^n$ can be expressed in the form $a_1(e_1 \otimes v_1) + a_2(e_2 \otimes v_2)$ for some unit vectors $v_1, v_2 \in \mathbf{C}^n$ and complex coefficients a_1, a_2.

Now suppose we are given an initial elementary tensor product state $(a_1 e_1 + a_2 e_2) \otimes v$, where the two subsystems are each in well-defined individual states. Here the observer is oblivious to the state of the photon. As long as there is no physical interaction between the two subsystems, time evolution will preserve the property that the composite system is in an elementary state. However, if the subsystems do interact then after some passage of time the composite system will be in a general state of the form $b_1(e_1 \otimes v_1) + b_2(e_2 \otimes v_2)$. This overall state is now "entangled" in the sense that it is not possible to determine the state of one subsystem independently of the state of the other subsystem. The remarkable fact at this point is that the total state is a composite of the states $e_1 \otimes v_1$ and $e_2 \otimes v_2$, which are orthogonal *and will remain so at all future times* due to the unitarity of time evolution. Thus the system can be decomposed into two orthogonal states in which the state of the observer is respectively correlated with the two different possible polarizations of the photon. In each of the two states the observer apparently has the subjective experience of being in a universe in which the photon has the corresponding polarization.

This analysis suggests the surprising possibility that there is in some sense more than one copy of the observer after the measurement has been made. In general, after any measurement of a finite state system we can decompose the state of the composite system into an orthogonal sum of states in each of which the observer sees the measured system in a particular eigenstate. Now in most elementary treatments of quantum mechanics it is assumed that one of the terms in this sum is "real" and the rest are not; this transition of the state of the system to a single term in the sum is the "collapse" referred to in Section 1.2. One then has to deal with the problem of what mechanism causes the system to collapse and how such a mechanism might be triggered. Moreover, the infinite dimensional case, in which there generally is no decomposition of the Hilbert space into a discrete family of eigenspaces, presents the separate problem of what the end result of the collapse should even be. Alternatively, it has been suggested that no collapse takes place and an image of parallel universes has been put forward in its place.

Thus, one is supposedly faced with the problem of deciding whether every one of a family of possible universes really exists, or if not how and when the process of selecting the real one occurs. Arguments on either side of this question can be given in terms of the parsimony either of not requiring a collapse or of not invoking the existence of unobservable universes.

However, it is possible that the whole issue is really what some philosophers might call a meaningless pseudo-problem. One could argue that any assertion that a given universe does or does not exist, or that other terms of the sum are or are not "really there", is in principle unverifiable, and hence metaphysical in the most literal sense. For example, suppose it was asserted after some particular measurement that only the even terms of the decomposition had given rise to existing universes, and that the odd terms had vanished. What actual content would such an assertion have?

Ordinary assertions about the existence of ordinary objects can be understood as indicating the appearance of the objects *in the speaker's universe.* In principle, they can always be verified or falsified. But this is exactly what one cannot do with statements about the existence of other universes. It may be that one is simply tricked into thinking that the latter sort of assertion is meaningful by a false analogy with the former sort.

For reasons like these, one may want to question whether statements about when and how a "collapse" takes place have any genuine content at all.

1.6 Quantum computation

In theory, the fact that quantum states can be superimposed should enable us to perform some computations faster than is classically possible. (Actually building a "quantum computer" which could do this presents serious engineering obstacles, but it seems likely that these will be overcome in time.) One can simulate any quantum computation on a classical computer, and vice versa, so there is no absolute problem which can be solved on one but not the other; the issue is solely one of speed.

To illustrate the way superposition can be used to speed up a computation, we will describe an algorithm that searches n data points in $O(\sqrt{n})$ time. The problem is formulated as follows. Let $\mathbf{C}^n$, $n \geq 3$, model the phase space of some quantum system; its physical constitution is irrelevant. Let V be a unitary $n \times n$ matrix which is diagonal with respect to the standard basis $\{e_1, \ldots, e_n\}$, so $Ve_i = \lambda_i e_i$, and assume there exists i_0 such that $\lambda_i = 1$ for $i \neq i_0$ and $\lambda_{i_0} = -1$. Suppose also that the transformation $v \mapsto Vv$ on $\mathbf{C}^n$ is physically realizable, so that in the course of our computation we can apply V to the state of the system at will.

We wish to find the eigenvector e_{i_0} whose eigenvalue is -1. Classically, there is no way to find this -1 eigenvector with certainty in fewer than n steps, and weakening the problem to ask only that we find the eigenvector with probability p shortens the computation to pn steps, which is still $O(n)$. In contrast, the quantum mechanical algorithm that we are about to present finds the -1 eigenvector with probability p in $O(\sqrt{n})$ steps, for any $p < 1$. (Finding it with certainty still cannot be done in fewer than n steps.)

The computation also requires that we be able to apply the matrix $U = [u_{ij}]$ to the system, where $u_{ij} = 2n^{-1}$ for $i \neq j$ and $u_{ii} = -1+2n^{-1}$. Note that this matrix is also unitary. The algorithm is then to prepare the system in the state $v = (n^{-1/2}, \ldots, n^{-1/2})$ and apply the operators U and V alternately k times, that is, apply the operator $(UV)^k$. The claim is that after this process the system will be in the -1 eigenstate with probability p such that p increases with k.

To see this, let $w = (1, \ldots, 1) - e_{i_0}$ and write the initial state of the system as $v = n^{-1/2}w + n^{-1/2}e_{i_0}$. It is easy to check that a state of the form $aw + be_{i_0}$ is taken into another state of this form by the operator UV. So say $(UV)^k v = a_k w + b_k e_{i_0}$. Then $a_0 = b_0 = n^{-1/2}$ and we have the recursion relations $a_{k+1} = 2c_k - a_k$ and $b_{k+1} = 2c_k + b_k$, where $c_k = \frac{n-1}{n}a_k - \frac{1}{n}b_k$.

Since $(n-1)a_k^2 + b_k^2 = 1$, if b_k lies between $n^{-1/2}$ and $2^{-1/2}$ then a_k

lies between $(2n-2)^{-1/2}$ and $n^{-1/2}$. In this case c_k is bounded below by $2^{-1/2}((n-1)^{1/2}-1)n^{-1}$. This shows that as long as $n^{-1/2} \leq b_k \leq 2^{-1/2}$ holds, b_k will increase at each step by at least $2^{1/2}((n-1)^{1/2}-1)n^{-1} \approx 2^{1/2}n^{-1/2}$. So for large n, on the order of $n^{1/2}$ steps are needed to ensure $b_k \geq 2^{-1/2}$. At this point a measurement of the system will place in it the state e_{i_0} with probability at least $1/2$. Running the same procedure m times yields a probability of at least $1-2^{-m}$ of reaching the state e_{i_0} in at least one trial, and for any fixed value of m this still takes $O(\sqrt{n})$ steps. This shows that the algorithm has the property that was claimed.

One is sometimes told that the reason quantum algorithms can be faster than classical algorithms is because they are able to carry out multiple computations simultaneously, due to superposition. However, examination of the above algorithm reveals no precise sense in which simultaneous computations are being performed. Perhaps it is clearer to say that the reason quantum algorithms can be faster than classical algorithms lies in the phemomenon of superposition together with the possibility of applying arbitrary unitary matrices. Using only permutation matrices is equivalent to computing with a classical finite state machine.

1.7 Notes

The classic introductions to quantum mechanics are [18] and [52]. A modern approach is given in [8].

Our explanation of the correspondence between real-valued functions and self-adjoint operators given in Section 1.3 is due in a slightly different form to Mackey [45]. We will discuss this further in Chapter 3.

For more on path integrals see [27].

The so-called "many-worlds" interpretation discussed in Section 1.5 is expounded in detail in [25]. In particular, a derivation of the probability interpretation of Section 1.2 is given there. Our suggestion that it is meaningless to ask whether parallel universes really exist is based on philosophical grounds articulated in [5].

The quantum search algorithm given in Section 1.6 is due to Grover [31]. There is also a quantum algorithm for factorization which is presumably (provided $P \neq NP$) exponentially faster than any classical algorithm; it is due to Shor [65].

Chapter 2

Hilbert Spaces

2.1 Definitions and examples

In the last chapter we considered only finite dimensional Hilbert spaces. We now present the infinite dimensional theory, paying particular attention to the analogies between Hilbert spaces and sets. As we saw in Chapter 1, this analogy is grounded in physical reality, and it will be the basis of all of our subsequent work.

We begin with the formal definition of Hilbert spaces. (It should be made clear, however, that readers who are not already familiar with other functional analysis topics at a comparable level, which we do not review, will find later chapters difficult to follow.) We adopt the mathematical convention of making inner products antilinear in the second variable; in the physics literature they are usually linear in the second and antilinear in the first.

DEFINITION 2.1.1 *Let $\mathcal{H}$ be a complex vector space. A pseudo inner product on $\mathcal{H}$ is a map $\langle \cdot, \cdot \rangle : \mathcal{H} \times \mathcal{H} \to \mathbf{C}$ satisfying*

(a) $\langle av_1 + bv_2, w\rangle = a\langle v_1, w\rangle + b\langle v_2, w\rangle$;
(b) $\overline{\langle v, w\rangle} = \langle w, v\rangle$; *and*
(c) $\langle v, v\rangle \geq 0$

for all $a, b \in \mathbf{C}$ and $v_1, v_2, v, w \in \mathcal{H}$. We write $\|v\| = \langle v, v\rangle^{1/2}$. If $\|v\| = 0$ implies $v = 0$ then $\langle \cdot, \cdot \rangle$ is an inner product.

A Hilbert space is a complex vector space equipped with an inner product whose corresponding norm $\|\cdot\|$ is complete.

This definition of Hilbert spaces is contingent on us verifying that $\|\cdot\|$ really is a norm, which we do immediately below, in Corollary 2.1.3. First we need the Cauchy-Schwarz inequality.

PROPOSITION 2.1.2
Let $\mathcal{H}$ be a complex vector space, let $\langle \cdot, \cdot \rangle$ be a pseudo inner product on $\mathcal{H}$, and let $v, w \in \mathcal{H}$. Then $|\langle v, w \rangle| \leq \|v\|\|w\|$.

PROOF If $\|v\| = \|w\| = 0$ then

$$0 \leq \langle v - \langle v, w\rangle w, v - \langle v, w\rangle w\rangle = -2|\langle v, w\rangle|^2,$$

hence $\langle v, w \rangle = 0$. Otherwise, without loss of generality $\|w\| \neq 0$; then

$$0 \leq \left\langle \|w\|v - \frac{\langle v, w\rangle}{\|w\|}w, \|w\|v - \frac{\langle v, w\rangle}{\|w\|}w \right\rangle = \|w\|^2\|v\|^2 - |\langle v, w\rangle|^2,$$

so again $|\langle v, w\rangle| \leq \|v\|\|w\|$. ∎

COROLLARY 2.1.3
If $\langle \cdot, \cdot \rangle$ is a pseudo inner product then $\| \cdot \|$ is a pseudonorm, and if $\langle \cdot, \cdot \rangle$ is an inner product then $\| \cdot \|$ is a norm.

PROOF Suppose $\langle \cdot, \cdot \rangle$ is a pseudo inner product. It is immediate from the definition of $\| \cdot \|$ that $\|v\| \geq 0$ for all v. Next, we have

$$\|av\| = \langle av, av\rangle^{1/2} = (|a|^2\langle v, v\rangle)^{1/2} = |a|\|v\|$$

for all $a \in \mathbf{C}$. Finally, by the Cauchy-Schwarz inequality we have $\mathrm{Re}\langle v, w\rangle \leq |\langle v, w\rangle| \leq \|v\|\|w\|$, and hence

$$\begin{aligned} \|v + w\|^2 &= \langle v + w, v + w\rangle \\ &= \|v\|^2 + 2\mathrm{Re}\langle v, w\rangle + \|w\|^2 \\ &\leq \|v\|^2 + 2\|v\|\|w\| + \|w\|^2 \\ &= (\|v\| + \|w\|)^2, \end{aligned}$$

yielding the triangle inequality.

If $\langle \cdot, \cdot \rangle$ is an inner product then $v \neq 0$ implies $\|v\|^2 = \langle v, v\rangle \neq 0$, so $\| \cdot \|$ is a norm. ∎

Our definition of Hilbert spaces is now logically sound.

Pseudo inner products can be converted into inner products using the following technique. We call this construction factoring out null vectors.

PROPOSITION 2.1.4
Let $\langle \cdot, \cdot \rangle$ be a pseudo inner product on a complex vector space $\mathcal{H}$. Then the set $\mathcal{H}_0 = \{v \in \mathcal{H} : \langle v, v\rangle = 0\}$ is a linear subspace of $\mathcal{H}$, and $\langle \cdot, \cdot \rangle$ descends to an inner product on $\mathcal{H}/\mathcal{H}_0$.

PROOF The fact that $\mathcal{H}_0 = \{v : \|v\| = 0\}$ is a linear subspace follows from the fact that $\|\cdot\|$ is a pseudonorm (Corollary 2.1.3). In particular, $v, w \in \mathcal{H}_0$ implies $v + w \in \mathcal{H}_0$ by the triangle inequality.

To show that $\langle \cdot, \cdot \rangle$ descends to $\mathcal{H}/\mathcal{H}_0$, let $v, w \in \mathcal{H}$ and $v_0, w_0 \in \mathcal{H}_0$. We must verify that $\langle v + v_0, w + w_0 \rangle = \langle v, w \rangle$. But

$$\langle v_0, w \rangle = \langle v, w_0 \rangle = \langle v_0, w_0 \rangle = 0$$

by the Cauchy-Schwarz inequality, so the desired equality holds.

It is now routine to check that $\langle \cdot, \cdot \rangle$ defines a pseudo inner product on $\mathcal{H}/\mathcal{H}_0$, and this is actually an inner product because $v \notin \mathcal{H}_0$ implies $\langle v, v \rangle \neq 0$. ∎

Having turned a pseudo inner product into an inner product in this way, we may still need to complete the space. There is no obstacle to doing this; the proof is routine but tedious, so we omit it.

PROPOSITION 2.1.5

Let $\mathcal{H}$ be a complex vector space and let $\langle \cdot, \cdot \rangle$ be an inner product on $\mathcal{H}$. Then the formula

$$\langle \lim v_n, \lim w_n \rangle = \lim \langle v_n, w_n \rangle$$

defines an inner product on the Cauchy completion of $\mathcal{H}$ which makes it a Hilbert space.

For the remainder of this section we discuss examples. None of these actually uses the preceding techniques of factoring out null vectors and completing; these will be needed later. In particular, all of the examples we consider now are complete from the start, which is good because it means that we have a concrete representation of every vector in the Hilbert space.

Example 2.1.6

$\mathbf{C}^n$ is a Hilbert space, with inner product

$$\langle v, w \rangle = \sum_{i=1}^{n} a_i \bar{b}_i,$$

where $v = (a_1, \ldots, a_n)$ and $w = (b_1, \ldots, b_n)$.

The space $\mathbf{C}^n$ of complex n-tuples can be identified with the set of functions from $\{1, \ldots, n\}$ into $\mathbf{C}$. In this way, the preceding example can be seen as a special case of the following more general construction.

Example 2.1.7

Let X be a set. For any function $f : X \to \mathbf{C}$, we write $\sum_X f(x) = a$ ($a \in \mathbf{C}$) if for every $\epsilon > 0$ there exists a finite subset S of X such that $|a - \sum_{S'} f(x)| < \epsilon$ for any finite $S' \subset X$ which contains S.

Define $l^2(X)$ to be the collection of all functions $f : X \to \mathbf{C}$ such that $\|f\|^2 = \sum_X |f(x)|^2$ is finite. For $f, g \in l^2(X)$ we define $\langle f, g\rangle$ by $\langle f, g\rangle = \sum f(x)\overline{g(x)}$.

To see that $\sum f(x)\overline{g(x)}$ exists, given $\epsilon > 0$ find a finite set $S \subset X$ such that $\|f|_S\| \geq \|f\| - \epsilon$ and $\|g|_S\| \geq \|g\| - \epsilon$. Then for any finite subset $S' \subset X$ which contains S, the Cauchy-Schwarz inequality in $l^2(S')$ yields

$$\left|\sum_{S'-S} f(x)\overline{g(x)}\right| \leq \left\|f|_{S'-S}\right\|\left\|g|_{S'-S}\right\| \leq \epsilon^2.$$

So $\langle f, g\rangle$ is indeed well-defined.

Finally, to show $l^2(X)$ is complete, let (f_n) be a Cauchy sequence. Then for any $x \in X$ the sequence $(f_n(x))$ satisfies $|f_m(x) - f_n(x)| \leq \|f_m - f_n\|$ and hence is Cauchy in $\mathbf{C}$. So we can define a limit function $f = \lim f_n : X \to \mathbf{C}$ pointwise. This function belongs to $l^2(X)$ because for any finite $S \subset X$ we have $\|f|_S\| \leq \sup \|f_n\|$. Finally, given $\epsilon > 0$ we can find an integer N such that $m, n \geq N$ implies $\|f_m - f_n\| \leq \epsilon$; then for any $n \geq N$ and any finite $S \subset X$ we have

$$\left\|(f - f_n)|_S\right\| = \lim_m \left\|(f_m - f_n)|_S\right\| \leq \epsilon,$$

so $f_n \to f$.

Next we consider L^2 spaces. Any $l^2(X)$ can be regarded as an L^2 space by treating the set X as a measure space with counting measure. However, this fact is of little use to us because much of what we do measure theoretically in later chapters requires σ-finiteness and if X is uncountable then counting measure does not have this property. So in general we have to treat sets and measure spaces separately.

To simplify the presentation we are going to assume throughout that all measures are σ-finite.

We need the following lemma, which gives a useful criterion for determining whether a normed vector space is complete.

LEMMA 2.1.8

A normed vector space $\mathcal{V}$ is complete if and only if the sum $\sum_1^\infty v_n$ converges whenever (v_n) is a sequence of vectors such that $\sum_1^\infty \|v_n\|$ converges.

PROOF If $\mathcal{V}$ is complete and $\sum_1^\infty \|v_n\| < \infty$ then the sum $\sum_1^\infty v_n$ converges because the sequence of partial sums $\sum_1^N v_n$ is Cauchy.

To prove the converse, let (w_n) be a Cauchy sequence in $\mathcal{V}$; we must show it converges. Choose a subsequence (w_{k_n}) with the property that

$$\|w_{k_n} - w_{k_{n+1}}\| \leq 2^{-n}$$

for all n. Then define $v_1 = w_{k_1}$ and $v_n = w_{k_n} - w_{k_{n-1}}$ for $n > 1$. By hypothesis $\sum v_n$ converges, but the nth partial sum of $\sum v_n$ equals w_{k_n}, so the sequence (w_{k_n}) converges. This implies that (w_n) converges. ∎

Example 2.1.9

Let μ be a σ-finite measure on a set X and let $L^2(X) = L^2(X, \mu)$ be the space of measurable functions $f : X \to \mathbf{C}$, modulo functions supported on null sets, which satisfy $\|f\|^2 = \int |f|^2 d\mu < \infty$. Then for any $f, g \in L^2(X)$ the inner product $\langle f, g\rangle = \int f\bar{g}d\mu$ is well-defined because the inequality $2|f\bar{g}| \leq |f|^2 + |g|^2$ implies that $f\bar{g}$ is integrable. This is a genuine inner product because $\int |f|^2\, d\mu = 0$ implies $f = 0$ almost everywhere.

To see that $L^2(X)$ is complete, let $(f_n) \subset L^2(X)$ satisfy $\sum \|f_n\| = C < \infty$. Let $g_N = \sum_1^N |f_n|$ and $g = \lim g_N$. Since $\|g_N\| \leq C$ for all N, the monotone convergence theorem implies that $\|g\| \leq C$ as well. So $g < \infty$ almost everywhere, which implies that $\sum f_n$ converges almost everywhere to a measurable function f. Finally, $|f| \leq g$ so $f \in L^2(X)$, and $\|f - \sum_1^N f_n\| \to 0$ by the dominated convergence theorem since $f - \sum_1^N f_n \to 0$ almost everywhere. Thus $L^2(X)$ is complete by Lemma 2.1.8.

For the sake of clarity we occasionally denote the L^2 norm by $\|\cdot\|_2$.

A similar argument shows that $L^p(X)$ is complete for $1 \leq p \leq \infty$. After $p = 2$ we will be most interested in the cases $p = \infty$ and $p = 1$ (but we will not formally introduce them).

2.2 Subspaces

We want to develop an analogy between sets and Hilbert spaces. The analog of an element of a set is a unit vector in a Hilbert space — for instance, in $l^2(X)$ (Example 2.1.7) the element $x \in X$ corresponds to the function $\chi_x = \chi_{\{x\}}$ which takes the value 1 at x and is 0 elsewhere. Likewise, closed subspaces of a Hilbert space are analogous to subsets of a set. We now introduce the operations $\vee$, $\wedge$, and $\perp$ on subspaces, which correspond to union, intersection, and complementation of subsets. In the finite dimensional case these reduce to the operations defined in Section 1.2.

PROPOSITION 2.2.1
Let $\{E_\kappa : \kappa \in J\}$ be a family of subspaces of a Hilbert space $\mathcal{H}$. Then the closure of span$\{E_\kappa\}$ *is the smallest closed subspace of $\mathcal{H}$ which contains every E_κ, and $\bigcap E_\kappa$ is the largest closed subspace of $\mathcal{H}$ which is contained in every E_κ. If E is a closed subspace of $\mathcal{H}$ then the set*

$$\{v \in \mathcal{H} : \langle v, w\rangle = 0 \text{ for all } w \in E\}$$

is a closed subspace of $\mathcal{H}$.

For example, the last statement is proven as follows. It is clear that the given set of vectors is a subspace; to prove closure, suppose that $v_n \to v$ and that each v_n belongs to the set. Then we have $\langle v, w\rangle = \langle v_n - v, w\rangle \to 0$ for any $w \in E$ by the Cauchy-Schwarz inequality, so v is also in the set.

DEFINITION 2.2.2 *Let $\mathcal{H}$ be a Hilbert space. The* join *of a family $\{E_\kappa : \kappa \in J\}$ of closed subspaces of $\mathcal{H}$, denoted $\bigvee E_\kappa$, is their closed span $\overline{\text{span}}\{E_\kappa\}$ and their* meet *is their intersection $\bigwedge E_\kappa = \bigcap E_\kappa$.*

The orthocomplement *of a subspace E of $\mathcal{H}$ is the closed subspace*

$$E^\perp = \{v \in \mathcal{H} : \langle v, w\rangle = 0 \text{ for all } w \in E\}.$$

We write $v \perp w$ if $\langle v, w\rangle = 0$, $v \perp E$ if $v \in E^\perp$, and $E_1 \perp E_2$ if $E_1 \subset E_2^\perp$; and we say in the respective cases that v and w, v and E, or E_1 and E_2, are orthogonal.

The operations $\vee$, $\wedge$, and $\perp$ actually behave very much like $\cup$, $\cap$, and c (complementation): the set theoretic laws $S \cup S^c = X$, $S \cap S^c = \emptyset$, $S^{cc} = S$, $(S_1 \cup S_2)^c = S_1^c \cap S_2^c$, and $(S_1 \cap S_2)^c = S_1^c \cup S_2^c$, for any subsets $S, S_1, S_2 \subset X$, all transfer to subspaces of a Hilbert space. Before proving this we require two preliminary lemmas which are both of independent interest. The first is the parallelogram law and the second says that vectors of minimal norm exist in any closed convex set.

LEMMA 2.2.3
Let $\mathcal{H}$ be a Hilbert space. Then

$$\|v + w\|^2 + \|v - w\|^2 = 2\|v\|^2 + 2\|w\|^2$$

for all $v, w \in \mathcal{H}$. For any $\epsilon > 0$ there exists $\delta > 0$ such that $\|v\| \le 1$, $\|w\| \le 1$, and $\|\frac{1}{2}(v + w)\| \ge 1 - \delta$ imply $\|v - w\| \le \epsilon$.

PROOF The parallelogram law is a straightforward calculation with inner products. For the second assertion, given $\epsilon > 0$ let $\delta = \epsilon/8$. Then

$||v|| \leq 1$, $||w|| \leq 1$, and $||\frac{1}{2}(v+w)|| \geq 1-\delta$ together imply

$$||v-w||^2 = 2||v||^2 + 2||w||^2 - ||v+w||^2 \leq 4 - 4(1-\delta)^2 \leq \epsilon,$$

as desired. ∎

The second assertion of Lemma 2.2.3 is called uniform convexity. It and completeness are the only properties needed to prove the next lemma.

LEMMA 2.2.4
Let K be a closed, convex subset of a Hilbert space. Then there exists a unique vector $v \in K$ of minimal norm.

PROOF Let $a = \inf\{||v|| : v \in K\}$. If $a = 0$ then there is a sequence of vectors in K whose norms converge to zero, hence $0 \in K$, and we are done. Otherwise let v_n be a sequence in K such that $||v_n|| \to a$. Given $\epsilon > 0$, choose δ as in Lemma 2.2.3 and find $N \in \mathbf{N}$ such that $n \geq N$ implies $||v_n|| \leq a(1+\delta) \equiv b$. Then for $m, n \geq N$ we have $||v_m||, ||v_n|| \leq b$ and $\frac{1}{2}(v_m + v_n) \in K$, and hence $||\frac{1}{2}(v_m+v_n)|| \geq a \geq b(1-\delta)$; so uniform convexity implies $||v_m - v_n|| \leq b\epsilon$. This shows that the sequence (v_n) is Cauchy, so it converges to some $v \in K$, and we have $||v|| = \lim ||v_n|| = a$.

For uniqueness, suppose $w \in K$ and $||w|| = ||v||$. Then $\frac{1}{2}(v+w)$ also belongs to K, so by minimality of $||v||$ we have $||\frac{1}{2}(v+w)|| \geq ||v|| = ||w||$. Uniform convexity then implies that $||v-w|| = 0$, i.e., $v = w$. ∎

THEOREM 2.2.5
Let E be a closed subspace of a Hilbert space $\mathcal{H}$. Then $E + E^\perp = E \vee E^\perp = \mathcal{H}$ and $E \wedge E^\perp = \{0\}$. Every $v \in \mathcal{H}$ can be expressed uniquely in the form $v = v_1 + v_2$ with $v_1 \in E$ and $v_2 \in E^\perp$.

PROOF Let $v \in \mathcal{H}$. By Lemma 2.2.4 there exists a vector $v_0 \in v + E$ of minimal norm. We claim that $v_0 \in E^\perp$. To verify this, choose $w \in E$, $w \neq 0$. Then $||v_0 + aw|| \geq ||v_0||$ for all $a \in \mathbf{C}$, by minimality of $||v_0||$. In particular, taking $a = -\langle v_0, w\rangle/||w||^2$ and performing a short computation, we get

$$||v_0||^2 - \frac{|\langle v_0, w\rangle|^2}{||w||^2} = \left\| v_0 - \frac{\langle v_0, w\rangle}{||w||^2} w \right\|^2 \geq ||v_0||^2.$$

Thus $\langle v_0, w\rangle = 0$, as claimed. We conclude that $v_0 \in E^\perp$, and $v = (v - v_0) + v_0$ is the desired decomposition of v. Thus $E + E^\perp = \mathcal{H}$.

$E \wedge E^\perp = \{0\}$ because any vector simultaneously in E and $E^\perp$ is orthogonal to itself.

If $v = v_1 + v_2 = v_1' + v_2'$ with $v_1, v_1' \in E$ and $v_2, v_2' \in E^\perp$ then $v_1 - v_1' = v_2' - v_2$. Since the left side belongs to E and the right side belongs to $E^\perp$, and we know that $E \wedge E^\perp = \{0\}$, we conclude that both sides are zero. Thus $v_1 = v_1'$ and $v_2 = v_2'$. This proves that the decomposition $v = v_1 + v_2$ is unique. ∎

The trick used to prove $v_0 \in E^\perp$ in the preceding theorem also appeared in the proof of the Cauchy-Schwarz inequality (Proposition 2.1.2). The point is that if $\langle v, w\rangle = \langle w, w\rangle$ then $\|v - w\|^2 = \|v\|^2 - \|w\|^2$. That is, if $v - w$ is orthogonal to w then $\|w\|^2 + \|v - w\|^2 = \|v\|^2$, which is just the Pythagorean theorem. Furthermore, given any v and w, $w \neq 0$, we can ensure the hypothesis by scaling w by the factor $\langle v, w\rangle/\|w\|^2$, i.e., replacing w by the projection of v onto w.

COROLLARY 2.2.6
Let E and E_κ ($\kappa \in J$) be closed subspaces of a Hilbert space $\mathcal{H}$. Then

(a) $E^{\perp\perp} = E$;
(b) $(\bigvee E_\kappa)^\perp = \bigwedge E_\kappa^\perp$; and
(c) $(\bigwedge E_\kappa)^\perp = \bigvee E_\kappa^\perp$.

PROOF

(a) Let $v \in E$. Then $v \perp E^\perp$, so $v \in E^{\perp\perp}$. This shows that $E \subset E^{\perp\perp}$. Suppose the containment is proper. Then by Theorem 2.2.5 there exists a nonzero $w \in E^{\perp\perp}$ orthogonal to E. But w is also orthogonal to $E^\perp$, and hence $w = 0$, a contradiction. So $E^{\perp\perp} = E$.

(b) Suppose $v \in (\bigvee E_\kappa)^\perp$. Then $v \perp E_\kappa$ for all κ, so $v \in \bigwedge E_\kappa^\perp$. This shows one containment.

Conversely, suppose $v \in \bigwedge E_\kappa^\perp$. Then v is orthogonal to every element of every E_κ, and hence it is orthogonal to the closed span of all of the E_κ. This verifies the reverse containment.

(c) By (a) and (b),

$$\bigvee E_\kappa^\perp = \left(\bigvee E_\kappa^\perp\right)^{\perp\perp} = \left(\bigwedge E_\kappa^{\perp\perp}\right)^\perp = \left(\bigwedge E_\kappa\right)^\perp,$$

as desired. ∎

We have now shown that the basic identities obeyed by subsets of a set are also obeyed by subspaces of a Hilbert space. However, there is a slightly more complicated set theoretic identity which does not transfer to Hilbert spaces. It is called the distributive law, and it asserts that

$S_1 \cup (S_2 \cap S_3) = (S_1 \cup S_2) \cap (S_1 \cup S_3)$ for any subsets S_1, S_2, S_3 of a set. The following example falsifies this law for Hilbert subspaces. It also falsifies the dual law obtained by interchanging $\cup$ and $\cap$, which is true for sets as well.

Example 2.2.7

Take $\mathcal{H} = \mathbf{C}^2$ and

$$\begin{aligned} E_1 &= \{(a,a) : a \in \mathbf{C}\} \\ E_2 &= \{(a,0) : a \in \mathbf{C}\} \\ E_3 &= \{(0,a) : a \in \mathbf{C}\}. \end{aligned}$$

Then $E_1 \vee (E_2 \wedge E_3) = E_1$ and $(E_1 \vee E_2) \wedge (E_1 \vee E_3) = \mathcal{H}$.

In the case of $l^2(X)$, each subset S of X gives rise to a closed subspace $l^2(S)$ of $l^2(X)$ consisting of those functions whose support is contained in S. It is easy to see that this correspondence between subsets and subspaces takes $\cup$, $\cap$, and c into $\vee$, $\wedge$, and $\perp$. (Thus, Example 2.2.7 hinges on the fact that there are other subspaces in $l^2(\{0,1\})$ besides those of the form $l^2(S)$ for $S \subset \{0,1\}$.) This observation can also be made in the context of L^2 spaces, but in order to state it we need a lemma. This is our first result that uses σ-finiteness.

If S and S' are measurable sets, we say that S essentially contains S' if $S' - S$ is null.

LEMMA 2.2.8

Let (X, μ) be a σ-finite measure space and let $\{S_\kappa : \kappa \in J\}$ be a family of measurable subsets of X. Then there is a measurable set S such that

(a) S essentially contains each S_κ and
(b) any measurable set that essentially contains each S_κ also essentially contains S

and a measurable set S' such that

(a′) S' is essentially contained in each S_κ and
(b′) any measurable set that is essentially contained in each S_κ is also essentially contained in S'.

The sets S and S' are unique up to null sets.

PROOF We prove the first assertion. Assume first that $\mu(X)$ is finite. Since the family $\{S_\kappa\}$ is arbitrary, without loss of generality we can assume it is closed under finite unions. Now define $a = \sup_{\kappa \in J} \mu(S_\kappa)$. Since $\mu(X)$ is finite, so is a.

Fix a sequence (S_{κ_n}) such that $\mu(S_{\kappa_n}) \to a$ and let $S = \bigcup S_{\kappa_n}$. Then S essentially contains every S_κ because if $\mu(S_{\tilde\kappa} - S) > 0$ for some $\tilde\kappa$ then $\mu(S_{\kappa_n} \cup S_{\tilde\kappa})$ converges to $\mu(S \cup S_{\tilde\kappa}) > a$, which contradicts the maximality of a. Also, if T is any set which essentially contains each S_{κ_n} then it essentially contains S.

Now let μ be any σ-finite measure and let (X_n) be a partition of X into finite measure subsets. By the preceding argument, for each n we can find a set $T_n \subset X_n$ which essentially contains $S_\kappa \cap X_n$ for all $\kappa \in J$, and is essentially contained in any other set with this property. Then let $S = \bigcup T_n$. Properties (a) and (b) hold because they hold on each X_n. Furthermore, S is unique up to null sets because any other set T with the same properties would both essentially contain and be essentially contained in S.

The existence and uniqueness of S' can either be proven by analogy with the above, or as a consequence of it via complementation. ∎

We call the sets S and S' of the preceding proposition the essential union and the essential intersection of the family $\{S_\kappa\}$. A more sophisticated proof of their existence involves taking the weak* limit in $L^\infty(X) \cong L^1(X)^*$ of the characteristic functions χ_{S_κ}, assuming the family $\{S_\kappa\}$ is closed under finite unions or intersections and is directed by inclusion or reverse inclusion, respectively.

Example 2.2.9

(a) Let X be a set. Then $l^2(S)$ is a closed subspace of $l^2(X)$ and we have $l^2(S)^\perp = l^2(S^c)$ for every $S \subset X$. Also $\bigvee l^2(S_\kappa) = l^2(\bigcup S_\kappa)$ and $\bigwedge l^2(S_\kappa) = l^2(\bigcap S_\kappa)$ for any family of sets $S_\kappa \subset X$.
(b) Let (X, μ) be a σ-finite measure space. Then $L^2(S)$ is a closed subspace of $L^2(X)$ and we have $L^2(S)^\perp = L^2(S^c)$, for every measurable $S \subset X$. Also $\bigvee L^2(S_\kappa) = L^2(S)$ and $\bigwedge L^2(S_\kappa) = L^2(S')$ for any family of measurable sets $S_\kappa \subset X$, where S and S' are respectively the essential union and the essential intersection of the S_κ.

2.3 *Orthonormal bases*

In Example 2.1.7 we observed that $l^2(X)$ is a Hilbert space. The point of orthonormal bases is to allow us to reverse this construction and realize any Hilbert space in the form $l^2(X)$. We start with the definition of orthonormal bases and a proof that they always exist.

DEFINITION 2.3.1 *A subset Λ of a Hilbert space $\mathcal{H}$ is orthonormal if $\|v\| = 1$ for all $v \in \Lambda$ and $v \perp w$ for all distinct $v, w \in \Lambda$. It is an orthornormal basis if in addition $\overline{\text{span}}(\Lambda) = \mathcal{H}$.*

We will say that Λ generates $\mathcal{H}$ if $\overline{\text{span}}(\Lambda) = \mathcal{H}$. This is generally more useful than the concept of algebraically spanning $\mathcal{H}$.

Observe that orthonormal sets are always linearly independent. To see this suppose Λ is orthonormal and $v_1, \ldots, v_n \in \Lambda$ obey a linear dependence $\sum_1^n a_i v_i = 0$. Then for any $1 \leq j \leq n$ we have

$$a_j = \left\langle \sum_{i=1}^{n} a_i v_i, v_j \right\rangle = 0.$$

So Λ must be independent.

PROPOSITION 2.3.2

Let Λ be an orthonormal set in a Hilbert space $\mathcal{H}$. Then Λ extends to an orthonormal basis.

PROOF By Zorn's lemma, we can find a maximal orthonormal set Λ' containing Λ. We claim that Λ' generates $\mathcal{H}$ (meaning that $\overline{\text{span}}(\Lambda') = \mathcal{H}$; see above). Suppose not and let $E = \overline{\text{span}}(\Lambda')$. Theorem 2.2.5 implies the existence of a nonzero vector v orthogonal to E, and $v/\|v\|$ is then a vector of norm one that is orthogonal to every vector in Λ'. This contradicts maximality of Λ', so we conclude that $\overline{\text{span}}(\Lambda') = \mathcal{H}$. Thus Λ' is an orthonormal basis of $\mathcal{H}$ which contains Λ. ∎

In particular, applying this proposition to the set $\Lambda = \emptyset$ shows that every Hilbert space has an orthonormal basis. Now we can reverse the construction of Example 2.1.7 and show that every Hilbert space is isometrically isomorphic to an l^2 space.

THEOREM 2.3.3

Let $\mathcal{H}$ be a Hilbert space with orthonormal basis $\{e_x : x \in X\}$. Then $\mathcal{H}$ is isometrically isomorphic to $l^2(X)$ by a map which takes e_x to χ_x.

PROOF Define $U : \text{span}\{e_x\} \to l^2(X)$ by

$$U\left(\sum a_x e_x\right) = \sum a_x \chi_x.$$

(Both sums are finite.) This map satisfies $U(e_x) = \chi_x$, and it is clearly linear. Moreover, it preserves inner products because

$$\left\langle \sum a_x e_x, \sum b_x e_x \right\rangle = \sum a_x \bar{b}_x = \left\langle \sum a_x \chi_x, \sum b_x \chi_x \right\rangle.$$

In particular, it preserves norms, so it is an isometry.

U uniquely extends to $\overline{\text{span}}\{e_x\} = \mathcal{H}$ by continuity. It is surjective because its range is closed and contains the orthonormal basis $\{\chi_x\}$ of $l^2(X)$. ∎

COROLLARY 2.3.4

Let $\{e_\kappa : \kappa \in J\}$ be an orthonormal basis of a Hilbert space $\mathcal{H}$ and let $v \in \mathcal{H}$. Then

$$v = \sum_{\kappa \in J} a_\kappa e_\kappa \qquad \text{and} \qquad \|v\|^2 = \sum_{\kappa \in J} |a_\kappa|^2$$

where the coefficients a_κ are given by $a_\kappa = \langle v, e_\kappa \rangle$.

This corollary is an immediate consequence of Theorem 2.3.3 and the fact that it is true of the standard basis in $l^2(J)$.

We define the dimension of a Hilbert space $\mathcal{H}$ to be the cardinality of an orthonormal basis of $\mathcal{H}$. This depends on the fact that any two bases have the same cardinality, which is clear if either one is finite. Otherwise, let Λ and Λ' be orthonormal bases of an infinite dimensional Hilbert space $\mathcal{H}$ and consider the set D of all finite linear combinations $\sum a_\kappa e_\kappa$ where each a_κ belongs to the countable set $\mathbf{Q}+i\mathbf{Q} \subset \mathbf{C}$ and each e_κ belongs to Λ. Now D is dense in $\mathcal{H}$ and the open balls of radius $1/\sqrt{2}$ centered at the elements of Λ' are disjoint, so $\text{card}(D) \geq \text{card}(\Lambda')$. But $\text{card}(D) = \aleph_0 \cdot \text{card}(\Lambda) = \text{card}(\Lambda)$, so we conclude $\text{card}(\Lambda) \geq \text{card}(\Lambda')$. By symmetry, the two are equal.

The following remarkable fact is now a consequence of Theorem 2.3.3.

COROLLARY 2.3.5

Any two Hilbert spaces of the same dimension are isometrically isomorphic.

In particular, any countably infinite dimensional Hilbert space is isometrically isomorphic to $l^2(\mathbf{N})$. As an example we consider L^2 of the unit circle.

Example 2.3.6

Let $\mathbf{T} = \mathbf{R}/2\pi\mathbf{Z}$. The functions $\tilde{e}_n = (2\pi)^{-1/2}e^{inx}$ $(n \in \mathbf{Z})$ form an orthonormal basis of $L^2(\mathbf{T})$: orthonormality is easy to check, and the fact that their span is dense follows from its density in $C(\mathbf{T})$ — a consequence of the Stone-Weierstrass theorem — and the density of $C(\mathbf{T})$ in $L^2(\mathbf{T})$, which is standard measure theory.

Consider the standard orthonormal basis $e_n = \chi_n$ $(n \in \mathbf{Z})$ of $l^2(\mathbf{Z})$. By Theorem 2.3.3 the map $\tilde{\mathcal{F}} : L^2(\mathbf{T}) \to l^2(\mathbf{Z})$ which takes $\tilde{e}_n$ to e_n

is an isometry. In general, if $f \in L^2(\mathbf{T})$ then the nth entry of $\tilde{\mathcal{F}}f$ is $\langle f, \tilde{e}_n \rangle = (2\pi)^{-1/2} \int_0^{2\pi} f e^{-inx} dx$.

The map $\tilde{\mathcal{F}}$ of the preceding example is the Fourier transform on the circle and the values $\langle f, \tilde{e}_n \rangle$ are the Fourier coefficients of f.

In the case of $L^2(\mathbf{R})$ it is a little harder to find a nice orthonormal basis. There is one, however; see Section 4.3.

2.4 Duals and direct sums

So far we have been working inside a single Hilbert space. In this section and the next we will look at three constructions which start with one or more Hilbert spaces and produce a new space as a result. The first construction, dualization, has no set theoretic analog, but the other two have.

DEFINITION 2.4.1 *Let $\mathcal{H}$ be a Hilbert space. Denote its dual space, the set of bounded linear functionals (i.e., bounded linear maps from $\mathcal{H}$ into $\mathbf{C}$) by $\mathcal{H}^*$. For each $v \in \mathcal{H}$ the map $\hat{v} : w \mapsto \langle w, v \rangle$ is bounded by the Cauchy-Schwarz inequality, and hence belongs to $\mathcal{H}^*$. This defines a map $v \mapsto \hat{v}$ from $\mathcal{H}$ to $\mathcal{H}^*$.*

A map $T : \mathcal{V} \to \mathcal{W}$ between vector spaces is antilinear if

$$T(av + bw) = \bar{a}Tv + \bar{b}Tw$$

for all $v, w \in \mathcal{V}$ and $a, b \in \mathbf{C}$. It is an anti-isomorphism if it is antilinear and one-to-one.

PROPOSITION 2.4.2
Any Hilbert space $\mathcal{H}$ is isometrically anti-isomorphic to its dual via the map $v \mapsto \hat{v}$.

PROOF Verifying antilinearity is routine. To see that $\|\hat{v}\| = \|v\|$, observe that for any $w \in \mathcal{H}$ we have

$$|\hat{v}(w)| = |\langle w, v \rangle| \leq \|w\|\|v\|,$$

and hence $\|\hat{v}\| \leq \|v\|$, while

$$|\hat{v}(v)| = |\langle v, v \rangle| = \|v\|^2,$$

and hence $\|\hat{v}\| \geq \|v\|$. This also shows that $v \mapsto \hat{v}$ is isometric.

For surjectivity, let $\omega : \mathcal{H} \to \mathbf{C}$ be any linear functional. The kernel of ω is a closed, codimension one subspace of $\mathcal{H}$, so by Theorem 2.2.5 we have $\mathcal{H} = \ker(\omega) + E$ where E is the one-dimensional orthocomplement of $\ker(\omega)$. Say $E = \mathrm{span}(v_0)$.

Let $v_1 = \bar{a}v_0$ where $a = \omega(v_0)/\|v_0\|^2$; we will show that $\hat{v}_1 = \omega$. For any $w \in \ker(\omega) = E^\perp$ we have $\langle w, v_1\rangle = 0$, so that $\hat{v}_1(w) = 0 = \omega(w)$. Also, a direct computation shows that $\hat{v}_1(v_0) = \omega(v_0)$. Since $\mathcal{H}$ is spanned by $\ker(\omega)$ and v_0 it follows that $\hat{v}_1 = \omega$, as claimed. This proves surjectivity. ∎

It follows that $\mathcal{H}^*$ is a Hilbert space in the sense that its norm is compatible with an inner product. Namely, for $v, w \in \mathcal{H}$ define $\langle \hat{v}, \hat{w}\rangle = \langle w, v\rangle$. Since $v \mapsto \hat{v}$ is a bijection this defines an inner product on $\mathcal{H}^*$. By the preceding result we have in particular that

$$\langle \hat{v}, \hat{v}\rangle = \langle v, v\rangle = \|v\|^2 = \|\hat{v}\|^2$$

for all $v \in \mathcal{H}$, so this inner product does give rise to the original norm on $\mathcal{H}^*$.

Next we consider the direct sum construction.

DEFINITION 2.4.3 *Let $\{\mathcal{H}_\kappa : \kappa \in J\}$ be a family of Hilbert spaces. Define the* <u>*direct sum*</u> *$\bigoplus \mathcal{H}_\kappa$ to be the set of all sequences (v_κ), also denoted $\bigoplus v_\kappa$, such that $v_\kappa \in \mathcal{H}_\kappa$ for all κ and $\sum \|v_\kappa\|^2 < \infty$. Give it the inner product*

$$\left\langle \bigoplus v_\kappa, \bigoplus w_\kappa \right\rangle = \sum_{\kappa\in J} \langle v_\kappa, w_\kappa\rangle.$$

This sum converges because $|\langle v, w\rangle| \leq \|v\|\|w\| \leq \frac{1}{2}(\|v\|^2 + \|w\|^2)$, and an easy calculation shows that it does define an inner product.

The structure of direct sums is exhibited in the next result.

PROPOSITION 2.4.4
Let $\{\mathcal{H}_\kappa : \kappa \in J\}$ be a family of Hilbert spaces. Then $\bigoplus \mathcal{H}_\kappa$ is a Hilbert space. Each $\mathcal{H}_\kappa$ naturally embeds in the direct sum, and if Λ_κ is an orthonormal basis of $\mathcal{H}_\kappa$ for each $\kappa \in J$ then $\bigcup \Lambda_\kappa$ is an orthonormal basis of $\bigoplus \mathcal{H}_\kappa$.

Completeness of the direct sum is verified by an argument similar to the one used in Example 2.1.7. This is no accident; the direct sum construction reduces to $l^2(J)$ when each $\mathcal{H}_\kappa$ is one-dimensional. The rest of the proposition is an easy consequence of Theorem 2.3.3 in the following way. For each κ let Λ_κ be an orthonormal basis of $\mathcal{H}_\kappa$. Then we may identify $\mathcal{H}_\kappa$ with $l^2(\Lambda_\kappa)$, and an elementary calculation shows that $\bigoplus \mathcal{H}_\kappa$ is correspondingly identified with $l^2(\bigcup \Lambda_\kappa)$.

Evidently direct sums of Hilbert spaces correspond to disjoint unions of sets. The same is true for measure spaces:

Example 2.4.5

(a) Let (X_κ) be a family of sets and let X be their disjoint union. Then $l^2(X) \cong \bigoplus l^2(X_\kappa)$.

(b) Let (X_n, μ_n) be a countable family of σ-finite measure spaces and let (X, μ) be their disjoint union. Then μ is also σ-finite, and $L^2(X) \cong \bigoplus L^2(X_n)$.

There is also a measurable version of direct summation. The situation here involves a family of Hilbert spaces $\{\mathcal{H}_x : x \in X\}$ where X is a σ-finite measure space. If each $\mathcal{H}_x$ is separable (i.e., of either finite or countably infinite dimension) then we can partition X into a countable family of subsets X_n $(0 \leq n \leq \infty)$ on which $\dim(\mathcal{H}_x) = n$, perform the construction on each X_n separately, and direct sum the result.

This reduces the problem to the case where the dimension of the spaces $\mathcal{H}_x$ is constant. Since Hilbert spaces of the same dimension are isomorphic, we need only consider the problem of taking a "measurable direct sum" of a single Hilbert space $\mathcal{H}$ over a measure space X. The following definition indicates how this is done.

DEFINITION 2.4.6 *Let μ be a σ-finite measure on a set X and let $\mathcal{H}$ be a separable Hilbert space. A function $f : X \to \mathcal{H}$ is* <u>weakly measurable</u> *if the function $x \mapsto \langle f(x), v\rangle$ is measurable for each $v \in \mathcal{H}$. Then $L^2(X; \mathcal{H})$ is the set of all weakly measurable functions $f : X \to \mathcal{H}$ such that*

$$\|f\|^2 = \int_X \|f(x)\|^2 < \infty,$$

modulo functions which are zero almost everywhere. An inner product on $L^2(X; \mathcal{H})$ is given by

$$\langle f, g\rangle = \int_X \langle f(x), g(x)\rangle.$$

The preceding integral exists because

$$2|\langle f(x), g(x)\rangle| \leq 2\|f(x)\|\|g(x)\| \leq \|f(x)\|^2 + \|g(x)\|^2$$

pointwise almost everywhere, and completeness of $L^2(X; \mathcal{H})$ can be verified using Lemma 2.1.8, just as for ordinary L^2 spaces (cf. Example 2.1.9).

The following proposition gives an alternative characterization of the space $L^2(X;\mathcal{H})$. Yet another characterization will be given in Example 2.5.4 (c).

PROPOSITION 2.4.7
Let (X,μ) be a σ-finite measure space and let $\mathcal{H}$ be a separable Hilbert space of dimension n ($1 \le n \le \infty$). Then $L^2(X;\mathcal{H})$ is isomorphic to $\bigoplus_{i=1}^{n} L^2(X)$.

PROOF We assume $n = \infty$; the finite dimensional case is similar but easier.

First we define a linear isometry $U : \bigoplus_1^\infty L^2(X) \to L^2(X;\mathcal{H})$. Let $\{e_i : i \in \mathbf{N}\}$ be an orthonormal basis of $\mathcal{H}$. For any sequence of functions $f_i \in L^2(X)$, all but finitely many of which are zero, define

$$U\left(\bigoplus f_i\right)(x) = \sum_{i=1}^{\infty} f_i(x) \cdot e_i.$$

The right side is clearly measurable, and we have

$$\left\|\sum f_i \cdot e_i\right\|^2 = \sum \|f_i\|^2 = \left\|\bigoplus f_i\right\|^2,$$

so U is an isometry on its domain of definition. But this domain is dense in $\bigoplus L^2(X)$, so U extends by continuity to the entire space.

For surjectivity, let $f \in L^2(X;\mathcal{H})$ and define $f_i(x) = \langle f(x), e_i\rangle$ for each i; then $f = \sum f_i \cdot e_i$ and each term of the sum belongs to the range of U. But since U is an isometry, its range is closed. This shows that U is surjective. ∎

We can now formalize the notion of a measurable direct sum. This construction will be central to Chapter 3, and it will remain important in Chapters 5 and 6. We will develop it further in Section 9.2.

DEFINITION 2.4.8 *Let X be a σ-finite measure space. A* <u>(separable) measurable Hilbert bundle</u> *over X is a disjoint union*

$$\mathcal{X} = \bigcup (X_n \times \mathcal{H}_n)$$

where $\{X_n\}$ is a measurable partition of X and $\mathcal{H}_n$ is a Hilbert space of dimension n, $0 \le n \le \infty$.

The <u>Hilbert space of L^2 sections</u> *of $\mathcal{X}$ is the direct sum $L^2(X;\mathcal{X}) = \bigoplus L^2(X_n;\mathcal{H}_n)$. Equivalently, it is the set of weakly measurable functions*

$f : X \to \bigcup \mathcal{H}_n$ *with the properties that* $f(x) \in \mathcal{H}_n$ *for all* $x \in X_n$ *and* $\int \|f(x)\|^2 < \infty$.

We will discuss continuous Hilbert bundles in Section 9.1.

If the fiber spaces $\mathcal{H}_n$ in Definition 2.4.8 are not assumed to be separable, then the measure theoretic issues become somewhat murky. In this setting the bundle construction seems to break down, but there is a corresponding module approach which is still workable; see Chapter 9, especially Theorem 9.4.11, which has no separability assumptions.

In any case, we will always take our measurable Hilbert bundles to be separable.

2.5 Tensor products

Our last Hilbert space construction, the tensor product, is the most sophisticated. Tensor products of two (or finitely many) Hilbert spaces work like this. For any $v \in \mathcal{H}$ and $w \in \mathcal{K}$ there is a corresponding vector $v \otimes w \in \mathcal{H} \otimes \mathcal{K}$, and vectors of this form generate $\mathcal{H} \otimes \mathcal{K}$. The inner product of two such vectors is given by $\langle v \otimes w, v' \otimes w' \rangle = \langle v, v' \rangle \langle w, w' \rangle$. In particular, the norm of $v \otimes w$ is $\|v\|\|w\|$. If $\{e_\kappa\}$ and $\{\tilde{e}_{\tilde{\kappa}}\}$ are orthonormal bases of $\mathcal{H}$ and $\mathcal{K}$, respectively, then $\{e_\kappa \otimes \tilde{e}_{\tilde{\kappa}}\}$ is an orthonormal basis of $\mathcal{H} \otimes \mathcal{K}$.

In the case of infinitely many factors there are two competing definitions of tensor products, both of which reduce to the preceding construction when the number of factors is finite. The first takes $\bigotimes \mathcal{H}_\kappa$ to be generated by vectors of the form $\bigotimes v_\kappa$ where each $v_\kappa \in \mathcal{H}_\kappa$ and the product $\prod \|v_\kappa\|^2$ — which will be the norm of $\bigotimes v_\kappa$ — converges. The result is a Hilbert space for which the vectors $\bigotimes v_\kappa$ with $v_\kappa \in \Lambda_\kappa$ for all κ form an orthonormal basis, if each Λ_κ is an orthonormal basis of $\mathcal{H}_\kappa$.

This tensor product has had little application, probably because it is essentially never separable. The other definition, which we adopt, requires the choice of a distinguished unit vector u_κ in each $\mathcal{H}_\kappa$; this vector plays the role of a sort of zero or "ground state."

DEFINITION 2.5.1 *Let* $\{\mathcal{H}_\kappa : \kappa \in J\}$ *be a family of Hilbert spaces and for each* κ *fix a unit vector* $u_\kappa \in \mathcal{H}_\kappa$. *Consider the sequences* (v_κ) *such that each* v_κ *belongs to* $\mathcal{H}_\kappa$ *and* $v_\kappa = u_\kappa$ *for all but finitely many values of* κ.

Let $\mathcal{V}$ *be a vector space with a basis* $\{e_{(v_\kappa)}\}$ *indexed by all such sequences* (v_κ). *Give* $\mathcal{V}$ *the pseudo inner product defined by*

$$\langle e_{(v_\kappa)}, e_{(w_\kappa)} \rangle = \prod \langle v_\kappa, w_\kappa \rangle,$$

extending linearly. (All but finitely many terms of this product are 1, so it is well-defined.) Then the tensor product $\bigotimes \mathcal{H}_\kappa$ is the Hilbert space formed by factoring out null vectors and completing, as in Propositions 2.1.4 and 2.1.5.

We write $\bigotimes v_\kappa$ for the equivalence class of the element $e_{(v_\kappa)}$ in $\bigotimes \mathcal{H}_\kappa$. Observe that $\langle \bigotimes v_\kappa, \bigotimes w_\kappa \rangle = \prod \langle v_\kappa, w_\kappa \rangle$. Also notice that if the index set J is finite, the unit vectors u_κ are irrelevant.

The following two propositions give the basic properties and essential structure of tensor products.

PROPOSITION 2.5.2
Let $\mathcal{H}$ and $\mathcal{K}$ be Hilbert spaces. Then

(a) $av \otimes w = v \otimes aw = a(v \otimes w)$ and
(b) $(v_1 + v_2) \otimes w = v_1 \otimes w + v_2 \otimes w$

for all $a \in \mathbf{C}$, $v, v_1, v_2 \in \mathcal{H}$, and $w \in \mathcal{K}$.

PROOF
(a) By direct computation,

$$\|av \otimes w - a(v \otimes w)\|^2 = 2\|av\|^2\|w\|^2 - 2\mathrm{Re}\,\bar{a}\langle av, v\rangle\langle w, w\rangle = 0.$$

Thus $av \otimes w = a(v \otimes w)$, and $v \otimes aw = a(v \otimes w)$ similarly.

(b) Taking the inner product of $(v_1 + v_2) \otimes w - (v_1 \otimes w + v_2 \otimes w)$ with itself yields

$$\langle (v_1 + v_2) - v_1 - v_2, (v_1 + v_2) - v_1 - v_2\rangle\langle w, w\rangle = 0.$$

So $(v_1 + v_2) \otimes w = v_1 \otimes w + v_2 \otimes w$. ∎

PROPOSITION 2.5.3
Let $\{\mathcal{H}_\kappa : \kappa \in J\}$ be a family of Hilbert spaces with distinguished unit vectors u_κ and let $\mathcal{H} = \bigotimes \mathcal{H}_\kappa$ be their tensor product. For each κ let Λ_κ be an orthonormal basis of $\mathcal{H}_\kappa$ which contains u_κ. Then the set

$$\Lambda = \{\bigotimes e_\kappa : \text{ each } e_\kappa \in \Lambda_\kappa, \text{ and } e_\kappa = u_\kappa \text{ for all but finitely many } \kappa\}$$

is an orthonormal basis of $\bigotimes \mathcal{H}_\kappa$.

PROOF It is easy to check that Λ is orthonormal. To show that it generates $\bigotimes \mathcal{H}_\kappa$, we must verify that $\bigotimes v_\kappa$ is in the closure of its span whenever (v_κ) is a sequence such that $v_\kappa \in \mathcal{H}_\kappa$ for all κ and $v_\kappa = u_\kappa$ for all but finitely many κ. This reduces the problem to the case of a tensor product of finitely many Hilbert spaces, and by induction we reduce to the case of two Hilbert spaces.

Thus let $\mathcal{H}$ and $\mathcal{K}$ be Hilbert spaces and let $\{e_\kappa\}$ and $\{\tilde{e}_{\tilde{\kappa}}\}$ be respective orthonormal bases. Let $v \otimes w \in \mathcal{H} \otimes \mathcal{K}$ and given $\epsilon > 0$ choose $v' \in \operatorname{span}\{e_\kappa\}$ and $w' \in \operatorname{span}\{\tilde{e}_{\tilde{\kappa}}\}$ such that $\|v - v'\|, \|w - w'\| \leq \epsilon$. Then

$$\begin{aligned} \|v \otimes w - v' \otimes w'\| &\leq \|v \otimes w - v' \otimes w\| + \|v' \otimes w - v' \otimes w'\| \\ &= \|v - v'\|\|w\| + \|v'\|\|w - w'\| \\ &\leq (\|v'\| + \|w\|)\,\epsilon \\ &\leq (\|v\| + \|w\| + \epsilon)\,\epsilon. \end{aligned}$$

This shows that $v \otimes w$ is in the closure of $\operatorname{span}\{e_\kappa \otimes \tilde{e}_{\tilde{\kappa}}\}$, and it follows that $\operatorname{span}\{e_\kappa \otimes \tilde{e}_{\tilde{\kappa}}\}$ is dense in $\mathcal{H} \otimes \mathcal{K}$, as desired. ∎

As with the direct sum (Proposition 2.4.4), we can embed each factor $\mathcal{H}_{\kappa_0}$ into the tensor product $\bigotimes \mathcal{H}_\kappa$. In this case the natural embedding is the map $v \mapsto \bigotimes v_\kappa$ where $v_{\kappa_0} = v$ and $v_\kappa = u_\kappa$ for $\kappa \neq \kappa_0$.

The set theoretic analog of a tensor product of two Hilbert spaces is a cartesian product of two sets. Indeed, $l^2(X) \otimes l^2(Y) \cong l^2(X \times Y)$. This follows from the last proposition since the natural basis of $l^2(X \times Y)$ can be identified with the product of the natural bases of $l^2(X)$ and $l^2(Y)$. We can also compute some more complicated examples.

Example 2.5.4

(a) Let X and Y be sets. Then $l^2(X \times Y) \cong l^2(X) \otimes l^2(Y)$.
(b) Let X and Y be σ-finite measure spaces. Then $L^2(X \times Y)$ is isometrically isomorphic to $L^2(X) \otimes L^2(Y)$ via the identification of $f \otimes g \in L^2(X) \otimes L^2(Y)$ with the function $f(x)g(y) \in L^2(X \times Y)$.
(c) Let X be a σ-finite measure space and let $\mathcal{H}$ be a Hilbert space. Then $L^2(X; \mathcal{H})$ is naturally isomorphic to $L^2(X) \otimes \mathcal{H}$. This follows from Propositions 2.4.7 and 2.5.3.

Tensor powers — tensor products in which every factor is the same — can be symmetrized or antisymmetrized, and we turn to this topic now. First we give the abstract definitions, and then we describe their interpretation in terms of an orthonormal basis.

DEFINITION 2.5.5 *Let $\mathcal{H}$ be a Hilbert space, let $n \geq 1$, and let $\mathcal{H}^{\otimes n} = \mathcal{H} \otimes \cdots \otimes \mathcal{H}$ be the n-fold tensor power of $\mathcal{H}$.*

(a) Define the <u>symmetrization</u> of $v_1 \otimes \cdots \otimes v_n \in \mathcal{H}^{\otimes n}$ to be the vector

$$[v_1 \otimes \cdots \otimes v_n]_s = \frac{1}{n!} \sum_\sigma v_{\sigma(1) \otimes \cdots \otimes \sigma(n)},$$

where the sum is taken over all permutations σ of the set $\{1,\ldots,n\}$. The n-fold symmetric tensor power *of $\mathcal{H}$ is the closed linear span, in $\mathcal{H}^{\otimes n}$, of the vectors $[v_1 \otimes \cdots \otimes v_n]_s$ with $v_1,\ldots,v_n \in \mathcal{H}$. It is denoted $\mathcal{H}_s^{\otimes n}$ or $\mathcal{H} \otimes_s \cdots \otimes_s \mathcal{H}$.*

(b) Define the antisymmetrization *of $v_1 \otimes \cdots \otimes v_n \in \mathcal{H}^{\otimes n}$ to be the vector*

$$[v_1 \otimes \cdots \otimes v_n]_a = \frac{1}{n!} \sum_{\sigma} (-1)^{|\sigma|} v_{\sigma(1)\otimes\cdots\otimes\sigma(n)},$$

where $|\sigma|$ is the parity of σ and the sum is again taken over all permutations of the set $\{1,\ldots,n\}$. The n-fold antisymmetric tensor power *of $\mathcal{H}$ is the closed linear span, in $\mathcal{H}^{\otimes n}$, of the vectors $[v_1 \otimes \cdots \otimes v_n]_a$ with $v_1,\ldots,v_n \in \mathcal{H}$. It is denoted $\mathcal{H}_a^{\otimes n}$ or $\mathcal{H} \otimes_a \cdots \otimes_a \mathcal{H}$.*

PROPOSITION 2.5.6

Let $\mathcal{H}$ be a separable Hilbert space, let $\{e_i\}$ be an orthonormal basis of $\mathcal{H}$, and let $n \geq 1$. Then

(a) the vectors

$$[e_{i_1} \otimes \cdots \otimes e_{i_n}]_s$$

with $i_1 \leq \cdots \leq i_n$ form an orthonormal basis of $\mathcal{H}_s^{\otimes n}$ without repetition, and

(b) the vectors

$$[e_{i_1} \otimes \cdots \otimes e_{i_n}]_a$$

with $i_1 < \cdots < i_n$ form an orthonormal basis of $\mathcal{H}_a^{\otimes n}$ without repetition.

For example, if $\mathcal{H} = \mathbf{C}^2$ and $\{e_1, e_2\}$ is the standard basis, then $\mathcal{H}\otimes_s\mathcal{H}$ has a basis consisting of the three vectors $[e_1 \otimes e_1]_s$, $[e_1 \otimes e_2]_s$, and $[e_2 \otimes e_2]_s$, while $\mathcal{H}\otimes_a\mathcal{H}$ has a basis consisting of the single vector $[e_1 \otimes e_2]_a$. (Note: in this case $\mathcal{H}^{\otimes 2} = \mathcal{H}_s^{\otimes 2} \oplus \mathcal{H}_a^{\otimes 2}$, but the analogous statement is not true if $\dim(\mathcal{H}) > 2$.)

The proof of Proposition 2.5.6 is left to the reader. If $\dim(\mathcal{H}) = m$ is finite, a little combinatorics shows that the dimension of $\mathcal{H}_s^{\otimes n}$ is $\binom{m+n-1}{n}$ and the dimension of $\mathcal{H}_a^{\otimes n}$ is $\binom{m}{n}$.

In general there is no good way to define measurable tensor products. But in the special cases $\mathcal{H} = l^2(\mathbf{N})$ and $\mathcal{H} = \mathbf{C}^2$, we can use the preceding constructions to express discrete tensor powers in a way which suggests a definition of measurable tensor powers.

Example 2.5.7

(a) Let X be a countable set, let $\mathcal{H}_x = l^2(\mathbf{N})$ for $x \in X$, and take $u_x = e_0$ for all x, where $\{e_k\}$ is the canonical basis of $l^2(\mathbf{N})$. Then the tensor power $\bigotimes \mathcal{H}_x$ has a natural basis consisting of vectors of the form $\bigotimes e_{k_x}$ with each $k_x \in \mathbf{N}$ and $k_x = 0$ for all but finitely many x.

For such a sequence $k = (k_x)$, write $|k| = \sum k_x$. Then the set of all sequences with $|k| = n$ may be identified with the set of all n-element subsets of X, allowing multiplicity. These are naturally in one-to-one correspondence with the orthonormal basis of the n-fold symmetric tensor power of $l^2(X)$ described in Proposition 2.5.6, so we have a natural isomorphism between $\bigotimes_{x \in X} \mathcal{H}_x$ and the direct sum, over $n \in \mathbf{N}$, of the n-fold symmetric tensor powers of $l^2(X)$.

(b) Let X be a countable set, let $\mathcal{K}_x = \mathbf{C}^2$ for $x \in X$, and let $\{e_1, e_2\}$ be the canonical basis of $\mathbf{C}^2$. Taking $u_x = e_1$ for all x, the tensor power $\bigotimes \mathcal{K}_x$ has a basis consisting of vectors of the form $\bigotimes v_x$ with $v_x = e_2$ for finitely many $x \in X$ and $v_x = e_1$ for all other x. There is a natural bijection between this basis of $\bigotimes \mathcal{K}_x$ and the collection of all finite subsets of X, not allowing multiplicity.

But the set of n-element subsets of X is naturally in one-to-one correspondence with the orthonormal basis of the n-fold antisymmetric tensor power of $l^2(X)$ described in Proposition 2.5.6. Thus, there is a natural isomorphism between $\bigotimes_{x \in X} \mathcal{K}_x$ and the direct sum, over $n \in \mathbf{N}$, of the n-fold antisymmetric tensor powers of $l^2(X)$.

The requirement that X be countable in the preceding example is not really necessary.

This example motivates the following definition.

DEFINITION 2.5.8 *Let $\mathcal{H}$ be a Hilbert space.*

(a) The <u>*symmetric (Boson) Fock space*</u> *over $\mathcal{H}$ is the direct sum*

$$\mathcal{F}_s\mathcal{H} = \mathbf{C} \oplus \mathcal{H} \oplus \mathcal{H}_s^{\otimes 2} \oplus \cdots$$

of the n-fold symmetric tensor powers of $\mathcal{H}$, for $n \in \mathbf{N}$.

(b) The <u>*antisymmetric (Fermion) Fock space*</u> *over $\mathcal{H}$ is the direct sum*

$$\mathcal{F}_a\mathcal{H} = \mathbf{C} \oplus \mathcal{H} \oplus \mathcal{H}_a^{\otimes 2} \oplus \cdots$$

of the n-fold antisymmetric tensor powers of $\mathcal{H}$, for $n \in \mathbf{N}$.

In light of Example 2.5.7, if X is a σ-finite measure space then we regard $\mathcal{F}_s L^2(X)$ as a measurable tensor product of the Hilbert spaces $l^2(\mathbf{N})$ over the index set X, and we regard $\mathcal{F}_a L^2(X)$ as a measurable tensor product of the Hilbert spaces $\mathbf{C}^2$ over the index set X. These spaces will be used in Section 7.1. "Boson" and "Fermion" are physical terms; their significance was indicated in Section 1.5.

2.6 Quantum logic

Quantum logic was the first fully articulated quantum version of a classical mathematical subject.

Classical propositional logic can be understood in the following way. Let X be a set, which we think of as the set of all possible states of some (not necessarily physical) system. Any definite proposition p about the system will be true in some states and not in others, so we have a one-to-one correspondence between the subsets of X and the propositions about the system, up to equivalent propositions.

Under this correspondence the logical operators $\vee$, $\wedge$, and $\neg$ (and, or, and not) correspond to the set theoretic operators $\cup$, $\cap$, and c (union, intersection, complement). For instance, the states in which $p \wedge q$ holds are precisely the states in which p and q both hold, which is to say the intersection of the set of states which satisfy p and the set of states which satisfy q. In this way we can convert propositional logic — the logic of propositions — into set theory.

The quantum mechanical analog of this formulation of propositional logic replaces the set X with a Hilbert space $\mathcal{H}$ and models propositions by closed subspaces of $\mathcal{H}$. This purely formal analog has a natural physical interpretation if $\mathcal{H}$ actually is the Hilbert space of some quantum system. Then any closed subspace E of $\mathcal{H}$ represents a proposition about the system, namely, the proposition that the state vector lies in E. In contrast to the classical case, one cannot in general say with certainty whether a given state vector satisfies a given proposition (as one can say whether an element lies in a subset), but can only give the probability that it will do so. Conversely, to any proposition p about the system one can associate the closed subspace generated by the state vectors which satisfy p *with full probability.*

The logical interpretation of $\neg$ and $\wedge$ is straightforward. A unit vector v belongs to $E^{\perp}$ precisely if, with full probability, v fails to satisfy the proposition associated to E. And $v \in E_1 \wedge E_2$ precisely if v, with full probability, satisfies both the proposition associated to E_1 and the proposition associated to E_2. The disjunction $\vee$ is not so easily interpreted, and is best understood as a non-primitive symbol which is defined in terms of $\neg$ and $\wedge$ by the law $p \vee q = \neg(\neg p \wedge \neg q)$.

Quantum logic, in the above sense, is usually understood as a many-valued logic, with the possible truth-values being the closed subspaces of $\mathcal{H}$. But the intuition behind this interpretation is unclear. It makes more sense to think of quantum logic as a many-valued logic for which the truth-values are probabilities, values in the interval $[0, 1]$. As in the classical case, although it is the case that every proposition p corresponds to a subspace E of $\mathcal{H}$, the truth-value of p only makes sense *relative to a given state v of the underlying system*, and the only sensible value this can be is the probability that an experiment testing p will produce a positive result for the state v.

On the basis of this point of view it is possible to formulate a quantum version of the predicate calculus. This is the logical setting that involves variables, relations, and quantification, in which all ordinary mathematics is done.

In the predicate calculus, a well-formed formula (wf.) is built up from atomic wfs, which are of the form $r(x_1, \ldots, x_n)$. Here r stands for some relation and $x_1, \ldots, x_n$ are variables (not necessarily distinct). If A and B are wfs then so are $\neg A$, $A \vee B$, $A \wedge B$, and $(\exists x)A$ and $(\forall x)A$ for any variable x. Every wf. can be built up from atomic wfs in this way.

Classically, the truth of a given wf. can be determined only when the relation and variable symbols are given some concrete interpretation. This means that a set X is fixed, and to each variable symbol x we must assign an element of X and to each relation symbol r which takes n variables we must assign a subset of X^n. The truth-value of the wf. can then be determined inductively in a straightforward way.

The quantum analog of the predicate calculus involves the same language, and the same notion of a well-formed formula, but interprets these formulas in a Hilbert space $\mathcal{H}$ rather than a set X. Now the variable symbols x are interpreted as unit vectors in $\mathcal{H}$, and the n-place relation symbols r are interpreted as closed subspaces of $\mathcal{H}^{\otimes n}$.

In order to determine the truth-value of a wf. in a given Hilbert space interpretation, we initially ignore the unit vectors corresponding to the variables and inductively associate a closed subspace of a tensor power of $\mathcal{H}$ to each subformula of the wf. At the end of this inductive process, when we have a closed subspace E of $\mathcal{H}^{\otimes n}$ which corresponds to the entire wf., we define its truth-value in the given interpretation to be (in the language of Section 1.2) the probability that $v_1 \otimes \cdots \otimes v_n$ satisfies the event E, where v_i is the unit vector in $\mathcal{H}$ which corresponds to the variable x_i.

For the most part this procedure can be carried out without difficulty. There are two problematic points, however. The first is the issue of repeated variables. If $r(x, y)$ is a two-place relation then it is interpreted as a closed subspace E of $\mathcal{H} \otimes \mathcal{H}$. How then to interpret $r(x, x)$? The most natural solution is to interpret it as $E \cap (\mathcal{H} \otimes_s \mathcal{H})$. Thus, we treat repeated variables as distinct, but use the symmetric tensor product rather than the full tensor product.

The second issue is how to interpret quantification. Suppose r corresponds to the closed subspace E of $\mathcal{H}^{\otimes n} = \mathcal{H} \otimes \mathcal{H}'$, where $\mathcal{H}' = \mathcal{H}^{\otimes(n-1)}$, and the variable x corresponds to the first copy of $\mathcal{H}$. Then we let $(\forall x)r$ correspond to the set of vectors $w \in \mathcal{H}'$ such that $v \otimes w \in E$ for all $v \in \mathcal{H}$. This is indeed a closed subspace. Having dealt with universal quantification, we can then handle existential quantification by means of the equation $(\exists x)r = \neg(\forall x)\neg r$.

Alternatively, we can interpret $(\exists x)r$ independently and prove the preceding equality as a theorem. This is worth discussing because intuitively, existential quantification involves projection from a tensor product onto one of its factors. In general there is no natural projection map from $\mathcal{H} \otimes \mathcal{K}$ onto $\mathcal{H}$; however, given a subspace E of $\mathcal{H} \otimes \mathcal{K}$ there is a subspace $\pi(E)$ of $\mathcal{K}$ which can naturally be considered its projection. Namely, there exists a closed subspace of $\mathcal{K}$ whose tensor product with $\mathcal{H}$ contains E and which is the smallest subspace with this property. Then if r corresponds to the closed subspace E of $\mathcal{H}^{\otimes n}$, we let $(\exists x)r$ correspond to $\pi(E)$.

To prove the preceding assertion, for each $u \in \mathcal{K}$ define a map $\pi_u : \mathcal{H} \otimes \mathcal{K} \to \mathcal{H}$ by setting $\pi_u(v \otimes w) = \langle w, u\rangle v$, and for each $u \in \mathcal{H}$ define $\pi'_u : \mathcal{H} \otimes \mathcal{K} \to \mathcal{K}$ similarly.

LEMMA 2.6.1

Let $\mathcal{H}$ and $\mathcal{K}$ be Hilbert spaces and let $v_0 \in \mathcal{H} \otimes \mathcal{K}$. Then there exist closed subspaces E_1 of $\mathcal{H}$ and E_1' of $\mathcal{K}$ with $v_0 \in E_1 \otimes E_1'$ and such that $E_1 \subset E_2$ and $E_1' \subset E_2'$ whenever E_2 and E_2' are closed subspaces of $\mathcal{H}$ and $\mathcal{K}$ with $v_0 \in E_2 \otimes E_2'$.

PROOF Let E_1 be the closure of $\{\pi_u(v_0) : u \in \mathcal{K}\}$ and let E_1' be the closure of $\{\pi'_u(v_0) : u \in \mathcal{H}\}$. Then E_1 and E_1' are closed subspaces of $\mathcal{H}$ and $\mathcal{K}$, respectively.

Choose an orthonormal basis (e_κ) of $\mathcal{H}$ which contains a basis of E_1 and an orthonormal basis $(\tilde{e}_{\tilde{\kappa}})$ of $\mathcal{K}$ which contains a basis of E_1'. Write $v_0 = \sum_{\kappa,\tilde{\kappa}} a_{\kappa\tilde{\kappa}}(e_\kappa \otimes \tilde{e}_{\tilde{\kappa}})$. Then, considering the map $\pi_{\tilde{e}_{\tilde{\kappa}}} : v_0 \mapsto \sum_\kappa a_{\kappa\tilde{\kappa}} e_\kappa$, we see that $a_{\kappa\tilde{\kappa}} = 0$ whenever $e_\kappa \notin E_1$. Similarly, $a_{\kappa\tilde{\kappa}} = 0$ whenever $\tilde{e}_{\tilde{\kappa}} \notin E_1'$. Thus $v_0 \in E_1 \otimes E_1'$.

Now suppose E_2 and E_2' are closed subspaces of $\mathcal{H}$ and $\mathcal{K}$ such that $v_0 \in E_2 \otimes E_2'$. For any $u \in \mathcal{K}$ we claim that $\pi_u(E_2 \otimes E_2') \subset E_2$. It is clear that $\pi_u(v \otimes w) = \langle w, u\rangle v \in E_2$ for all $v \in E_2$ and $w \in E_2'$, and since vectors of the form $v \otimes w$ generate $E_2 \otimes E_2'$, the claim follows by linearity and continuity of π_u. Thus since $v_0 \in E_2 \otimes E_2'$, it follows that $\pi_u(v_0) \in E_2$ for all $u \in \mathcal{K}$, and hence $E_1 \subset E_2$. Similarly, $E_1' \subset E_2'$. ∎

PROPOSITION 2.6.2

Let $\mathcal{H}$ and $\mathcal{K}$ be Hilbert spaces and let E be a closed subspace of $\mathcal{H} \otimes \mathcal{K}$. Then there exist closed subspaces E_1 of $\mathcal{H}$ and E_1' of $\mathcal{K}$ with $E \subset E_1 \otimes E_1'$ and such that $E_1 \subset E_2$ and $E_1' \subset E_2'$ whenever E_2 and E_2' are closed subspaces of $\mathcal{H}$ and $\mathcal{K}$ with $E \subset E_2 \otimes E_2'$.

PROOF For each $v_0 \in E$ let E_{v_0} and E'_{v_0} be the subspaces described by the lemma, i.e., which are minimal among those whose tensor product contains v_0. Then simply let E_1 and E'_1 be the joins of the E_{v_0} and E'_{v_0}, respectively, as v_0 ranges over all elements of E. ∎

We observe that if $\mathcal{H} = l^2(X)$ and $\mathcal{K} = l^2(Y)$ for some sets X and Y, and if $R \subset X \times Y$, then we have $\pi(l^2(R)) = l^2(\tilde{\pi}(R))$, where π is the projection of subspaces into $\mathcal{H}$ defined above and $\tilde{\pi} : X \times Y \to X$ is the usual projection of sets. To verify $\subset$, observe that $l^2(\tilde{\pi}(R)) \otimes \mathcal{K}$ contains $l^2(R)$. To verify $\supset$ note that for any $(x, y) \in R$ we have $\pi_{e_y}(e_x \otimes e_y) = e_x$, i.e., every element in a basis for $l^2(\tilde{\pi}(R))$ is of the form $\pi_u(v)$ for some $u \in \mathcal{K}$ and $v \in l^2(R)$. Thus we get $\supset$ by the construction of the projection of subspaces. This shows that our quantum mechanical interpretation of existential quantification is a valid analog of classical existential quantification.

The above is only a sketch of how to determine the truth-value of a wf. relative to a Hilbert space interpretation, but it contains all of the essential ideas. One can show that the l^2 version of the classical interpretation of any wf. will give rise, by this procedure, to the same truth-value as in the classical case. However, there will generally be other Hilbert space interpretations which do not arise from classical structures in this way.

2.7 Notes

Most of the material in this chapter can be found in many standard texts; [8], [13], [34], and [58] are all good for further reading.

The quantum propositional calculus was introduced in [7]. See also [37]. The quantum predicate calculus will be treated more thoroughly in [75].

Chapter 3

Operators

3.1 Unitaries and projections

In this chapter we discuss linear operators on Hilbert spaces. The central result is the spectral theorem, which completely describes the structure of the most important kinds of operators. Throughout the chapter, $\mathcal{H}$ will be a Hilbert space; in Sections 3.5 – 3.8 it will be separable.

DEFINITION 3.1.1 *Let $\mathcal{H}$ be a Hilbert space. $B(\mathcal{H})$ will denote the space of all bounded linear operators $A : \mathcal{H} \to \mathcal{H}$.*

The norm of an operator is given by $\|A\| = \sup\{\|Av\| : \|v\| \leq 1\}$. It is easy to check that the operator norm is complete; thus $B(\mathcal{H})$ is a Banach space. If $\dim(\mathcal{H}) = n$ is finite, then we can identify $B(\mathcal{H})$ with the space $M_n(\mathbf{C})$ of $n \times n$ complex matrices, as in Chapter 1. However, this singles out a preferred basis, which may only complicate matters.

In addition to adding and scaling operators, we can multiply them in the sense of composition and also take their adjoints. In the context of $M_n(\mathbf{C})$ the adjoint is the Hermitian transpose matrix, but it can be defined for general Hilbert spaces in a basis-independent manner. We do this first, and then prove it agrees with the Hermitian transpose of matrices in Corollary 3.1.4.

Any bounded linear map $A : \mathcal{V} \to \mathcal{W}$ between Banach spaces has a natural "adjoint" map going from $\mathcal{W}^*$ to $\mathcal{V}^*$. The adjoint of a Hilbert space operator is essentially this map, but we also identify $\mathcal{H}$ with $\mathcal{H}^*$ via the map $v \mapsto \hat{v}$ of Definition 2.4.1. For clarity we use the notation $A^\#$ for the Banach space adjoint and A^* for the Hilbert space adjoint.

DEFINITION 3.1.2 *Let $A \in B(\mathcal{H})$. Define a bounded linear map $A^\# : \mathcal{H}^* \to \mathcal{H}^*$ by setting $A^\#(\omega) = \omega \circ A$ for $\omega \in \mathcal{H}^*$. Then define $A^* : \mathcal{H} \to \mathcal{H}$ by $A^*v = (A^\#\hat{v})\check{}$, where $\check{}$ is the inverse of $\hat{}$. A^* is the* adjoint of A.

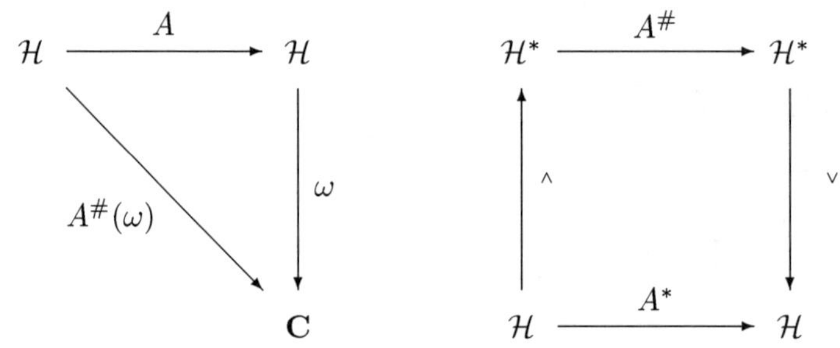

Figure 3.1 $A^{\#}$ and A^*

More generally, if $\mathcal{H}$ and $\mathcal{K}$ are Hilbert spaces, one can define the adjoint of any bounded map $A : \mathcal{H} \to \mathcal{K}$ by the same method. It will be a bounded map from $\mathcal{K}$ to $\mathcal{H}$.

The fundamental property of the adjoint is given in the next proposition. This can also be used as a definition.

PROPOSITION 3.1.3
Let $A \in B(\mathcal{H})$. Then $A^* \in B(\mathcal{H})$ and

$$\langle Av, w\rangle = \langle v, A^*w\rangle$$

for all $v, w \in \mathcal{H}$.

PROOF A^* is linear because it is a composition of two antilinear maps and one linear map, and it is bounded because $v \mapsto \hat{v}$ is isometric and $\|A^{\#}\| = \|A\|$. For the second assertion, let $v, w \in \mathcal{H}$; then

$$\langle v, A^*w\rangle = \langle v, (A^{\#}\hat{w})\check{}\,\rangle = (A^{\#}\hat{w})(v) = \hat{w}(Av) = \langle Av, w\rangle,$$

as desired. ∎

It is generally easier to compute adjoints using the preceding result rather than Definition 3.1.2. This is possible because the inner products $\langle v, A^*w\rangle$ completely determine $(A^*w)\hat{}$, and hence they completely determine A^*w, for all $w \in \mathcal{H}$.

COROLLARY 3.1.4
Let $A \in B(\mathcal{H})$ and suppose $\mathcal{H} \cong \mathbf{C}^n$ is a finite dimensional Hilbert space. Then $B(\mathcal{H}) \cong M_n(\mathbf{C})$. If A is represented by the matrix $[a_{ij}]$ then A^* is represented by its Hermitian transpose $[\bar{a}_{ji}]$.

PROOF The first assertion is trivial. For the second, let $\{e_1, \ldots, e_n\}$ be the standard orthonormal basis of $\mathbf{C}^n$. Then

$$\langle A^* e_i, e_j \rangle = \overline{\langle e_j, A^* e_i \rangle} = \overline{\langle A e_j, e_i \rangle} = \bar{a}_{ji}.$$

Thus the (i, j) entry of the matrix which represents A^* is $\bar{a}_{ji}$. ∎

Now we list the basic laws satisfied by adjoints.

PROPOSITION 3.1.5
Let $A, B \in B(\mathcal{H})$ and $a, b \in \mathbf{C}$. Then

(a) $(aA + bB)^* = \bar{a}A^* + \bar{b}B^*$;
(b) $(AB)^* = B^* A^*$;
(c) $A^{**} = A$;
(d) $\|A^*\| = \|A\|$; and
(e) $\|A^* A\| = \|A\|^2$.

Properties (a) – (c) follow from similar laws for $A^\#$, and property (d) is a consequence of the fact that the map $v \mapsto \hat{v}$ is an isometry. To verify (e), note first that $\|A^* A\| \leq \|A^*\| \|A\| = \|A\|^2$ (using property (d)). Conversely, for any unit vector $v \in \mathcal{H}$ we have

$$\|Av\|^2 = \langle Av, Av \rangle = \langle A^* Av, v \rangle \leq \|A^* A\|;$$

taking the supremum over all unit vectors v yields $\|A\|^2 \leq \|A^* A\|$. This proves (e).

The most important classes of operators on a Hilbert space are defined in terms of adjoints; we identify them now.

DEFINITION 3.1.6 *Denote the identity map on $\mathcal{H}$ by I.*

(a) $A \in B(\mathcal{H})$ is <u>self-adjoint</u> if $A = A^*$.
(b) $N \in B(\mathcal{H})$ is <u>normal</u> if $NN^* = N^* N$.
(c) $U \in B(\mathcal{H})$ is <u>unitary</u> if $UU^* = U^* U = I$.
(d) $P \in B(\mathcal{H})$ is a <u>projection</u> if $P^2 = P^* = P$.

Projections are sometimes called "orthogonal projections" to distinguish them from operators which only satisfy $P^2 = P$. But we will not need to consider the latter type of operators.

Observe that normality is implied by the other three conditions. Normal operators are important because they are precisely the operators for which one can prove a spectral theorem (see Theorem 3.5.3). Unitaries and projections have simple geometric properties, as we will see shortly, and self-adjoint operators are conceptually fundamental because they are the Hilbert space analogs of real-valued functions on a set.

Before describing the structure of unitaries and projections, we first give some examples.

Example 3.1.7

Let (X,μ) be a σ-finite measure space and let $\mathcal{H} = L^2(X)$. For $f \in L^\infty(X)$ define the multiplication operator $M_f \in B(\mathcal{H})$ by setting $M_f g = fg$ for $g \in L^2(X)$.

The norm of M_f is $\|f\|_\infty$. Its adjoint is $M_{\bar{f}}$, the operator of multiplication by the pointwise complex conjugate of f. It follows that M_f is normal. Furthermore, if f is real-valued then M_f is self-adjoint; if $|f(x)| = 1$ for almost every x then M_f is unitary; and if $f(x) = 0$ or 1 almost everywhere then M_f is a projection.

Exactly the same statements hold if $\mathcal{H} = L^2(X;\mathcal{X})$, where $\mathcal{X}$ is a measurable Hilbert bundle over X (Definition 2.4.8). Multiplication by $f \in L^\infty(X)$ still makes sense.

The preceding example is general; see Theorem 3.5.3. But there is also a very different intuition for unitary operators, illustrated in the following example.

Example 3.1.8

Let (X,μ) be a σ-finite measure space and let $f : X \to X$ be a measure preserving bijection. Define the composition operator $C_f : L^2(X) \to L^2(X)$ by $C_f g = g \circ f$. Then $C_f^* g = g \circ f^{-1}$, and so C_f is unitary.

Now we proceed to a general geometric description of unitaries and projections. The only tool we need is the following polarization identity, whose proof is a routine computation. It is useful because it allows us to recover any inner product $\langle Av, w\rangle$, and hence complete knowledge of the operator A, from inner products of the form $\langle Av, v\rangle$.

LEMMA 3.1.9
Let $A \in B(\mathcal{H})$. Then

$$\langle Av, w\rangle = \frac{1}{4}\sum_{k=0}^{3} i^k \langle A(v + i^k w), (v + i^k w)\rangle$$

for all $v, w \in \mathcal{H}$.

PROPOSITION 3.1.10
Let $U \in B(\mathcal{H})$. The following are equivalent:

(a) U is unitary;
(b) U is invertible and $U^{-1} = U^*$; and
(c) U is an surjective isometry.

PROOF

(a) $\Rightarrow$ (b). Trivial.

(b) $\Rightarrow$ (c). Suppose $U^{-1} = U^*$. Then

$$\langle Uv, Uw\rangle = \langle U^*Uv, w\rangle = \langle v, w\rangle$$

for all $v, w \in \mathcal{H}$. In particular, $\|Uv\|^2 = \|v\|^2$ for all $v \in \mathcal{H}$, so U is an isometry. It is surjective because it is invertible.

(c) $\Rightarrow$ (a). Suppose U is an isometry. Then

$$\langle U^*Uv, v\rangle = \langle Uv, Uv\rangle = \|Uv\|^2 = \|v\|^2 = \langle v, v\rangle$$

for all $v \in \mathcal{H}$. Thus $\langle (U^*U - I)v, v\rangle = 0$ for all v, and it follows by polarization that $U^*U - I = 0$, i.e. $U^*U = I$.

If U is surjective, then U^{-1} exists and is also an isometry. Since $U^*U = I$, it follows that $U^{-1} = U^*$; so $UU^* = I$ as well. ∎

In contrast to the finite dimensional case, isometries need not be surjective in general. The simplest counterexample is the unilateral shift S^+ on $l^2(\mathbf{N})$ defined by

$$S^+(a_0, a_1, \ldots) = (0, a_0, \ldots).$$

Its adjoint, the backward shift S^-, satisfies

$$S^-(a_0, a_1, \ldots) = (a_1, a_2, \ldots),$$

and from this one can verify that $S^-S^+ = I$ but $S^+S^- \neq I$.

Now we turn to projections.

DEFINITION 3.1.11 *Let E be a closed subspace of $\mathcal{H}$. For any $v \in \mathcal{H}$ let $v = v_1 + v_2$ be the unique decomposition with $v_1 \in E$ and $v_2 \in E^\perp$, as in Theorem 2.2.5. Then define $P_E : \mathcal{H} \to \mathcal{H}$, the orthogonal projection onto E, by $P_E v = v_1$.*

PROPOSITION 3.1.12
The projections in $B(\mathcal{H})$ are precisely the orthogonal projections onto closed subspaces of $\mathcal{H}$.

PROOF First, fix a closed subspace $E \subset \mathcal{H}$. We verify that P_E is a projection in the sense of Definition 3.1.6. It is linear because if

$v = v_1 + v_2$ and $w = w_1 + w_2$ with $v_1, w_1 \in E$ and $v_2, w_2 \in E^\perp$ then $av + bw = (av_1 + bw_1) + (av_2 + bw_2)$, so

$$P_E(av + bw) = av_1 + bw_1 = aP_E v + bP_E w.$$

It is bounded because $\|v_1\|^2 \leq \|v_1\|^2 + \|v_2\|^2 = \|v\|^2$. Also

$$\langle P_E v, w\rangle = \langle v_1, w_1 + w_2\rangle = \langle v_1, w_1\rangle = \langle v_1 + v_2, w_1\rangle = \langle v, P_E w\rangle,$$

so that $P_E = P_E^*$, and $P_E^2 = P_E$ is clear. Thus every P_E is a projection in $B(\mathcal{H})$.

Conversely, suppose P is a projection and let $E = P(\mathcal{H})$ be its range. E is closed because if $(v_n) \subset E$ and $v_n \to v$ then $v_n = Pv_n \to Pv$ implies $Pv = v$, and thus $v \in E$. Now choose $v \in \mathcal{H}$ and write $v = v_1 + v_2$ where $v_1 \in E$ and $v_2 \in E^\perp$. Then $Pv_1 = v_1$ and

$$\|Pv_2\|^2 = \langle Pv_2, Pv_2\rangle = \langle P^*Pv_2, v_2\rangle = \langle Pv_2, v_2\rangle = 0,$$

so $Pv_2 = 0$. Thus $Pv = v_1 = P_E v$, and we conclude $P = P_E$. So P is an orthogonal projection onto a closed subspace of $\mathcal{H}$. ∎

3.2 Continuous functional calculus

In this and the following two sections we develop the tools needed for the spectral theorem. We begin with the spectrum.

DEFINITION 3.2.1 *Let $A \in B(\mathcal{H})$. The* <u>spectrum</u> *of A, $\mathrm{sp}(A)$, is the set of $\lambda \in \mathbf{C}$ such that $A - \lambda I$ is not invertible in $B(\mathcal{H})$.*

Example 3.2.2

(a) Let A be a complex $n \times n$ matrix. Then $A - \lambda I$ is invertible if and only if its kernel is zero. That is, it fails to be invertible if and only if there exists a nonzero vector $v \in \mathbf{C}^n$ such that $(A - \lambda I)v = 0$, i.e., $Av = \lambda v$. So $\mathrm{sp}(A)$ is exactly the set of eigenvalues of A.

(b) Now let $\mathcal{H} = L^2(X)$ for X a σ-finite measure space and let $f \in L^\infty(X)$. The <u>essential range</u> of f is the set of $\lambda \in \mathbf{C}$ such that $f^{-1}(\mathcal{O})$ has positive measure for every open neighborhood $\mathcal{O}$ of λ.

Let M_f be the operator of multiplication by f, as in Example 3.1.7. If λ does not belong to the essential range of f, then for some $\epsilon > 0$ we have $|f(x) - \lambda| \geq \epsilon$ for almost every $x \in X$. Thus $\|(f - \lambda)^{-1}\|_\infty \leq 1/\epsilon$, and multiplication by $(f - \lambda)^{-1}$ is an inverse of $M_f - \lambda I = M_{f-\lambda}$. So $\mathrm{sp}(M_f)$ is contained in the essential range of f.

Conversely, suppose λ belongs to the essential range of f and let $\epsilon > 0$. Find a function $g \in L^2(X)$ of norm one which is supported on $f^{-1}(\mathcal{O})$ where $\mathcal{O}$ is the ball of radius ϵ about λ. Then

$$\|(M_f - \lambda I)(g)\|^2 = \int |(f - \lambda)g|^2 \leq \epsilon^2 \int |g|^2 = \epsilon^2$$

since $|f - \lambda| \leq \epsilon$ on the support of g. Suppose $A \in B(\mathcal{H})$ were an inverse of $M_f - \lambda I$. Then the above implies that $\|A\| \geq 1/\epsilon$ for all positive ϵ, which is absurd, so we conclude that $M_f - \lambda I$ has no inverse in $B(\mathcal{H})$. Thus $\mathrm{sp}(M_f)$ is exactly the essential range of f.

The same conclusions hold if $L^2(X)$ is replaced by $L^2(X; \mathcal{X})$ for any measurable Hilbert bundle $\mathcal{X}$ over X (cf. Example 3.1.7).

If A is a diagonal $n \times n$ complex matrix, a short computation shows that the norm of A, as an operator, equals the modulus of its largest eigenvalue. In light of the previous example, we can say that $\|A\| = \max\{\lambda : \lambda \in \mathrm{sp}(A)\}$. This is not true for all matrices, but it does still hold for self-adjoint operators in infinite dimensions. This is the main fact we need about the spectrum, and we now proceed to prove it.

The following lemma gives a simple sufficient condition for invertibility in $B(\mathcal{H})$.

LEMMA 3.2.3

Let $A \in B(\mathcal{H})$ and suppose $\|A\| < 1$. Then $I - A$ is invertible.

PROOF Note that $\|A^n\| \leq \|A\|^n$. It follows that the infinite series

$$I + A + A^2 + \cdots$$

converges in $B(\mathcal{H})$. A trivial calculation shows that the product of this operator and $I - A$, in either order, equals the identity. Therefore $I - A$ is invertible. ∎

PROPOSITION 3.2.4

Let $A \in B(\mathcal{H})$ be self-adjoint. Then

$$\begin{aligned} \|A\| &= \sup\{|\langle Av, v\rangle| : \|v\| = 1\} \\ &= \sup\{|\lambda| : \lambda \in \mathrm{sp}(A)\}. \end{aligned}$$

PROOF For the first equality, let $C = \sup\{|\langle Av, v\rangle| : \|v\| = 1\}$. Then $|\langle Av, v\rangle| \leq C\|v\|^2$ for all $v \in \mathcal{H}$. Also note that $\langle Av, v\rangle = \langle v, Av\rangle = \overline{\langle Av, v\rangle}$, so that $\langle Av, v\rangle$ is always real. So if v and w are any unit vectors in $\mathcal{H}$, then polarization and the parallelogram law (Lemmas 3.1.9 and 2.2.3) yield

$$\begin{aligned} |\mathrm{Re}\langle Av, w\rangle| &= \frac{1}{4}\,|\langle A(v+w), (v+w)\rangle - \langle A(v-w), (v-w)\rangle| \\ &\leq \frac{C}{4}\left(\|v+w\|^2 + \|v-w\|^2\right) \end{aligned}$$

$$= \frac{C}{4}(2\|v\|^2 + 2\|w\|^2)$$
$$= C.$$

More generally, we have

$$|\mathrm{Re}\, a\langle Av, w\rangle| = |\mathrm{Re}\langle Av, \bar{a}w\rangle| \leq C$$

for any $a \in \mathbf{C}$ with $|a| = 1$; in particular, taking $a = |\langle Av, w\rangle|/\langle Av, w\rangle$ yields $|\langle Av, w\rangle| \leq C$, and as this is true for all unit vectors v and w we conclude that $\|A\| \leq C$. Conversely, $C \leq \|A\|$ by the Cauchy-Schwarz inequality. So we have proven the first equality.

Now let $C' = \sup\{|\lambda| : \lambda \in \mathrm{sp}(A)\}$. Choose $\lambda \in \mathbf{C}$ such that $|\lambda| > \|A\|$. Then $\|\frac{1}{\lambda}A\| < 1$, so $I - \frac{1}{\lambda}A$ is invertible by the lemma. Hence $A - \lambda I$ is invertible, and so $\lambda \notin \mathrm{sp}(A)$. This shows that $C' \leq \|A\|$.

To prove $\|A\| \leq C'$, we use the first equality and the fact that $\langle Av, v\rangle$ is real for all v. It follows that we can find a sequence of unit vectors v_n such that $\langle Av_n, v_n\rangle \to \pm\|A\|$. Let $\lambda = \lim \langle Av_n, v_n\rangle = \pm\|A\|$; then

$$\begin{aligned}\|Av_n - \lambda v_n\|^2 &= \|Av_n\|^2 - 2\langle Av_n, \lambda v_n\rangle + \|\lambda v_n\|^2 \\ &\leq 2\|A\|^2 - 2\lambda\langle Av_n, v_n\rangle.\end{aligned}$$

The last expression converges to zero as $n \to \infty$, so $(A - \lambda I)v_n$ also converges to zero. But $\|v_n\| = 1$ for all n, so $A - \lambda I$ cannot be invertible. We conclude that $\lambda = \pm\|A\| \in \mathrm{sp}(A)$, so that $\|A\| \leq C'$. ∎

COROLLARY 3.2.5

The spectrum of any self-adjoint operator $A \in B(\mathcal{H})$ is a compact subset of the real line.

PROOF Boundedness of the spectrum follows immediately from the theorem. To prove compactness suppose $\lambda \notin \mathrm{sp}(A)$ and let $B = (A - \lambda I)^{-1}$. Then for any $\mu \in \mathbf{C}$ with $|\lambda - \mu| < \|B\|^{-1}$ we have

$$\begin{aligned}\|I - (A - \mu I)B\| &= \|\,[(A - \lambda I) - (A - \mu I)]\, B\| \\ &= \|(\mu - \lambda)B\| < 1;\end{aligned}$$

by Lemma 3.2.3 this implies that $(A - \mu I)B$ is invertible, so $A - \mu I$ is right invertible, and a similar argument shows that it is left invertible. Thus $\mu \notin \mathrm{sp}(A)$. This shows that the complement of $\mathrm{sp}(A)$ is open. So the spectrum of A is compact.

It remains to prove that $\mathrm{sp}(A) \subset \mathbf{R}$. Let $a + ib \in \mathbf{C}$ with a and b real and $b \neq 0$. We must show that $A - (a + ib)I$ is invertible. Let $A' = \frac{1}{b}(A - aI)$; then A' is also self-adjoint and it will suffice to show that $A' - iI$ is invertible.

Observe that

$$\begin{aligned}\|(A' \pm iI)v\|^2 &= \langle A'v, A'v\rangle \pm \langle A'v, iv\rangle \pm \langle iv, A'v\rangle + \langle v, v\rangle \\ &= \|A'v\|^2 + \|v\|^2.\end{aligned}$$

In particular, $\|(A' - iI)v\| \geq \|v\|$ and hence $A' - iI$ is an isomorphism onto its range. So we need only show that its range is dense in $\mathcal{H}$. But if $w \perp \operatorname{ran}(A' - iI)$ then

$$0 = \langle w, (A' - iI)(A' + iI)w\rangle = \|(A' + iI)w\|^2 = \|A'w\|^2 + \|w\|^2,$$

so $w = 0$. Thus $\operatorname{ran}(A' - iI)$ is indeed dense in $\mathcal{H}$, and we conclude that $A' - iI$ is invertible in $B(\mathcal{H})$, as desired. ∎

Now we turn to functional calculi for a self-adjoint operator. For any $A \in B(\mathcal{H})$ we can form polynomials in A just using basic algebra of operators. The idea of functional calculus is to generalize this to functions other than polynomials. We first treat continuous functions, and then in the next section we will pass to bounded Borel functions.

DEFINITION 3.2.6 *Let $p(z) = a_0 + a_1 z + \cdots + a_n z^n$ be a polynomial with complex coefficients and let $A \in B(\mathcal{H})$. We define $p(A)$ to be the operator $a_0 I + a_1 A + \cdots + a_n A^n$.*

LEMMA 3.2.7
Let $A \in B(\mathcal{H})$ be self-adjoint and let $p(z)$ be a polynomial. Then $\operatorname{sp}(p(A)) = p(\operatorname{sp}(A))$.

PROOF Let $\lambda \in \mathbf{C}$ and write $p(z) - \lambda = a(z - b_1)\cdots(z - b_n)$. The complex numbers $b_1, \ldots, b_n$ are the roots of the equation $p(z) = \lambda$. Now

$$p(A) - \lambda I = a(A - b_1 I)\cdots(A - b_n I)$$

is a product of commuting operators, so it is invertible if and only if each factor $A - b_i I$ is invertible, i.e., each $b_i \notin \operatorname{sp}(A)$, or equivalently, $p(z) \neq \lambda$ for all $z \in \operatorname{sp}(A)$. Thus λ is not in $\operatorname{sp}(p(A))$ if and only if it is not in $p(\operatorname{sp}(A))$. ∎

The following theorem describes the continuous functional calculus for a self-adjoint operator. Here we use the notation $C(X)$ for the Banach space of continuous complex-valued functions on the compact Hausdorff space X, with norm $\|f\|_\infty = \sup\{|f(x)| : x \in X\}$. We also use the following terminology. A linear map T from a conjugation-closed algebra of functions into $B(\mathcal{H})$ is a <u>$*$-homomorphism</u> if $T(\bar{f}) = T(f)^*$ and

$T(fg) = T(f)T(g)$ for all f and g. It is unital if $T(1_X) = I$, and it is a $*$-isomorphism if it is isometric. (We will generalize this terminology in Definition 5.1.1.) Here 1_X is the function which is constantly 1 on X.

THEOREM 3.2.8

Let $A \in B(\mathcal{H})$ be self-adjoint. Then the map which takes the polynomial $p(x) \in C(\mathrm{sp}(A))$ to the operator $p(A) \in B(\mathcal{H})$ extends uniquely to a unital $*$-isomorphism from $C(\mathrm{sp}(A))$ into $B(\mathcal{H})$.

PROOF First suppose the coefficients of p are real. Then $p(A)$ is self-adjoint, and it follows from Proposition 3.2.4 and Lemma 3.2.7 that

$$\|p(A)\| = \sup\{|p(\lambda)| : \lambda \in \mathrm{sp}(A)\} = \|p\|_\infty,$$

where the sup norm of p is taken in $C(\mathrm{sp}(A))$. For an arbitrary complex polynomial p, apply the above to the real polynomial $|p|^2 = p\bar{p}$; this yields $\|p(A)p(A)^*\| = \|p\|_\infty^2$. By Proposition 3.1.5 (e), it follows that $\|p(A)\| = \|p\|_\infty$. Thus the map taking $p(x)$ to $p(A)$ is an isometry. By the Stone-Weierstrass theorem, the polynomials are dense in $C(\mathrm{sp}(A))$, so this map uniquely extends to an isometry from $C(\mathrm{sp}(A))$ into $B(\mathcal{H})$ which is clearly unital.

It is trivial that $(ap + bq)(A) = ap(A) + bq(A)$, $(pq)(A) = p(A)q(A)$, and $\bar{p}(A) = p(A)^*$ for any polynomials $p(x)$ and $q(x)$ and any $a, b \in \mathbf{C}$. The same properties hold on all of $C(\mathrm{sp}(A))$ by continuity. ∎

For any $f \in C(\mathrm{sp}(A))$, we write $f(A)$ for the corresponding operator. Thus $f(A) = \lim p_n(A)$ where (p_n) is any sequence of polynomials which converges to f uniformly on $\mathrm{sp}(A)$.

Example 3.2.9

Let $f \in L^\infty(X)$ and let M_f be the corresponding multiplication operator on $L^2(X)$ (or on $L^2(X; \mathcal{X})$), as in Example 3.1.7. Then for any polynomial $p(x)$ it is easy to see that $p(M_f) = M_{p\circ f}$. By continuity it follows that for any $g \in C(\mathrm{sp}(M_f))$ we have $g(M_f) = M_{g\circ f}$.

3.3 Borel functional calculus

We now want to extend the continuous functional calculus of the previous section to Borel functions. In particular, this will allow us to define operators of the form $\chi_S(A)$ where χ_S is the characteristic function of a Borel subset of $\mathrm{sp}(A)$. These operators are crucial to understanding the structure of the operator A.

The method of defining $f(A)$ for a Borel function f is again extension by continuity, but now with respect to different topologies.

DEFINITION 3.3.1 *The* <u>weak operator topology</u> *on* $B(\mathcal{H})$ *is the weakest topology such that for every* $v, w \in \mathcal{H}$ *the map* $A \mapsto \langle Av, w\rangle$ *is continuous. In terms of nets, we have* $A_\kappa \to A$ *weak operator if and only if* $\langle A_\kappa v, w\rangle \to \langle Av, w\rangle$ *for all* $v, w \in \mathcal{H}$.

We also need the following generally useful fact. A <u>sesquilinear form</u> is a map $\{\cdot,\cdot\} : \mathcal{H} \times \mathcal{H} \to \mathbf{C}$ which is linear in the first variable and antilinear in the second. It is <u>bounded</u> if there exists $C > 0$ such that $|\{v, w\}| \leq C\|v\|\|w\|$ for all $v, w \in \mathcal{H}$.

LEMMA 3.3.2

Let $\{\cdot,\cdot\} : \mathcal{H} \times \mathcal{H} \to \mathbf{C}$ *be a bounded sesquilinear form. Then there is a unique operator* $A \in B(\mathcal{H})$ *such that* $\{v, w\} = \langle Av, w\rangle$ *for all* $v, w \in \mathcal{H}$. *Conversely, any bounded operator* A *defines a bounded sesquilinear form by this equation.*

PROOF Fix $v \in \mathcal{H}$ and consider the map $w \mapsto \overline{\{v, w\}}$. This is a bounded linear functional, so by Proposition 2.4.2 there exists a unique $v' \in \mathcal{H}$ such that $\overline{\{v, w\}} = \langle w, v'\rangle$; that is, $\{v, w\} = \langle v', w\rangle$ for all $w \in \mathcal{H}$. Define $Av = v'$. Then A is linear because

$$\begin{aligned}
\langle A(av_1 + bv_2), w\rangle &= \{av_1 + bv_2, w\} \\
&= a\{v_1, w\} + b\{v_2, w\} \\
&= a\langle Av_1, w\rangle + b\langle Av_2, w\rangle \\
&= \langle aAv_1 + bAv_2, w\rangle
\end{aligned}$$

for all w, and it is bounded because

$$|\langle Av, w\rangle| = |\{v, w\}| \leq C\|v\|\|w\|.$$

Uniqueness is clear, as is the converse. ∎

We can now define the Borel functional calculus. Recall the Riesz representation theorem: for any compact Hausdorff space X, the dual space $C(X)^*$ can be identified with the space $M(X)$ of regular complex Borel measures on X. This identification associates the measure μ with the linear functional $f \mapsto \int f\,d\mu$. We write $\mathrm{Bor}(X)$ for the space of bounded

Borel functions on X, and we define the μ-topology on $\mathrm{Bor}(X)$ to be the weakest topology for which integration against any regular Borel function is continuous. Thus $f_\kappa \to f\ (\mu)$ if and only if $\int f_\kappa\, d\mu \to \int f\, d\mu$ for all $\mu \in M(X)$.

THEOREM 3.3.3

Let $A \in B(\mathcal{H})$ be self-adjoint. There is a unique unital $*$-homomorphism $\mathrm{Bor}(\mathrm{sp}(A))$ *into $B(\mathcal{H})$ which extends the continuous functional calculus and is continuous from the μ-topology to the weak operator topology.*

PROOF Let $T : C(\mathrm{sp}(A)) \to B(\mathcal{H})$ be the $*$-homomorphism given by the continuous functional calculus for A. Then the double (Banach space) adjoint map $T^{\#\#}$ takes $C(\mathrm{sp}(A))^{**} \cong M(\mathrm{sp}(A))^*$ weak* continuously into $B(\mathcal{H})^{**}$.

Now for any bounded Borel function f on $\mathrm{sp}(A)$, the map $\omega_f : \mu \mapsto \int f\, d\mu$ is a bounded linear functional on $M(\mathrm{sp}(A))$ and hence belongs to $C(\mathrm{sp}(A))^{**}$. For $v, w \in \mathcal{H}$ let $\rho_{v,w} \in B(H)^*$ be the linear functional $A \mapsto \langle Av, w\rangle$. Then $\{v, w\} = T^{\#\#}\omega_f(\rho_{v,w})$ is a bounded sesquilinear form and so the lemma implies that $\{v, w\} = \langle f(A)v, w\rangle$ for some bounded operator $f(A)$. Note that if $f \in C(\mathrm{sp}(A))$ then this definition agrees with the continuous functional calculus because then

$$\underset{\text{Borel}}{\langle f(A)v, w\rangle} = T^{\#\#}\omega_f(\rho_{v,w}) = \rho_{v,w}(Tf) = \underset{\text{continuous}}{\langle f(A)v, w\rangle}.$$

Clearly, for any net (f_κ) of Borel functions on $\mathrm{sp}(A)$ we have $f_\kappa \to f$ (μ) if and only if $\omega_{f_\kappa} \to \omega_f$ weak*. It then follows from weak* continuity of $T^{\#\#}$ that $T^{\#\#}\omega_{f_\kappa}(\rho_{v,w}) \to T^{\#\#}\omega_f(\rho_{v,w})$ for all $v, w \in \mathcal{H}$, and hence that $f_\kappa(A) \to f(A)$ weak operator.

The map $f \mapsto f(A)$ is a $*$-homomorphism on $C(\mathrm{sp}(A))$, and the unit ball of $C(\mathrm{sp}(A))$ is weak* dense in the unit ball of $C(\mathrm{sp}(A))^{**}$. The fact that $\bar{f}(A) = f(A)^*$ therefore follows from continuity, and $fg(A) = f(A)g(A)$ can be proven by a two step continuity argument, where first we assume f is continuous and approximate g by continuous functions, and then we let f be arbitrary and approximate it by continuous functions. Thus the map $f \mapsto f(A)$ is a $*$-homomorphism on the set of all bounded Borel functions. It is unique by weak* density of $C(\mathrm{sp}(A))$ in $C(\mathrm{sp}(A))^{**}$. ∎

The passage from $C(\mathrm{sp}(A))$ to $C(\mathrm{sp}(A))^{**}$ in this theorem is an instance of a more general fact: if $\mathcal{V}$ is any Banach space then any bounded linear map from $\mathcal{V}$ into $B(\mathcal{H})$ has an extension to $\mathcal{V}^{**}$ which is weak* to weak operator continuous. The proof of this fact is substantially the same as the proof for $C(\text{sp}(A))$.

The real reason why this works is that $B(\mathcal{H})$ is a dual Banach space, and its weak* topology agrees with the weak operator topology on bounded sets (see Theorem 6.3.8). This is why the weak operator topology is so important, at least on bounded sets.

Example 3.3.4

Let $f \in L^\infty(X)$ be real-valued and let M_f be the corresponding multiplication operator on $L^2(X)$ (or $L^2(X;\mathcal{X})$). Then for any bounded Borel function g on $\text{sp}(M_f)$ we have $g(M_f) = M_{g\circ f}$ (cf. Example 3.2.9). In particular, if S is a Borel subset of $\mathbf{R}$ then $\chi_S(M_f) = M_{\chi_S\circ f} = M_{\chi_{f^{-1}(S)}}$.

3.4 Spectral measures

Let $A \in B(\mathcal{H})$ be self-adjoint and let S be a Borel subset of $\mathbf{R}$. Consider the characteristic function χ_S: it is bounded and its restriction to $\text{sp}(A)$ is Borel measurable. So we can form the operator $\chi_S(A)$. Since $\chi_S = \bar{\chi}_S = \chi_S^2$, the operator $\chi_S(A)$ also has these properties, and therefore it is an orthogonal projection onto a closed subspace of $\mathcal{H}$ by Proposition 3.1.12. It is called a spectral projection of A. Example 3.3.4 shows that if A is a multiplication operator then its spectral projections are also multiplication operators.

Spectral projections help explain why self-adjoint operators are the Hilbert space analog of real-valued functions. Let $f : X \to \mathbf{R}$ be a measurable function on a σ-finite measure space (X, μ). In Section 1.3 we viewed real-valued functions on a finite set as partitioning the set into subsets and tagging these subsets with real numbers; for general X we need some measurable version of a partition. If f is only defined up to null sets, the sets $f^{-1}(a)$ $(a \in \mathbf{R})$ may well all be null and hence individually carry no information about f. Instead, consider the nested family of sets $S_{(-\infty,a]} = f^{-1}(-\infty, a]$ for $a \in \mathbf{R}$.

Since the intervals $(-\infty, a]$ generate the Borel sets, and

$$f^{-1}(S^c) = \left(f^{-1}(S)\right)^c$$
$$f^{-1}\left(\bigcup S_n\right) = \bigcup f^{-1}(S_n)$$
$$f^{-1}\left(\bigcap S_n\right) = \bigcap f^{-1}(S_n)$$

for any Borel sets $S, S_n \subset \mathbf{R}$, it follows that the sets $S_{(-\infty,a]}$ determine $f^{-1}(S)$ for every Borel subset S of $\mathbf{R}$, up to null sets.

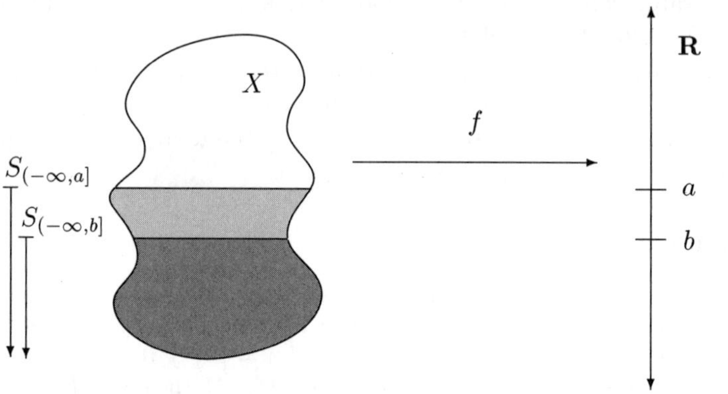

Figure 3.2 A "measurable partition"

The analogous Hilbert space construction is a family of closed subspaces $E_{(-\infty,a]}$ such that $E_{(-\infty,a]} \subset E_{(-\infty,b]}$ if $a \leq b$ and $E_{(-\infty,a]} = \bigcap_{b>a} E_{(-\infty,b]}$. The $E_{(-\infty,a]}$ determine a closed subspace $E(S)$ of $\mathcal{H}$ for every Borel subset S of $\mathbf{R}$ in the same way as in the set theoretic case, replacing $\cup$, $\cap$, and c with $\vee$, $\wedge$, and $\perp$. Such a map E from Borel subsets of $\mathbf{R}$ to closed subspaces of $\mathcal{H}$ is called a "spectral measure" on $\mathbf{R}$, and this is the Hilbert space version of a real-valued function. The spectral theorem states that spectral measures on $\mathbf{R}$ correspond precisely to self-adjoint operators on $\mathcal{H}$.

The following is the formal definition of a spectral measure. Let a measurable space be a set X together with a σ-algebra of subsets Ω.

DEFINITION 3.4.1 *Let (X, Ω) be a measurable space. An $\mathcal{H}$-valued spectral measure on X is a map E from Ω into the set of closed subspaces of $\mathcal{H}$ such that the projections $P_{E(S)}$ and $P_{E(T)}$ commute for all $S, T \in \Omega$ and*

(a) $E(\emptyset) = 0$ and $E(X) = \mathcal{H}$;
(b) $E(S^c) = E(S)^\perp$; and
(c) $E(\bigcup S_n) = \bigvee E(S_n)$ and $E(\bigcap S_n) = \bigwedge E(S_n)$

for all $S, S_n \in \Omega$.

There is a simple geometric interpretation of the condition that $P_{E(S)}$ and $P_{E(T)}$ commute. This is the case if and only if $E(S) = E_1 \oplus E_2$ and $E(T) = E_1 \oplus E_3$ for some mutually orthogonal subspaces E_1, E_2, and E_3. We leave the proof of this fact as an exercise. We also omit the proof that if $X = \mathbf{R}$ then Definition 3.4.1 is equivalent to the previous description of spectral measures in terms of the sets $E_{(-\infty,a]}$.

Spectral measures can be difficult to work with if one has no mental picture of them. This is the value of the following result; in the separable case, the picture becomes especially simple (Corollary 3.4.3).

THEOREM 3.4.2

Let E be an $\mathcal{H}$-valued spectral measure on a measurable space X. Then there is a family of probability measures $\{\mu_\kappa\}$ on X and an isometric isomorphism U from $\bigoplus_\kappa L^2(X, \mu_\kappa)$ onto $\mathcal{H}$ such that

$$E(S) = U\left(\bigoplus_\kappa L^2(S, \mu_\kappa|_S)\right)$$

for every measurable $S \subset X$.

PROOF For any $v \in \mathcal{H}$ let E_v be the closed subspace generated by the vectors $P_{E(S)}v$ as S ranges over all measurable subsets of X. Let $\{v_\kappa\}$ be a maximal family of unit vectors in $\mathcal{H}$ such that the subspaces E_{v_κ} are orthogonal. Then the E_{v_κ} generate $\mathcal{H}$: if this were false, we could find a unit vector $v \in \mathcal{H}$ such that $v \perp E_{v_\kappa}$ for all κ. But then $\langle P_{E(S)}v, P_{E(T)}v_\kappa\rangle = \langle v, P_{E(S\cap T)}v_\kappa\rangle = 0$ for all $S, T \subset X$ and all κ, which shows that $E_v \perp E_{v_\kappa}$ for all κ, contradicting maximality.

For each κ define $\mu_\kappa(S) = \|P_{E(S)}v_\kappa\|^2$. It is straightforward to verify that these are probability measures on X. Then for any measurable set $S \subset X$, define $U_\kappa(\chi_S) = P_{E(S)}v_\kappa$. Extend linearly to the span of the characteristic functions in $L^2(X, \mu_\kappa)$; the result is an isometric map into $\mathcal{H}$ because

$$\begin{aligned}\left\langle U_\kappa\left(\sum a_i\chi_{S_i}\right), U_\kappa\left(\sum b_j\chi_{S_j}\right)\right\rangle &= \sum_{i,j} a_i\bar{b}_j\langle P_{E(S_i)}v_\kappa, P_{E(S_j)}v_\kappa\rangle\\ &= \sum_{i,j} a_i\bar{b}_j\mu(S_i\cap S_j)\\ &= \left\langle\sum a_i\chi_{S_i}, \sum b_j\chi_{S_j}\right\rangle.\end{aligned}$$

This computation also shows that U_κ is well-defined, since it implies that $U_\kappa(\sum a_i\chi_{S_i}) = 0$ if $\sum a_i\chi_{S_i} = 0$. Thus U_κ extends to an isometric embedding of $L^2(X, \mu_\kappa)$ into $\mathcal{H}$, and its range is E_{v_κ}. Since $\bigvee E_{v_\kappa} = \mathcal{H}$, it follows that the direct sum map $U = \bigoplus U_\kappa$ is an isometric isomorphism from $\bigoplus_\kappa L^2(X, \mu_\kappa)$ onto $\mathcal{H}$.

Fix $S \subset X$. For any κ and any $S' \subset S$ we have $U_\kappa(\chi_{S'}) = P_{E(S')}v_\kappa \in E(S)$; it follows by linearity and continuity that $U_\kappa(L^2(S, \mu_\kappa|_S)) \subset E(S)$ for all κ, and therefore $U(\bigoplus L^2(S, \mu_\kappa|_S)) \subset E(S)$. The reverse containment follows by taking complements. ∎

COROLLARY 3.4.3

Let E be an $\mathcal{H}$-valued spectral measure on a measurable space X and suppose $\mathcal{H}$ is separable. Then there is a probability measure μ on X, a measurable Hilbert bundle $\mathcal{X}$ over X, and an isometric isomorphism U from $L^2(X;\mathcal{X})$ onto $\mathcal{H}$ such that $U(L^2(S;\mathcal{X}|_S)) = E(S)$ for every measurable $S \subset X$.

PROOF Let $\{\mu_k\}$ be probability measures satisfying the conclusion of Theorem 3.4.2. Since $\mathcal{H}$ is separable, this family is countable.

Define $\mu = \sum_1^\infty 2^{-k}\mu_k$. Then μ is also a probability measure, and μ_k is absolutely continuous with respect to μ for each k. Thus $\mu_k = f_k\mu$ for some $f_k \in L^1(X,\mu)$, $f_k \geq 0$. For each k let $S_k = \{x \in X : f_k(x) > 0\}$, and for $0 \leq n \leq \infty$ let

$$X_n = \{x \in X : x \in S_k \text{ for exactly } n \text{ values of } k\}.$$

Then $\{X_n\}$ is a measurable partition of X. For each n let $\mathcal{H}_n$ be an n-dimensional Hilbert space, and let $\mathcal{X} = \bigcup(X_n \times \mathcal{H}_n)$ be the corresponding measurable Hilbert bundle. By Proposition 2.4.7 we can identify $L^2(X;\mathcal{X})$ with $\bigoplus L^2(S_k,\mu|_{S_k})$.

For each k the map $V_k : L^2(S_k,\mu|_{S_k}) \to L^2(S_k,\mu_k|_{S_k})$ defined by $V_k(g) = g/\sqrt{f_k}$ is an isometric isomorphism, and this gives rise to an isometric isomorphism

$$\begin{aligned} V : L^2(X;\mathcal{X}) &\cong \bigoplus L^2(S_k,\mu|_{S_k}) \\ &\to \bigoplus L^2(S_k,\mu_k|_{S_k}) \cong \bigoplus L^2(X;\mu_k). \end{aligned}$$

Composing V with the map U from Theorem 3.4.2 yields the desired result. ∎

Theorem 3.4.2 is highly noncanonical: a different choice of vectors v_κ would lead to a different family of measures. Some degree of canonicality is regained in Corollary 3.4.3; here μ could be replaced by any measure with which it is mutually absolutely continuous, but otherwise the construction is rigid. We say that the construction only depends on the measure class of μ.

The inability of Corollary 3.4.3 to generalize to the nonseparable case is illustrated by the following example. Let $\mathcal{H} = L^2[0,1] \oplus l^2[0,1]$, using Lebesgue measure on the first summand and counting measure on the second. Define an $\mathcal{H}$-valued spectral measure on $X = [0,1]$ by setting $E(S) = L^2(S) \oplus l^2(S)$ for any Borel set $S \subset [0,1]$. In some sense $\mathcal{H}$ is a bundle over $[0,1]$ with two-dimensional fibers, but it cannot be identified with $L^2[0,1]\otimes\mathbf{C}^2$ for any measure on $[0,1]$. Because of this difficulty, and the fact that spectral measures are such a central topic, it will generally be convenient for us to work only with separable Hilbert spaces (just as we work only with σ-finite measure spaces). Still, for most purposes Theorem 3.4.2 is sufficient for work in the nonseparable case, if one needs to do this.

3.5 The bounded spectral theorem

For the remainder of the chapter $\mathcal{H}$ will be a fixed separable Hilbert space.

We now have enough machinery to give a quick proof of a strong form of the spectral theorem for bounded self-adjoint operators. Given such an operator A, the idea is first to obtain a spectral measure on $\mathrm{sp}(A)$ by applying the Borel functional calculus to characteristic functions, and then to use Corollary 3.4.3 to realize $\mathcal{H}$ as the L^2 sections of a measurable Hilbert bundle. Finally, we observe that this identification associates A to the operator of multiplication by x.

THEOREM 3.5.1
Let $A \in B(\mathcal{H})$ be self-adjoint. Then there is a probability measure μ on $\mathrm{sp}(A)$, a measurable Hilbert bundle $\mathcal{X}$ over $\mathrm{sp}(A)$, and an isometric isomorphism $U : L^2(\mathrm{sp}(A);\mathcal{X}) \cong \mathcal{H}$ such that $A = UM_xU^{-1}$.

PROOF Recall that $\mathrm{sp}(A) \subset \mathbf{R}$ by Corollary 3.2.5. Now for any Borel set $S \subset \mathrm{sp}(A)$ let $E(S)$ be the range of the spectral projection $\chi_S(A)$ of A for S. The fact that E is a spectral measure follows immediately from the properties of the Borel functional calculus. Let μ, $\mathcal{X}$, and U be as in Corollary 3.4.3.

It is straightforward to verify that the map $f \mapsto UM_fU^{-1}$ is a unital $*$-homomorphism from $\mathrm{Bor}(\mathrm{sp}(A))$ into $B(\mathcal{H})$ which is continuous from the μ-topology to the weak operator topology. As it agrees with the Borel functional calculus on characteristic functions, continuity implies that the two must agree on all bounded Borel functions. In particular, taking $f(x) = x$, we obtain $A = f(A) = UM_xU^{-1}$. ∎

Evidently there is a one-to-one correspondence between bounded self-adjoint operators on $\mathcal{H}$ and spectral measures on compact subsets of $\mathbf{R}$. Given such an operator one can use its spectral projections to define a spectral measure on its spectrum, and conversely, given such a spectral measure one can use Corollary 3.4.3 to transfer M_x to $UM_xU^{-1} \in B(\mathcal{H})$.

This is still true when $\mathcal{H}$ is nonseparable. Although we cannot apply

Corollary 3.4.3 in this case, Theorem 3.4.2 still holds, and it can be used to establish a weaker version of the realization of A as a multiplication operator. Namely, let $\{\mu_\kappa\}$ be probability measures on $\mathrm{sp}(A)$ and U an isometric isomorphism from $\bigoplus L^2(\mathrm{sp}(A), \mu_\kappa)$ onto $\mathcal{H}$ as in Theorem 3.4.2. For each κ let X_κ be $\mathrm{sp}(A)$ equipped with the measure μ_κ, and let X be the disjoint union of the X_κ. Then $\bigoplus L^2(\mathrm{sp}(A), \mu_\kappa)$ can be identified with $L^2(X)$, and we have $A = UM_fU^{-1}$ where $f(x) = x$ on each X_κ. Of course, X is not σ-finite in this case.

It is worth noting the extent to which the construction of Theorem 3.5.1 is canonical. If A and A' are unitarily equivalent self-adjoint operators, meaning that there is a unitary U such that $A = U^{-1}A'U$, then $\mathrm{sp}(A) = \mathrm{sp}(A')$ and a continuity argument shows that $f(A) = U^{-1}f(A')U$ for all bounded Borel functions f. Thus their spectral measures are unitarily equivalent. Then the measures μ and μ' from Corollary 3.4.3 must be mutually absolutely continuous, and $\mathcal{X}$ and $\mathcal{X}'$ are characterized by the sequences (X_n) and (X_n') of Definition 2.4.8, which must be the same up to μ-null sets. Thus $\mathrm{sp}(A)$, μ (up to measure class), and (X_n) (up to null sets) constitute a set of invariants for A. Conversely, if B is any other self-adjoint operator with the same invariants then Theorem 3.5.1 gives rise to a unitary equivalence between A and B. So $\mathrm{sp}(A)$, μ, and (X_n) are a *complete* set of invariants.

Next we want to apply the self-adjoint theory to prove a version of the spectral theorem for normal operators. This can be done in the following way. If $N \in B(\mathcal{H})$ is normal, that is, $N^*N = NN^*$, then the self-adjoint operators $\mathrm{Re}\, N = \frac{1}{2}(N + N^*)$ and $\mathrm{Im}\, N = \frac{1}{2i}(N - N^*)$ commute. So $N = \mathrm{Re}\, N + i\mathrm{Im}\, N$ is a linear combination of commuting self-adjoint operators. Thus, the next result is all we need.

LEMMA 3.5.2

Let $A_1, A_2 \in B(\mathcal{H})$ be commuting self-adjoint operators and let E_1 and E_2 be the associated spectral measures on $\mathrm{sp}(A_1)$ and $\mathrm{sp}(A_2)$. Then there is a spectral measure E on $\mathrm{sp}(A_1) \times \mathrm{sp}(A_2)$ such that $E(S_1 \times \mathbf{R}) = E_1(S_1)$ and $E(\mathbf{R} \times S_2) = E_2(S_2)$ for all Borel sets $S_1 \subset \mathrm{sp}(A_1)$ and $S_2 \subset \mathrm{sp}(A_2)$.

This lemma can be proven by mimicking the construction of ordinary product measures: first define E on product sets by setting $E(S_1 \times S_2) = E_1(S_1) \cap E_2(S_2)$, then extend to finite unions of product sets, and finally extend to all Borel sets by taking limits. The proof that this indeed produces a spectral measure is not essentially different from the real-valued case. Alternatively, the lemma can be established in an algebraic manner; a more general result will be proven in this way in Lemma 5.3.4, so we omit the details here.

THEOREM 3.5.3

Let $N \in B(\mathcal{H})$ be normal. Then there is a probability measure μ on $\mathrm{sp}(N)$, a measurable Hilbert bundle $\mathcal{X}$ over $\mathrm{sp}(N)$, and an isometric isomorphism $U : L^2(\mathrm{sp}(N); \mathcal{X}) \cong \mathcal{H}$ such that $N = UM_zU^{-1}$.

PROOF Let $A_1 = \mathrm{Re}\, N$ and $A_2 = \mathrm{Im}\, N$, and let E be the spectral measure on $\mathrm{sp}(A_1) \times \mathrm{sp}(A_2) \subset \mathbf{R}^2$ given by the lemma. Let μ, $\mathcal{X}$, and U be as in Corollary 3.4.3.

We have $A_1 = UM_xU^{-1}$ and $A_2 = UM_yU^{-1}$ just as in the proof of Theorem 3.5.1. Thus $N = A_1 + iA_2 = UM_zU^{-1}$ where $z = x + iy$.

Evidently $M_{z-\lambda} = U^{-1}(A - \lambda I)U$ is invertible for any $\lambda \notin \mathrm{sp}(N)$. This implies that μ is supported on $\mathrm{sp}(N)$. ∎

Just as for self-adjoint operators, in the nonseparable case we can still prove that N is equivalent to a multiplication operator by using Theorem 3.4.2 in place of Corollary 3.4.3. See the comment following Theorem 3.5.1. Also, the observation made there that $\mathrm{sp}(A)$, μ, and (X_n) are a complete set of invariants applies here too.

Since unitary operators are normal, this theorem applies to them. But a multiplication operator M_f is unitary if and only if $|f| = 1$ almost everywhere. It follows that a normal operator is unitary if and only if its spectrum is contained in $\{z : |z| = 1\}$.

Essentially, Theorem 3.5.3 generalizes Theorem 3.5.1 to a pair of commuting self-adjoint operators. This will be subsumed under a far greater generalization when we discuss abelian C*-algebras in Section 5.3.

3.6 Unbounded operators

An operator is bounded if and only if it is continuous. For this reason, one generally prefers to work with bounded operators. However, there are cases where we must consider unbounded operators; in particular, some of the most basic operators in quantum mechanics are unbounded (see Section 4.1). This is related to the fact that the generators of one-parameter unitary groups are in general unbounded self-adjoint operators, as we will prove in Theorem 3.8.6.

Fortunately, the unbounded Hilbert space operators which arise in practice usually have a certain form which renders them moderately tractable. Namely, they are defined only on a dense (unclosed) subspace of the Hilbert space and on this domain they satisfy a kind of continuity property.

DEFINITION 3.6.1 *An unbounded operator on $\mathcal{H}$ is a linear map*

A from a dense subspace $D(A)$ of $\mathcal{H}$ into $\mathcal{H}$. It is closed if its graph $\Gamma(A) = \{v \oplus Av : v \in D(A)\}$ is a closed subspace of $\mathcal{H} \times \mathcal{H}$.

Occasionally one may have to deal with operators which are not densely defined. But since all of our unbounded operators will have dense domains, we incorporate this property into the definition.

Closed unbounded operators are still continuous in the sense that if (v_n) is a sequence in $D(A)$ and $v, w \in \mathcal{H}$, then $v_n \to v$ and $Av_n \to w$ imply that $v \in D(A)$ and $Av = w$. In particular, if A is closed and $D(A) = \mathcal{H}$ then A must be continuous and hence bounded. Thus, *not* requiring unbounded operators to be defined everywhere has the advantage that it allows us to formulate a reasonable continuity property, namely, graph closure. On the other hand, it has the disadvantage that in general one can neither add nor compose two unbounded operators (although one can always add or compose a bounded operator with an unbounded one). There is no general resolution of this difficulty; it must be dealt with case by case.

Often one first constructs an unbounded operator A which is not closed, and then defines $\bar{A}$ to be the closed unbounded operator whose graph is $\overline{\Gamma(A)}$. The operator $\bar{A}$ is called the closure of A. This procedure does not automatically work: one must check that $\overline{\Gamma(A)}$ really is the graph of an operator, which amounts to verifying that it contains no element of the form $0 \oplus v$ with $v \neq 0$.

We call a dense subspace $D \subset D(A)$ a core of A if $\Gamma(A) = \overline{\Gamma(A|_D)}$. Thus, in the last paragraph's construction $D(A)$ would be a core of $\bar{A}$. This is the unbounded version of a dense subspace of the domain. This concept will be useful in Chapter 4.

Next we define the adjoint of a closed unbounded operator. The slickest way to do this is in terms of the map $V : \mathcal{H} \oplus \mathcal{H} \to \mathcal{H} \oplus \mathcal{H}$ given by $V(v \oplus w) = -w \oplus v$. First we need to prove that the definition will make sense.

PROPOSITION 3.6.2

Let A be a closed unbounded operator on $\mathcal{H}$. Then $(V\Gamma(A))^{\perp}$ is the graph of a closed unbounded operator B. Its domain consists of precisely those vectors $w \in \mathcal{H}$ for which the map $v \mapsto \langle Av, w\rangle$ is bounded, and we have $\langle Av, w\rangle = \langle v, Bw\rangle$ for all $v \in D(A)$ and $w \in D(B)$.

PROOF Let $E = (V\Gamma(A))^{\perp}$. We begin by showing that the projection of E onto the first summand is one-to-one. It is enough to show that $0 \oplus w \in E$ implies $w = 0$. But if $(0 \oplus w) \perp (-Av \oplus v)$ for all $v \in D(A)$ then $w \perp v$ for all $v \in D(A)$, which implies $w = 0$ as desired.

For all $v \oplus w \in E$ define $Bv = w$; this is possible by the last paragraph. To show that $D(B)$ is dense, let $w \in D(B)^\perp$. Then $(0 \oplus w) \perp (-Bv \oplus v)$ for all $v \in D(B)$, so that $0 \oplus w \in (V(E))^\perp$. But since V is unitary it commutes with $\perp$, so

$$(V(E))^\perp = V((V\Gamma(A))^\perp)^\perp = (\Gamma(A))^{\perp\perp} = \Gamma(A).$$

Thus $w = 0$, and we conclude that $D(B)$ is dense in $\mathcal{H}$. As $E = \Gamma(B)$ is clearly closed, B is a closed unbounded operator.

If $w \in D(B)$ then $(w \oplus Bw) \perp (-Av \oplus v)$ for all $v \in D(A)$, and hence $\langle Av, w\rangle = \langle v, Bw\rangle$ for all $v \in D(A)$. This implies that the map $v \mapsto \langle Av, w\rangle = \langle v, Bw\rangle$ is bounded. Conversely, if the map $v \mapsto \langle Av, w\rangle$ is bounded then it is given by taking the inner product with some $w' \in \mathcal{H}$. Then $\langle Av, w\rangle = \langle v, w'\rangle$ for all $v \in D(A)$, and hence $(w \oplus w') \perp (-Av \oplus v)$. So $w \in D(B)$. This completes the proof. ∎

DEFINITION 3.6.3 *Let A be a closed unbounded operator on $\mathcal{H}$. Then A^* is the closed unbounded operator whose graph is $(V\Gamma(A))^\perp$. A is self-adjoint if $A = A^*$.*

It follows from Proposition 3.6.2 that Definition 3.6.3 generalizes the definition of adjoints of bounded operators. It is important to note that in the unbounded case the definition of self-adjoint operators includes the provision that $D(A) = D(A^*)$.

If A is closed, then $A^{**} = A$; this follows immediately from the definition of A^* and the fact that V and $\perp$ commute.

We can characterize A^* more simply as follows: $v_1 \in D(A^*)$ and $A^* v_1 = v_2$ if and only if $\langle v_1, Aw\rangle = \langle v_2, w\rangle$ for all $w \in D(A)$.

We indicate the unbounded version of multiplication operators.

Example 3.6.4

Let X be a σ-finite measure space and let $f : X \to \mathbf{C}$ be any measurable function. Define $M_f : L^2(X) \to L^2(X)$ by $M_f g = fg$, with domain the set of functions $g \in L^2(X)$ such that $fg \in L^2(X)$.

To see that $D(M_f)$ is dense define $X_k = \{x \in X : |f(x)| \le k\}$, so that $X = \bigcup_1^\infty X_k$. Then we have $fg \in L^2(X_k) \subset L^2(X)$ whenever $g \in L^2(X_k)$, so $L^2(X_k) \subset D(M_f)$ for all k. This is enough.

The graph of M_f is closed because if $g_n \to g$ and $fg_n \to h$, both in $L^2(X)$, then we can pass to a subsequence for which both converge pointwise almost everywhere. This implies $fg = h$, so $g \in D(M_f)$ and $M_f g = h$.

We have $M_f^* = M_{\bar{f}}$ where $\bar{f}$ is the pointwise complex conjugate of f; this can be verified directly from Proposition 3.6.2. In particular, $D(M_f^*) = D(M_f)$. Also, M_f is self-adjoint if and only if f is real almost everywhere.

Exactly the same results are true with $L^2(X)$ replaced by $L^2(X;\mathcal{X})$ for $\mathcal{X}$ a measurable Hilbert bundle over X.

3.7 The unbounded spectral theorem

From the point of view of spectral theory, unbounded self-adjoint operators are quite well-behaved. They arise from spectral measures on the real line in the same way as bounded self-adjoint operators, the only difference being that in the unbounded case the relevant spectral measure is not supported on a compact set.

We will prove this result by reducing to the bounded normal case via functional calculus: the function $f(x) = (x+i)^{-1}$ is bounded on the real line, so if A is an unbounded self-adjoint operator we expect $f(A) = (A+i)^{-1}$ to be a bounded normal operator. Having shown this, we will then invoke Theorem 3.5.3 to obtain a multiplication operator realization of $(A+i)^{-1}$, and finally we will recover the original operator A by applying the function $(x-i)^{-1}$. We proceed through a series of lemmas.

LEMMA 3.7.1
Let A be an unbounded self-adjoint operator. Then $A \pm iI$ has a bounded inverse in the sense that there exists $B \in B(\mathcal{H})$ such that $Bv \in D(A)$ and $(A \pm iI)Bv = v$ for all $v \in \mathcal{H}$, and $B(A \pm iI)v = v$ for all $v \in D(A)$.

PROOF Since A is closed, it is easy to check that $A \pm iI$ (with the same domain as A) is also closed. As in the proof of Corollary 3.2.5 we have

$$\|(A \pm iI)v\|^2 = \|Av\|^2 + \|v\|^2$$

for all $v \in D(A)$, so $\|(A \pm iI)v\| \geq \|v\|$. Thus projecting $\Gamma(A \pm iI)$ into the second summand yields an isomorphism between $\Gamma(A \pm iI)$ and the range of $A \pm iI$. So $\mathrm{ran}(A \pm iI)$ is closed, and we need only show that it is dense; this will imply that $\mathrm{ran}(A \pm iI) = \mathcal{H}$, so that $B = (A \pm iI)^{-1}$ is defined everywhere. B will then be bounded, and in fact a contraction, since $\|(A \pm iI)v\| \geq \|v\|$.

Thus, suppose $\langle w, (A \pm iI)v\rangle = 0$ for all $v \in D(A) = D(A \pm iI)$. Then $\langle w, Av\rangle = \pm\langle iw, v\rangle$ for all $v \in D(A)$, and hence $w \in D(A^*) = D(A)$ and $Aw = A^*w = \pm iw$. But then

$$\pm i\langle w, w\rangle = \langle Aw, w\rangle = \langle w, Aw\rangle = \mp i\langle w, w\rangle,$$

so that $w = 0$. We conclude that $\mathrm{ran}(A \pm iI)$ is indeed dense in $\mathcal{H}$. ∎

In the next lemma we let $D(A^2)$ be the set of vectors $v \in \mathcal{H}$ such that $v \in D(A)$ and $Av \in D(A)$. It is not *a priori* clear that A^2 is densely defined, but we do not need this fact here. (It will follow from Theorem 3.7.4, however.)

LEMMA 3.7.2
Let A be an unbounded self-adjoint operator. Then $A+iI$ carries $D(A^2)$ into $D(A-iI)$ and $A-iI$ carries $D(A^2)$ into $D(A+iI)$. We have $(A+iI)(A-iI)v = (A-iI)(A+iI)v$ for all $v \in D(A^2)$.

PROOF Suppose $v \in D(A^2)$. Then v and Av both belong to $D(A)$, so that $(A+iI)v = Av + iv \in D(A) = D(A-iI)$. Similarly, we have $(A-iI)v \in D(A+iI)$.

For $v \in D(A^2)$ we have

$$\begin{aligned}(A+iI)(A-iI)v &= (A+iI)(Av-iv) = A^2v + v\\ &= (A-iI)(Av+iv) = (A-iI)(A+iI)v,\end{aligned}$$

so $A+iI$ and $A-iI$ commute on $D(A^2)$. ∎

LEMMA 3.7.3
Let A be an unbounded self-adjoint operator. Then $(A+iI)^{-1} \in B(\mathcal{H})$ is normal and its adjoint is $(A-iI)^{-1}$.

PROOF The inverses exist by Lemma 3.7.1. Note that $\Gamma(B^{-1})$ is the transpose of $\Gamma(B)$ whenever B is invertible. This, together with the easy fact that $(A+iI)^* = A-iI$, implies that the adjoint of $(A+iI)^{-1}$ is $(A-iI)^{-1}$. To prove normality, let $u \in \mathcal{H}$. Then $u = (A+iI)v$ for some $v \in D(A)$, and $v = (A-iI)w$ for some $w \in D(A)$. Since v and w are both in $D(A)$ it follows that $Aw = v + iw \in D(A)$, so that $w \in D(A^2)$. By Lemma 3.7.2 we then have

$$u = (A+iI)(A-iI)w = (A-iI)(A+iI)w,$$

so

$$(A+iI)^{-1}(A-iI)^{-1}u = w = (A-iI)^{-1}(A+iI)^{-1}u.$$

Thus $(A+iI)^{-1}$ and $(A-iI)^{-1}$ commute. ∎

Now we can prove the spectral theorem for unbounded self-adjoint operators.

THEOREM 3.7.4
Let A be an unbounded self-adjoint operator on $\mathcal{H}$. Then there is a probability measure μ on $\mathbf{R}$, a measurable Hilbert bundle $\mathcal{X}$ over $\mathbf{R}$ and an isometric isomorphism $U : L^2(\mathbf{R}; \mathcal{X}) \cong \mathcal{H}$ such that $A = UM_xU^{-1}$.

PROOF By Lemma 3.7.3 the bounded operator $(A+iI)^{-1}$ is normal. Let $X = \mathrm{sp}(A+iI)^{-1}$ and find μ, $\mathcal{X}$, and U as in Theorem 3.5.3 such that $(A + iI)^{-1} = UM_zU^{-1}$. Then $A = UM_fU^{-1}$ where $f(z) = -i + 1/z$; since A is self-adjoint, it follows that $f(X) \subset \mathbf{R}$, so we can use f to identify X with a subset of $\mathbf{R}$. The result follows. ∎

As usual, in the nonseparable case we can still prove that A is equivalent to a multiplication operator by using Theorem 3.4.2 in place of Corollary 3.4.3.

3.8 Stone's theorem

The time evolution of a quantum mechanical system is modelled by a one-parameter group of unitary operators with a certain continuity property (cf. Section 1.4.) These groups also arise in purely mathematical settings, of course. Their general definition is as follows.

DEFINITION 3.8.1 *A weakly continuous one-parameter unitary group is a family $\{U_t : t \in \mathbf{R}\}$ of unitary operators with the properties that $U_0 = I$, $U_sU_t = U_{s+t}$ for all $s, t \in \mathbf{R}$, and $U_s \to U_t$ weak operator whenever $s \to t$ in $\mathbf{R}$.*

This continuity requirement is equivalent to the apparently stronger condition that $s \to t$ implies $U_sv \to U_tv$ for all $v \in \mathcal{H}$. That is because weak operator continuity implies $\langle U_{s-t}v, v\rangle \to \langle v, v\rangle = \|v\|^2$; so

$$\|U_sv - U_tv\|^2 = 2\|v\|^2 - 2\mathrm{Re}\langle U_sv, U_tv\rangle \to 0$$

as $s \to t$. The reverse implication is trivial.

Example 3.8.2

Let A be an unbounded self-adjoint operator on $\mathcal{H}$. Without loss of generality suppose $\mathcal{H} = L^2(\mathbf{R}; \mathcal{X})$ and $A = M_x$. For each $t \in \mathbf{R}$ write $e^{iAt} = M_{e^{itx}}$ and let $U_t = e^{itA}$. This is a weakly continuous one-parameter unitary group.

We now want to prove Stone's theorem, which states that conversely to Example 3.8.2, every weakly continuous one-parameter unitary group is of the form $U_t = e^{itA}$ for some unbounded self-adjoint operator A. The main tool we need is the Fourier transform on $\mathbf{R}$. Let $C_c^\infty(\mathbf{R})$ be the space of compactly supported C^∞ functions on $\mathbf{R}$. For $f \in C_c^\infty(\mathbf{R})$ we define $\hat{f}(t) = \frac{1}{\sqrt{2\pi}} \int f(x)e^{-itx}\,dx$. We will use the following properties of $\hat{f}$:

(a) $f \in C_c^\infty(\mathbf{R})$ implies $\hat{f} \in L^1(\mathbf{R})$,
(b) $(e^{-itx}f)\hat{}(s) = \hat{f}(s+t)$, and
(c) $\widehat{fg}(t) = \frac{1}{\sqrt{2\pi}} \int \hat{f}(s)\hat{g}(t-s)\,ds$.

We will also need to integrate operator-valued functions. This can be done in the following way.

DEFINITION 3.8.3 *Let X be a σ-finite measure space. Suppose $A : X \to B(\mathcal{H})$ has the property that the function $x \mapsto \langle A(x)v, w\rangle$ is measurable and satisfies $\int |\langle A(x)v, w\rangle| \leq C$ for all $v, w \in \mathcal{H}$. Then the <u>weak integral</u> of A is the bounded operator $A = \int A(x)$ which satisfies $\langle Av, w\rangle = \int \langle A(x)v, w\rangle$ for all $v, w \in \mathcal{H}$.*

Weak integrals exist by Lemma 3.3.2.

Now fix a weakly continuous one-parameter unitary group $\{U_t\}$ on $\mathcal{H}$. Define $T : C_c^\infty(\mathbf{R}) \to B(\mathcal{H})$ by

$$T(f) = \frac{1}{\sqrt{2\pi}} \int \hat{f}(t)U_t\,dt.$$

Here we are integrating the operator-valued function $t \mapsto \hat{f}(t)U_t$ according to Definition 3.8.3. This uses property (a) of the Fourier transform.

LEMMA 3.8.4
The map T is a $*$-homomorphism.

PROOF Let $f \in C_c^\infty(\mathbf{R})$. Then $\hat{\bar{f}}(t) = \bar{\hat{f}}(-t)$, so

$$\begin{aligned}\langle T(\bar{f})v, w\rangle &= \frac{1}{\sqrt{2\pi}} \int \hat{\bar{f}}(t)\langle U_t v, w\rangle\,dt \\ &= \frac{1}{\sqrt{2\pi}} \int \bar{\hat{f}}(t)\langle U_{-t} v, w\rangle\,dt \\ &= \langle v, T(f)w\rangle.\end{aligned}$$

Thus $T(\bar{f}) = T(f)^*$.

Now let $f, g \in C_c^\infty(\mathbf{R})$. Since $\widehat{fg}(t) = \frac{1}{\sqrt{2\pi}} \int \hat{f}(s)\hat{g}(t-s)\,ds$ (property (c) of the Fourier transform), we have

$$\begin{aligned}
\langle T(fg)v, w\rangle &= \frac{1}{\sqrt{2\pi}} \int \widehat{fg}(t)\langle U_t v, w\rangle\,dt \\
&= \frac{1}{2\pi} \int\int \hat{f}(s)\hat{g}(t-s)\langle U_{t-s}v, U_{-s}w\rangle\,dsdt \\
&= \frac{1}{2\pi} \int \hat{f}(s) \int \hat{g}(t)\langle U_t v, U_{-s}w\rangle\,dtds \\
&= \frac{1}{\sqrt{2\pi}} \int \hat{f}(s)\langle T(g)v, U_{-s}w\rangle\,dt \\
&= \langle T(f)T(g)v, w\rangle
\end{aligned}$$

for all $v, w \in \mathcal{H}$. The application of Fubini's theorem is legitimate because $\hat{f}(s)\hat{g}(t-s) \in L^1(\mathbf{R}^2)$. We conclude that $T(fg) = T(f)T(g)$. As T is clearly linear, this completes the proof. ∎

LEMMA 3.8.5

T extends to a unital $*$-homomorphism $\tilde{T}$ from Bor($\mathbf{R}$) *into $B(\mathcal{H})$ which is continuous from the μ-topology to the weak operator topology.*

PROOF We prove this by first extending T to $C_0(\mathbf{R})$ by norm continuity, and then passing to bounded Borel functions by the technique of Theorem 3.3.3. The main novelty is the verification that T is continuous in sup norm.

To see this, let $f \in C_c^\infty(\mathbf{R})$ be real-valued and choose $\lambda \in \mathbf{C}$ such that $|\lambda| > \|f\|_\infty$. Then $g = (f-\lambda)^{-1} + \lambda^{-1} \in C_c^\infty(\mathbf{R})$ and we have $fg - \lambda g - \lambda^{-1}f = 0$. Since T is multiplicative it follows that

$$(T(f) - \lambda I)(T(g) - \lambda^{-1}I) = T(fg - \lambda g - \lambda^{-1}f) + I = I.$$

Thus $\lambda \notin \mathrm{sp}(T(f))$. By Proposition 3.2.4 we conclude that $\|T(f)\| \leq \|f\|_\infty$. For complex-valued $f \in C_c(\mathbf{R})$ we then have

$$\|T(f)\|^2 = \|T(f)^*T(f)\| = \|T(|f|^2)\| \leq \||f|^2\|_\infty = \|f\|_\infty^2,$$

so T is a contraction.

Continuous extension of T to the set of bounded Borel functions on $\mathbf{R}$ is established by an argument similar to the one given in the proof of Theorem 3.3.3. The fact that $\tilde{T}$ is unital follows from the fact that the functions $f_n(x) = e^{-x^2/n}$ μ-converge to $1_\mathbf{R}$: we have $\hat{f}_n(t) = \sqrt{\frac{n}{2}}e^{-nt^2/4}$, and $T(f_n) \to I$ weak operator is a straightforward computation. This implies that $\tilde{T}(1_\mathbf{R}) = I$. ∎

We can now prove Stone's theorem.

THEOREM 3.8.6
Let $\{U_t\}$ be a weakly continuous one-parameter unitary group on $\mathcal{H}$. Then there is an unbounded self-adjoint operator A on $\mathcal{H}$ such that $U_t = e^{itA}$.

PROOF By Lemma 3.8.5, $\tilde{T}$ defines an $\mathcal{H}$-valued spectral measure E on $\mathbf{R}$ such that $P_{E(S)} = \tilde{T}(\chi_S)$ for $S \subset \mathbf{R}$. Find μ, $\mathcal{X}$, and U as in Corollary 3.4.3, identify $\mathcal{H}$ with $L^2(\mathbf{R};\mathcal{X})$, and let $A = M_x$. We must verify that $U_t = e^{itA}$.

It is enough to show that $\langle e^{itA}v, U_t w\rangle = \langle v, w\rangle$ for all $v, w \in L^2(K;\mathcal{X})$ where K is an arbitrary compact interval in $\mathbf{R}$. Fix K, v, and w and let $f \in C_c^\infty(\mathbf{R})$ satisfy $f(x) = e^{itx}$ for $x \in K$. Then $e^{-itx}f = 1$ on K and we have $T(e^{-itx}f)v = M_{e^{-itx}f}v = v$, so

$$\begin{aligned}\langle T(f)v, U_t w\rangle &= \frac{1}{\sqrt{2\pi}}\int \hat{f}(s)\langle U_s v, U_t w\rangle\, ds \\ &= \frac{1}{\sqrt{2\pi}}\int \hat{f}(s+t)\langle U_s v, w\rangle\, ds \\ &= \left\langle T(e^{-itx}f)v, w\right\rangle \\ &= \langle v, w\rangle.\end{aligned}$$

But $T(f)v = e^{itA}v$, so $\langle e^{itA}v, U_t w\rangle = \langle v, w\rangle$, as desired. ∎

The operator A is called the generator of the unitary group $\{U_t\}$ because iA is the derivative at $t = 0$ of the function $t \mapsto U_t$ in the following sense. By Theorem 3.7.4, assume $\mathcal{H} = L^2(\mathbf{R};\mathcal{X})$ and $A = M_x$. Then $g \in L^2(\mathbf{R};\mathcal{X})$ belongs to $D(A)$ if and only if $xg \in L^2(\mathbf{R};\mathcal{X})$. Now

$$\lim_{t\to 0}\frac{1}{t}(U_t g - U_0 g) = \lim_{t\to 0}\frac{1}{t}(e^{itx} - 1)g.$$

Since $\|xg\| < \infty$, for any $\epsilon > 0$ we can find $N > 0$ such that $g = g_1 + g_2$ where g_1 is supported on $[-N, N]$ and $\|xg_2\| \le \epsilon$. But $(e^{itx} - 1)/t \to ix$ uniformly on $[-N, N]$, while the elementary estimate $|(e^{itx} - 1)/t| \le |x|$ implies that $\|\frac{1}{t}(e^{itx} - 1)g_2\| \le \epsilon$. From this it follows that

$$iAg = ixg = \lim_{t\to 0}\frac{1}{t}(U_t g - U_0 g).$$

Thinking of U_t as a multiplication operator also helps explain our construction of A. Informally, for any Borel set $S \subset \mathbf{R}$ we have $T(\chi_S) =$

$\frac{1}{\sqrt{2\pi}}\int \hat{\chi}_S(t)U_t\,dt$, and $\frac{1}{\sqrt{2\pi}}\int \hat{\chi}_S(t)e^{ita}\,dt = \chi_S(a)$ by Fourier inversion. So if $U_t = M_{e^{itx}}$ then we ought to have $T(\chi_S) = M_{\chi_S}$. This nonrigorous argument shows why we expect T to give rise to the spectral measure associated with $A = M_x$.

3.9 Notes

Similar topics are covered in [13]. A shorter proof of a slightly weaker version of the spectral theorem for bounded operators is given in [58]. Our approach to spectral and multiplicity theory via Hilbert bundles seems to be new. Our proof of Stone's theorem is based on an idea of Ed Effros.

Chapter 4

The Quantum Plane

4.1 Position and momentum

We now have the necessary machinery to treat what is probably the most fundamental example in quantum mechanics: the one-dimensional particle. Its classical analog was mentioned in Section 1.1, and there we pointed out that the phase space of this classical system can be identified with the real plane $\mathbf{R}^2$. The phase space of the corresponding quantum system is modelled on the Hilbert space $L^2(\mathbf{R})$, and this space, together with some associated structure, plays the role of a "quantum" plane. As we will see in this and later chapters, the complex, topological, measure theoretic, metric, and differentiable structures of the ordinary plane all have quantum analogs. In this section we introduce "coordinates" and "translations" on the quantum plane.

We will begin with the position and momentum operators, which correspond to the coordinate functions on $\mathbf{R}^2$ (see Section 1.1). These operators are best understood in terms of the Fourier transform. Recall that for $f \in C_c^\infty(\mathbf{R})$ we define $\hat{f}(t) = \frac{1}{\sqrt{2\pi}} \int f(x)e^{-itx}\, dx$. We also require the Fourier inversion formula $f(x) = \frac{1}{\sqrt{2\pi}} \int \hat{f}(t)e^{itx}\, dt$.

PROPOSITION 4.1.1
Let $f, g \in C_c^\infty(\mathbf{R})$. Then $\int \hat{f}\hat{\bar{g}}\, dt = \int f\bar{g}\, dx$. The Fourier transform $\mathcal{F} : f \mapsto \hat{f}$ extends to a unitary operator on $L^2(\mathbf{R})$.

PROOF We have

$$\begin{aligned}\int \hat{f}(t)\bar{\hat{g}}(t)\, dt &= \frac{1}{\sqrt{2\pi}} \int\int \hat{f}(t)\overline{g(x)}e^{itx}\, dxdt \\ &= \int f(x)\overline{g(x)}\, dx.\end{aligned}$$

Fubini's theorem applies because the function $\hat{f}(t)\overline{g(x)}$ is in $L^1(\mathbf{R}^2)$.

This shows that $\mathcal{F}$ takes $C_c^\infty(\mathbf{R})$ isometrically into $L^2(\mathbf{R})$, and it extends to $L^2(\mathbf{R})$ by continuity.

To see that the extension is surjective, let $f \in C_c^\infty(\mathbf{R})$. Then differentiating under the integral sign (this is legitimate) yields

$$\begin{aligned}\hat{f}'(t) &= \frac{1}{\sqrt{2\pi}} \int -ixf(x)e^{-itx}\, dx \\ &= -i(xf)\hat{\ }(t);\end{aligned}$$

inductively, we obtain $d^n\hat{f}/dt^n = (-i)^n(x^n f)\hat{\ }$, and hence $\hat{f} \in C^\infty(\mathbf{R})$. As we also have $\hat{f} \in L^1(\mathbf{R}) \cap L^2(\mathbf{R})$, it follows that there is a sequence $(g_n) \subset C_c^\infty(\mathbf{R})$ such that $g_n \to \hat{f}$ in both L^1 and L^2 norms. Let $\tilde{g}_n(x) = g_n(-x)$. Then $(\tilde{g}_n)\hat{\ }(x) \to (\hat{f})\check{\ }(x) = f(x)$ pointwise (since $g_n \to \hat{f}$ in L^1 norm) and $(\tilde{g}_n)\hat{\ }$ converges in $L^2(\mathbf{R})$ (since $\tilde{g}_n$ converges in L^2 norm), so we must have $\mathcal{F}\tilde{g}_n \to f$ in $L^2(\mathbf{R})$. This shows that the range of the extension of $\mathcal{F}$ contains $C_c^\infty(\mathbf{R})$, and together with the fact that $\mathcal{F}$ is isometric this implies $\mathcal{F}$ is unitary. ∎

We use the same symbol $\mathcal{F}$ for the extension of the Fourier transform to $L^2(\mathbf{R})$. As we are now treating $\mathcal{F}$ as a map from $L^2(\mathbf{R})$ to itself, the variables x and t can be freely interchanged.

DEFINITION 4.1.2 *Fix a real number $\hbar > 0$. The* <u>position operator</u> *is the unbounded self-adjoint operator $\mathcal{Q} = M_x$ on $L^2(\mathbf{R})$. The* <u>momentum operator</u> *is the unbounded self-adjoint operator $\mathcal{P} = \hbar\mathcal{F}^{-1}M_x\mathcal{F}$.*

The number $\hbar$ is Planck's constant. For us its significance is that as it approaches zero the quantum system becomes classical. We will return to this point in the next section.

If $f \in C_c^\infty(\mathbf{R})$ then integration by parts yields

$$\begin{aligned}\mathcal{F}\left(-i\frac{df}{dx}\right)(t) &= \frac{1}{\sqrt{2\pi}} \int -i\frac{df}{dx}e^{-itx}\, dx \\ &= \frac{1}{\sqrt{2\pi}} \int tf(x)e^{-itx}\, dx \\ &= t\hat{f}(t),\end{aligned}$$

so $\mathcal{F}(-i(df/dx)) = M_x\mathcal{F}(f)$. Thus it makes sense to define $-i\frac{d}{dx}$ to be the unbounded self-adjoint operator $\mathcal{F}^{-1}M_x\mathcal{F}$. We can then write $\mathcal{P} = -i\hbar\frac{d}{dx}$.

Physically, $\mathcal{Q}$ and $\mathcal{P}$ are interpreted in the following way. If S is a Borel subset of $\mathbf{R}$ then $L^2(S) \subset L^2(\mathbf{R})$ models the event that the particle lies in the region S. That is, a state vector $f \in L^2(\mathbf{R})$ describes a particle which definitely belongs to S precisely if f is supported on S. Correspondingly, the spectral projection of $\mathcal{Q} = M_x$ for the function χ_S is the projection operator $\chi_S(M_x) = M_{\chi_S}$ with range $L^2(S)$, and on any state supported in $L^2(S)$ the observable $\mathcal{Q}$ will take values in S with certainty.

The momentum operator has exactly the same structure, up to the scaling factor $\hbar$, if one works in the Fourier transform picture. This means that the event that the particle's momentum lies in $\hbar S$ is modelled by the subspace $\mathcal{F}^{-1}(L^2(S))$, and hence that a state vector f describes a particle whose momentum definitely belongs to $\hbar S$ precisely if $\hat{f}$ is supported on S.

Recall that self-adjoint operators arise as generators of weakly continuous one-parameter unitary groups (Theorem 3.8.6). Since bounded operators are more tractable than unbounded operators, it is often easier to deal with the unitaries $e^{is\mathcal{Q}}$ and $e^{it\mathcal{P}}$ than with $\mathcal{Q}$ and $\mathcal{P}$ themselves. We have immediately that $e^{is\mathcal{Q}} = M_{e^{isx}}$ and $e^{it\mathcal{P}} = \mathcal{F}^{-1}M_{e^{i\hbar tx}}\mathcal{F}$. But also, if $f \in L^2(\mathbf{R})$ then

$$M_{e^{i\hbar tx}}\mathcal{F}(f) = e^{i\hbar tx}\mathcal{F}(f) = \mathcal{F}(T_{-\hbar t}f),$$

where T_t is the translation operator $T_t f(x) = f(x-t)$. Thus $e^{it\mathcal{P}} = T_{-\hbar t}$. Likewise, $e^{it\mathcal{Q}} = \mathcal{F}^{-1}T_t\mathcal{F}$.

Next we examine the failure of $\mathcal{Q}$ and $\mathcal{P}$, and $e^{is\mathcal{Q}}$ and $e^{it\mathcal{P}}$, to commute.

THEOREM 4.1.3
(a) Let $f \in C_c^\infty(\mathbf{R})$. Then $\mathcal{QP}f - \mathcal{PQ}f = i\hbar f$.
(b) $e^{is\mathcal{Q}}e^{it\mathcal{P}} = e^{-i\hbar st}e^{it\mathcal{P}}e^{is\mathcal{Q}}$ for all $s, t \in \mathbf{R}$.

PROOF
(a) We have $\mathcal{QP}f = -i\hbar x\frac{d}{dx}f$ and $\mathcal{PQ}f = -i\hbar\frac{d}{dx}(xf) = -i\hbar(f + x\frac{d}{dx}f)$. Thus $\mathcal{QP}f - \mathcal{PQ}f = i\hbar f$.
(b) If $f \in L^2(\mathbf{R})$ then $e^{is\mathcal{Q}}e^{it\mathcal{P}}f(x) = e^{isx}f(x+\hbar t)$ and $e^{it\mathcal{P}}e^{is\mathcal{Q}}f(x) = e^{is(x+\hbar t)}f(x+\hbar t)$. Thus $e^{is\mathcal{Q}}e^{it\mathcal{P}} = e^{-i\hbar st}e^{it\mathcal{P}}e^{is\mathcal{Q}}$. ∎

Usually in physics one just says that $\mathcal{QP} - \mathcal{PQ} = i\hbar I$. This and the fact that $\mathcal{Q}$ and $\mathcal{P}$ each commute with themselves are called the canonical commutation relations (CCRs). Now since $\mathcal{P}$ and $\mathcal{Q}$ are both unbounded, and are not defined on all of $L^2(\mathbf{R})$, there is a question of what is meant by the expression $\mathcal{QP} - \mathcal{PQ}$. Here we have dealt with this issue by restricting ourselves to vectors f which have the property that $f \in D(\mathcal{Q}) \cap D(\mathcal{P})$, $\mathcal{Q}f \in D(\mathcal{P})$, and $\mathcal{P}f \in D(\mathcal{Q})$. We will discuss commutators of unbounded operators further in Section 4.4.

The existence of this kind of question shows why it is generally preferrable to work with the bounded operators $e^{it\mathcal{Q}}$ and $e^{it\mathcal{P}}$ whenever possible. The equation in Theorem 4.1.3 (b), which can be thought of as an integral version of the equation in part (a), is called the Weyl form of the canonical commutation relations.

These exponentiated operators are significant in another way. Observe that if $f \in L^2(\mathbf{R})$ is supported on a set S — intuitively, its position "coordinate" lies in S — then $e^{-is\mathcal{P}/\hbar} f = T_s f$ is supported on $S + s$. Thus $e^{-is\mathcal{P}/\hbar}$ plays the quantum plane role of translation in the position variable. Likewise, $e^{it\mathcal{Q}/\hbar}$ plays the role of translation in the momentum variable.

Just as the operators $\mathcal{Q}$ and $\mathcal{P}$ act like coordinate functions on the quantum plane, polynomials in $\mathcal{Q}$ and $\mathcal{P}$ (that is, operators of the form $\sum a_{ij}\mathcal{Q}^i\mathcal{P}^j$) act like polynomials on the quantum plane. These expressions require some interpretation, since $\mathcal{Q}$ and $\mathcal{P}$ are unbounded.

PROPOSITION 4.1.4

Let $\sum a_{ij}x^iy^j$ be a complex polynomial in two variables, and for $f \in C_c^\infty(\mathbf{R})$ define $Af = \sum a_{ij}\mathcal{Q}^i\mathcal{P}^j f$ and $Bf = \sum \bar{a}_{ij}\mathcal{P}^j\mathcal{Q}^i f$. Let $\Gamma(A) = \{f \oplus Af : f \in C_c^\infty(\mathbf{R})\}$.

(a) We have $\langle Af, g\rangle = \langle f, Bg\rangle$ for all $f, g \in C_c^\infty(\mathbf{R})$.
(b) The closure of $\Gamma(A)$ in $L^2(\mathbf{R}) \oplus L^2(\mathbf{R})$ is the graph of an unbounded operator.

PROOF

(a) This follows from self-adjointness of $\mathcal{Q}$ and $\mathcal{P}$.

(b) The issue is whether the closure of $\Gamma(A)$ defines a single-valued operator on $L^2(\mathbf{R})$. It will suffice to show that $\overline{\Gamma(A)}$ contains no element of the form $0 \oplus f$ for $f \neq 0$. Thus, suppose that $(f_n) \subset L^2(\mathbf{R})$, $f_n \to 0$, and $Af_n \to f \in L^2(\mathbf{R})$. Then for any $g \in C_c^\infty(\mathbf{R})$ we have

$$\langle Af_n, g\rangle = \langle f_n, Bg\rangle \to 0.$$

Since $Af_n \to f$, this means that $\langle f, g\rangle = 0$ for all $g \in C_c^\infty(\mathbf{R})$, and hence $f = 0$. ∎

In light of part (b) of this proposition, it makes sense to define the operator $\sum a_{ij}\mathcal{Q}^i\mathcal{P}^j$ to be the closure of A. Thus, $C_c^\infty(\mathbf{R})$ is a core (in the sense of Section 3.6) for every operator of the form $\sum a_{ij}\mathcal{Q}^i\mathcal{P}^j$. Moreover, it has the pleasant property of being invariant in the sense that $\mathcal{Q}^i\mathcal{P}^j(C_c^\infty(\mathbf{R})) \subset C_c^\infty(\mathbf{R})$. Thus, polynomials in $\mathcal{Q}$ and $\mathcal{P}$ can be added and multiplied in a straightforward manner on $C_c^\infty(\mathbf{R})$. The product of any two polynomials can be computed using the commutation relation of Theorem 4.1.3 (a).

4.2 The tracial representation

The realization of $\mathcal{Q}$ and $\mathcal{P}$ as multiplication and differentiation operators on $L^2(\mathbf{R})$ is the most basic, but there are other models which bring out certain features of these operators more clearly. For instance, by choosing a suitable orthonormal basis of $L^2(\mathbf{R})$ we can use Theorem 2.3.3 to establish an isometric isomorphism between $L^2(\mathbf{R})$ and $l^2(\mathbf{N})$, in such a way that a certain aspect of the structure of $\mathcal{Q}$ and $\mathcal{P}$ becomes transparent. We will do this in Section 4.3.

In the present section we do something slightly different: we produce a unitary operator W on $L^2(\mathbf{R}^2) \cong L^2(\mathbf{R}) \otimes L^2(\mathbf{R})$ such that the operators $W^{-1}(\mathcal{Q} \otimes I)W$ and $W^{-1}(\mathcal{P} \otimes I)W$ have a special symmetrical form for which there is a reasonable limit as $\hbar \to 0$. These operators are equivalent not to $\mathcal{Q}$ and $\mathcal{P}$, but rather to $\mathcal{Q} \otimes I$ and $\mathcal{P} \otimes I$. There is no physical significance in this distinction; in effect we are treating a composite system where one subsystem is trivial. Our reasons for doing this are purely mathematical. Loosely speaking, this representation is a symmetrical combination of the ordinary representation of $\mathcal{Q}$ and $\mathcal{P}$ and the Fourier transform of the ordinary representation.

DEFINITION 4.2.1 *Define unitary operators U_s and V_t on $L^2(\mathbf{R}^2)$ $(s, t \in \mathbf{R})$ by*

$$U_s f(x_1, x_2) = e^{isx_1} f(x_1, x_2 - \frac{1}{2}\hbar s)$$

and

$$V_t f(x_1, x_2) = e^{itx_2} f(x_1 + \frac{1}{2}\hbar t, x_2).$$

For $(t_1, t_2) \in \mathbf{R}^2$ define the translation operator T_{t_1,t_2} on $L^2(\mathbf{R}^2)$ by

$$T_{t_1,t_2} f(x_1, x_2) = f(x_1 - t_1, x_2 - t_2).$$

Then in terms of multiplication and translation operators we can write $U_s = M_{e^{isx_1}} T_{\frac{1}{2}\hbar s\mathbf{j}}$ and $V_t = M_{e^{itx_2}} T_{-\frac{1}{2}\hbar t\mathbf{i}}$, where $\mathbf{i} = (1, 0)$ and $\mathbf{j} = (0, 1)$ are the standard basis vectors in $\mathbf{R}^2$.

PROPOSITION 4.2.2
There is a unitary operator W on $L^2(\mathbf{R}^2)$ such that for all $s, t \in \mathbf{R}$ we have $U_s = W^{-1}(e^{is\mathcal{Q}} \otimes I)W$ and $V_t = W^{-1}(e^{it\mathcal{P}} \otimes I)W$.

PROOF Note that $e^{is\mathcal{Q}} \otimes I$ and $e^{it\mathcal{P}} \otimes I$ are operators on $L^2(\mathbf{R}^2)$ which act trivially on the second variable. Define a unitary W_1 on $L^2(\mathbf{R}^2)$ by

$$W_1 f(x_1, x_2) = f\left(\frac{x_1 - x_2}{\sqrt{2}}, \frac{x_1 + x_2}{\sqrt{2}}\right).$$

Then

$$W_1^{-1}(e^{is\mathcal{Q}} \otimes I)W_1 = W_1^{-1} M_{e^{isx_1}} W_1 = M_{e^{is(x_1+x_2)/\sqrt{2}}}$$

and

$$W_1^{-1}(e^{it\mathcal{P}} \otimes I)W_1 = W_1^{-1} T_{-\hbar t\mathbf{i}} W_1 = T_{-\hbar t(\mathbf{i}+\mathbf{j})/\sqrt{2}}.$$

Next, let W_2 be the inverse Fourier transform in the second variable; that is, $W_2 = I \otimes \mathcal{F}^{-1}$. Then

$$W_2^{-1} M_{e^{is(x_1+x_2)/\sqrt{2}}} W_2 = M_{e^{isx_1\sqrt{2}}} T_{s\mathbf{j}/\sqrt{2}}$$

and

$$W_2^{-1} T_{-\hbar t(\mathbf{i}+\mathbf{j})/\sqrt{2}} W_2 = M_{e^{i\hbar t x_2/\sqrt{2}}} T_{-\hbar t\mathbf{i}/\sqrt{2}}.$$

Finally, define a unitary W_3 by

$$(W_3 f)(x_1, x_2) = \sqrt{\frac{\hbar}{2}} f\left(\frac{1}{\sqrt{2}} x_1, \frac{1}{\sqrt{2}} \hbar x_2\right);$$

then

$$W_3^{-1} M_{e^{isx_1/\sqrt{2}}} T_{s\mathbf{j}/\sqrt{2}} W_3 = M_{e^{isx_1}} T_{\frac{1}{2}\hbar s\mathbf{j}} = U_s$$

and

$$W_3^{-1} M_{e^{i\hbar t x_2/\sqrt{2}}} T_{-\hbar t\mathbf{i}/\sqrt{2}} W_3 = M_{e^{itx_2}} T_{-\frac{1}{2}\hbar t\mathbf{i}} = V_t.$$

So $W = W_1 W_2 W_3$ is the desired unitary. ∎

It follows that $W^{-1}(\mathcal{Q} \otimes I)W$ and $W^{-1}(\mathcal{P} \otimes I)W$ are the infinitesimal generators of the unitary groups $\{U_s\}$ and $\{V_t\}$, respectively. We can describe these operators more concretely: the first is $M_{x_1} + i\frac{\hbar}{2}\frac{\partial}{\partial x_2}$ and the second is $M_{x_2} - i\frac{\hbar}{2}\frac{\partial}{\partial x_1}$.

This, then, is the <u>tracial representation</u>, so-called for reasons to be explained in Section 5.6: the position operator is modelled by $M_{x_1} + i\frac{\hbar}{2}\frac{\partial}{\partial x_2}$, the momentum operator is modelled by $M_{x_2} - i\frac{\hbar}{2}\frac{\partial}{\partial x_1}$, and the unitary groups they generate are $\{U_s\}$ and $\{V_t\}$. We have assumed throughout that $\hbar > 0$, but let us now take $\hbar = 0$. Then the position and momentum operators become M_{x_1} and M_{x_2}, multiplication by the coordinate functions, which are the classical position and momentum observables. This is the sense in which $\mathcal{Q}$ and $\mathcal{P}$ approach the classical coordinate variables as $\hbar \to 0$. In the representation on $L^2(\mathbf{R})$ of Section 4.1 nothing similar happens and the $\hbar \to 0$ limit is rather mysterious.

Recall that the unitary operators $e^{-is\mathcal{P}/\hbar}$ and $e^{it\mathcal{Q}/\hbar}$ play the role of translation in the position and momentum variables. However, the operators $W^{-1}e^{-is\mathcal{P}/\hbar}W$ and $W^{-1}e^{it\mathcal{Q}/\hbar}W$ do not have limits as $\hbar \to 0$. We will resolve this difficulty in Section 5.4 by showing that there is a sense in which $e^{-is\mathcal{P}/\hbar}$ and $e^{it\mathcal{Q}/\hbar}$ are equivalent to the ordinary coordinate translations $T_{s\mathbf{i}}$ and $T_{t\mathbf{j}}$ on $L^2(\mathbf{R}^2)$.

4.3 Bargmann-Segal space

There is a nice orthonormal basis of $L^2(\mathbf{R})$ that we can use to establish an isometric isomorphism with $l^2(\mathbf{N})$ which puts $\mathcal{Q} + i\mathcal{P}$ and $\mathcal{Q} - i\mathcal{P}$, the operators analogous to the classical complex variables $z = x + iy$ and $\bar{z} = x - iy$, in a simple form. It is expressed in terms of the Hermite polynomials

$$H_n(x) = (-1)^n e^{x^2} \frac{d^n}{dx^n} e^{-x^2}.$$

The first few Hermite polynomials are $H_0(x) = 1$, $H_1(x) = 2x$, $H_2 = 4x^2 - 2$.

The Hermite functions are the functions

$$h_n(x) = N_n H_n(x/\sqrt{\hbar}) e^{-x^2/2\hbar},$$

where N_n is a normalizing factor chosen so that $||h_n|| = 1$. Specifically,

$$N_n = \frac{1}{\sqrt{(\pi\hbar)^{1/2} 2^n n!}}.$$

Moreover, the h_n are orthogonal; this can be proven in the following way. First, verify that the H_n satisfy

$$\frac{d}{dx}\left(e^{-x^2}\frac{dH_n}{dx}\right) = -2ne^{-x^2}H_n.$$

Then a short computation shows that

$$\frac{d}{dx}\left(e^{-x^2}(H_m H_n' - H_m' H_n)\right) = (2m - 2n)e^{-x^2}H_m H_n,$$

and from this it follows that

$$\begin{aligned}(2m - 2n)\int_{-\infty}^{\infty} h_m h_n \, dx &= \sqrt{\hbar} N_m N_n \int_{-\infty}^{\infty} (2m - 2n)e^{-x^2} H_m H_n \, dx \\ &= \sqrt{\hbar} N_m N_n e^{-x^2}(H_m H_n' - H_m' H_n)\Big|_{-\infty}^{\infty} = 0.\end{aligned}$$

So $\langle h_m, h_n \rangle = 0$ for $m \neq n$. (Note that h_n is real, hence the absence of a complex conjugate in our computation of the inner product.)

Since each H_n is a polynomial of degree exactly n, any polynomial can be written as a linear combination of the H_n. Therefore any $f \in L^2(\mathbf{R})$ which is orthogonal to every h_n is actually orthogonal to $p(x)e^{-x^2/2\hbar}$ for every polynomial $p(x)$. But then

$$\begin{aligned}(e^{-x^2/2\hbar}f)\hat{} &= \frac{1}{\sqrt{2\pi}}\int e^{-x^2/2\hbar}f(x)e^{-itx}\,dx\\ &= \frac{1}{\sqrt{2\pi}}\sum_n \int e^{-x^2/2\hbar}f(x)\frac{(-itx)^n}{n!}\,dx = 0.\end{aligned}$$

From this it follows that $e^{-x^2/2\hbar}f = 0$, and hence $f = 0$, almost everywhere. We conclude that the functions h_n constitute an orthonormal basis of $L^2(\mathbf{R})$. Thus, there is a natural isomorphism between $L^2(\mathbf{R})$ and $l^2(\mathbf{N})$ which takes (h_n) to the standard basis (e_n) of $l^2(\mathbf{N})$. We record this fact.

PROPOSITION 4.3.1

The Hermite functions (h_n) are an orthonormal basis of $L^2(\mathbf{R})$. The map $h_n \mapsto e_n$ extends to an isometric isomorphism from $L^2(\mathbf{R})$ onto $l^2(\mathbf{N})$.

The higher-dimensional case is worth mentioning here. In general, the $2n$-dimensional version of the quantum plane is simply obtained by taking nth tensor powers. For the $L^2(\mathbf{R})$ model this would result in a model on $L^2(\mathbf{R}^n)$. The nth tensor power of $l^2(\mathbf{N})$ was described in Example 2.5.7 (a): it is the direct sum of the k-fold symmetric tensor powers of $l^2(\{1,\ldots,n\}) \cong \mathbf{C}^n$. In other words, it is a symmetric Fock space over $\mathbf{C}^n$, as in Definition 2.5.8. For this reason, we call $l^2(\mathbf{N})$ the Fock space model of the quantum plane.

The Fock space model gives a particularly nice representation of the operators $\mathcal{Q} - i\mathcal{P}$ and $\mathcal{Q} + i\mathcal{P}$. Ignoring domain issues for the moment, we have

$$(\mathcal{Q} - i\mathcal{P})h_n = xh_n - \hbar\frac{d}{dx}h_n = \sqrt{2\hbar(n+1)}h_{n+1}$$

and

$$(\mathcal{Q} + i\mathcal{P})h_n = xh_n + \hbar\frac{d}{dx}h_n = \sqrt{2\hbar n}h_{n-1}$$

(with $(\mathcal{Q}+i\mathcal{P})h_0 = 0$). This can be shown using the recursion relations $H_n'(x) = 2nH_{n-1}(x) = 2xH_n(x) - H_{n+1}(x)$. Thus, $\mathcal{Q} - i\mathcal{P}$ and $\mathcal{Q} + i\mathcal{P}$ resemble the unilateral shift and the backward shift mentioned in Section 3.1; but they are unbounded because of the factors of $\sqrt{n+1}$ and $\sqrt{n}$.

Before discussing these operators further, we introduce another representation in which they become even simpler.

DEFINITION 4.3.2 *Let μ be $\frac{1}{2\pi\hbar}e^{-|z|^2/2\hbar}$ times Lebesgue measure on $\mathbf{C}$. This is a probability measure. The* <u>Bargmann-Segal space</u> *is the space $\mathcal{BS}$ of all analytic functions in $L^2(\mathbf{C},\mu)$. That is, $\mathcal{BS}$ consists of all entire analytic functions $f(z)$ on the complex plane such that*

$$\|f\|^2 = \frac{1}{2\pi\hbar}\int |f(z)|^2 e^{-|z|^2/2\hbar}\,dz$$

is finite.

The Bargmann-Segal space is also called Fischer space or Fock space, but the latter is a slight misnomer; as we noted earlier, $l^2(\mathbf{N})$ is the actual Fock space here (although the two are naturally isomorphic, as we will now show).

THEOREM 4.3.3
$\mathcal{BS}$ is a Hilbert space. It has an orthonormal basis consisting of the functions

$$\zeta_n(z) = \frac{z^n}{\sqrt{(2\hbar)^n n!}}$$

($n = 0, 1, \ldots$).

PROOF Since $\mathcal{BS}$ is a subspace of $L^2(\mathbf{C},\mu)$, to prove that it is a Hilbert space we need only show that it is closed.

Let f be an analytic function on $\mathbf{C}$. By standard complex analysis $f(z)$ is the average value of f on any disk centered at z, so using the Cauchy-Schwarz inequality we have

$$\begin{aligned}
|f(z)| &= \frac{1}{\pi R^2}\left|\int_{B(z;R)} f(w)\,dw\right| \\
&\le \frac{1}{\pi R^2}\left(\int_{B(z;R)} |f(w)|^2 e^{-|w|^2/2\hbar}\,dw\right)^{1/2}\left(\int_{B(z;R)} e^{|w|^2/2\hbar}\,dw\right)^{1/2} \\
&\le \frac{\sqrt{2\hbar}}{R} e^{(|z|+R)^2/4\hbar}\|f\|,
\end{aligned}$$

where $B(z;R)$ is the disk of radius R centered at z. It follows that any sequence of analytic functions which converges in L^2 norm also converges uniformly on compact sets, and therefore its limit (which we already know is in $L^2(\mathbf{C},\mu)$) is also analytic. So $\mathcal{BS}$ is complete.

The functions z^n are orthogonal, and a short computation shows that $\|z^n\|^2 = (2\hbar)^n n!$, so the functions ζ_n are orthonormal. In fact, for any $R > 0$ the restrictions of the functions z^n to $B(0;R)$ are orthogonal.

Now if $f \in \mathcal{BS}$ has a Taylor expansion $f(z) = \sum a_n z^n$, then this series converges to f uniformly on $B(0;R)$, and hence it also converges in $L^2(B(0;R), \mu|_{B(0;R)})$. Thus

$$\|f|_{B(0;R)}\|^2 = \sum |a_n|^2 \|z^n|_{B(0;R)}\|^2.$$

Taking the limit as $R \to \infty$ yields $\|f\|^2 = \sum |a_n|^2 \|z^n\|^2$. This shows that the sum $\sum a_n z^n$ converges in $\mathcal{BS}$, and since it converges uniformly on compact sets to f, it must converge to f in $\mathcal{BS}$. Thus, we have shown that the functions z^n generate $\mathcal{BS}$, and we conclude that (ζ_n) is an orthonormal basis. ∎

Having identified a natural orthonormal basis for $\mathcal{BS}$, we can now define an isomorphism between $\mathcal{BS}$ and $L^2(\mathbf{R})$ in the same way that we defined an isomorphism between $L^2(\mathbf{R})$ and $l^2(\mathbf{N})$ earlier. Namely, let $\mathcal{B} : L^2(\mathbf{R}) \to \mathcal{BS}$ be the surjective isometric isomorphism which takes h_n to ζ_n. This map is called the Bargmann transform.

Now we return to the operators $\mathcal{Q} - i\mathcal{P}$ and $\mathcal{Q} + i\mathcal{P}$. More precisely, we want to look at $A^+ = \frac{1}{\sqrt{2\hbar}}(\mathcal{Q} - i\mathcal{P})$ and $A^- = \frac{1}{\sqrt{2\hbar}}(\mathcal{Q} + i\mathcal{P})$, but first we need to define them rigorously. This could be done by the prescription given following Proposition 4.1.4, but it is more convenient for us to take a different (though equivalent) approach.

Our heuristic computation made earlier yielded $A^+ h_n = \sqrt{n+1} h_{n+1}$ and $A^- h_n = \sqrt{n} h_{n-1}$; thus the operators $\mathcal{B}A^+\mathcal{B}^{-1}$ and $\mathcal{B}A^-\mathcal{B}^{-1}$ on $\mathcal{BS}$ should satisfy

$$\begin{aligned}\mathcal{B}A^+\mathcal{B}^{-1} z^n &= \sqrt{(2\hbar)^n n!}\,\mathcal{B}A^+\mathcal{B}^{-1}\zeta_n \\ &= \sqrt{(2\hbar)^n (n+1)!}\,\zeta_{n+1} = \frac{1}{\sqrt{2\hbar}} z^{n+1}\end{aligned}$$

and

$$\begin{aligned}\mathcal{B}A^-\mathcal{B}^{-1} z^n &= \sqrt{(2\hbar)^n n!}\,\mathcal{B}A^-\mathcal{B}^{-1}\zeta_n \\ &= \sqrt{(2\hbar)^n n \cdot n!}\,\zeta_{n-1} = \sqrt{2\hbar}\, n z^{n-1}.\end{aligned}$$

That is, they should just be the operators $\frac{1}{\sqrt{2\hbar}} M_z$ and $\sqrt{2\hbar}\frac{d}{dz}$. It is easy to define these operators rigorously.

DEFINITION 4.3.4 *Let $M_z \in B(\mathcal{BS})$ be multiplication by z, with domain*

$$D(M_z) = \{f \in \mathcal{BS} : zf \in \mathcal{BS}\}$$

and let $\frac{d}{dz} \in B(\mathcal{BS})$ be differentiation, with domain

$$D\left(\frac{d}{dz}\right) = \{f \in \mathcal{BS} : f' \in \mathcal{BS}\}.$$

We also define $A^+ = \frac{1}{\sqrt{2\hbar}}\mathcal{B}^{-1}M_z\mathcal{B}$ *and* $A^- = \sqrt{2\hbar}\mathcal{B}^{-1}\frac{d}{dz}\mathcal{B}$. *These are, respectively, the* creation *and* annihilation *operators on* $L^2(\mathbf{R})$.

Clearly, $D(M_z)$ and $D(\frac{d}{dz})$ both contain all polynomials in z, so these operators are densely defined. Moreover, $C_c^\infty(\mathbf{R})$ is a core for both A^+ and A^-, so this definition is consistent with the comment made after Proposition 4.1.4. This can be seen by finding, for each n, a sequence $(f_i) \subset C_c^\infty(\mathbf{R})$ such that $f_i \to h_n$ and $A^\pm f_i \to A^\pm h_n$. By passing to the $l^2(\mathbf{N})$ model (see the proof of the next result) it is easy to see that the span of the h_n is a core for both A^+ and A^-, which implies that $C_c^\infty(\mathbf{R})$ is also a core.

Evidently we have $\mathcal{B}(\mathcal{Q}-i\mathcal{P})\mathcal{B}^{-1} = M_z$ and $\mathcal{B}(\mathcal{Q}+i\mathcal{P})\mathcal{B}^{-1} = 2\hbar\frac{d}{dz}$. It is also possible to write $\mathcal{B}(\mathcal{Q}+i\mathcal{P})\mathcal{B}^{-1} = PM_{\bar{z}}$, where P is the orthogonal projection of $L^2(\mathbf{C},\mu)$ onto $\mathcal{BS}$ and $\bar{z} = x - iy$ is the conjugate variable to z. Thus, the operators $\mathcal{Q} - i\mathcal{P}$ and $\mathcal{Q} + i\mathcal{P}$ have particularly elegant representations in the Bargmann-Segal picture. We will develop this point in the next section. Let us observe here, though, that there is a slight mismatch in that we would expect $\mathcal{Q} + i\mathcal{P}$ to correspond to $z = x + iy$ and $\mathcal{Q} - i\mathcal{P}$ to correspond to $\bar{z} = x - iy$. This would have happened if we defined $\mathcal{BS}$ to be the space of *antianalytic* functions in $L^2(\mathbf{C},\mu)$. Doing so leads to the formulas $\mathcal{B}(\mathcal{Q} + i\mathcal{P})\mathcal{B}^{-1} = P'M_z$ and $\mathcal{B}(\mathcal{Q}-i\mathcal{P})\mathcal{B}^{-1} = M_{\bar{z}}$ where $P' = I - P$. However, we have choosen instead to adopt the traditional definition of $\mathcal{BS}$ and accept the resulting minor disparity.

The meaning of the terms "creation" and "annihilation" will become more clear in Section 7.1, when we interpret the summands of Fock space as corresponding to different numbers of particles. The terms should not be taken too literally, however, as these are not physical operations which can actually be performed; they are not unitary.

PROPOSITION 4.3.5

Both M_z *and* $\frac{d}{dz}$ *are closed operators, they have the same domain, and* $\frac{1}{\sqrt{2\hbar}}M_z^* = \sqrt{2\hbar}\frac{d}{dz}$.

PROOF The easiest way to prove this is by passing to $l^2(\mathbf{N})$ via the correspondence $\zeta_n \leftrightarrow e_n$. The operator A on $l^2(\mathbf{N})$ corresponding to $\frac{1}{\sqrt{2\hbar}}M_z$ satisfies $Ae_n = \sqrt{n+1}e_{n+1}$, with domain

$$D(A) = \{(a_n) \in l^2(\mathbf{N}) : \sum(n+1)|a_n|^2 < \infty\},$$

and the operator B corresponding to $\sqrt{2\hbar}\frac{d}{dz}$ satisfies $Be_n = \sqrt{n}e_{n-1}$, with domain

$$D(B) = \{(a_n) \in l^2(\mathbf{N}) : \sum n|a_n|^2 < \infty\}.$$

From this it immediately follows that $D(A) = D(B)$, and therefore $D(M_z) = D(\frac{d}{dz})$. It is also easy to check that $\Gamma(A)$ and $\Gamma(B)$ are closed, so the same is true of M_z and $\frac{d}{dz}$. To compute A^*, let $b = (b_n) \in l^2(\mathbf{N})$; then $b \in D(A^*)$ if and only if the map

$$a \mapsto \langle Aa, b\rangle = \sum_{n=1}^{\infty} \sqrt{n} a_{n-1} \bar{b}_n$$

is bounded for $a \in D(A)$. This is certainly true if the sequence $(\sqrt{n} b_n)$ is in $l^2(\mathbf{N})$; conversely, if $(\sqrt{n} b_n)$ is not in $l^2(\mathbf{N})$, then taking $a^N_{n-1} = \sqrt{n} b_n$ for $1 \le n \le N$ and $a^N_{n-1} = 0$ for $n > N$ yields

$$\langle Aa^N, b\rangle = \sum_{n=1}^{\infty} \sqrt{n} a^N_{n-1} \bar{b}_n = \sum_{n=1}^{N} n|b_n|^2 = \|a^N\|^2,$$

and since $\|a^N\| \to \infty$ as $N \to \infty$, this shows that the map $a \mapsto \langle Aa, b\rangle$ is not bounded. Thus $D(A^*) = D(B)$, and for $b \in D(B)$ we have

$$\langle Aa, b\rangle = \sum_{n=1}^{\infty} \sqrt{n} a_{n-1} \bar{b}_n = \langle a, Bb\rangle.$$

So $A^* = B$, as claimed. It follows that $\frac{1}{\sqrt{2\hbar}} M_z^* = \sqrt{2\hbar}\frac{d}{dz}$. ∎

We conclude our treatment of Bargmann-Segal space with a brief discussion of the <u>reproducing kernels</u> $e_w(z) = e^{z\bar{w}/2\hbar}$. For any $w \in \mathbf{C}$, we have $|e_w(z)| \le e^{|w|\cdot|z|/2\hbar}$, so the integral $\frac{1}{2\pi\hbar}\int |e_w(z)|^2 e^{-|z|^2/2\hbar}\, dz$ converges, and hence $e_w \in \mathcal{BS}$. Now

$$e_w(z) = \sum \frac{(z\bar{w})^n}{(2\hbar)^n n!} = \sum \frac{\bar{w}^n}{\sqrt{(2\hbar)^n n!}} \zeta_n(z),$$

so for any $f = \sum a_n z^n = \sum \sqrt{(2\hbar)^n n!} a_n \zeta_n$ in $\mathcal{BS}$ we have

$$\langle f, e_w\rangle = \sum a_n w^n = f(w).$$

This is the meaning of the term "reproducing kernel": taking the inner product of a function in $\mathcal{BS}$ with e_w reproduces its value at w.

We have $2\hbar\frac{d}{dz} e_w = \bar{w} e_w$. So even though the vectors e_w are not orthogonal, they are eigenvectors of the (non-normal) operator $2\hbar\frac{d}{dz}$. Moreover, for any $f, g \in \mathcal{BS}$ we have

$$\int \langle f, e_w\rangle\langle e_w, g\rangle\, d\mu(w) = \int f(w)\overline{g(w)}\, d\mu(w) = \langle f, g\rangle.$$

That is,

$$f = \int \langle f, e_w \rangle e_w \, d\mu(w),$$

where the integral is taken in the weak sense (meaning just what was said in the previous line). So the family (e_w) behaves rather like a basis; cf. the first assertion of Corollary 2.3.4, replacing the sum with an integral. The functions e_w might be thought of as the "points" of the quantum plane, an idea that will gain more support in the next section.

We record these properties in the following result.

PROPOSITION 4.3.6
For each $w \in \mathbf{C}$ let $e_w(z) = e^{z\bar{w}/2\hbar}$. Then $e_w \in \mathcal{BS}$ and

(a) $f(w) = \langle f, e_w \rangle$ for all $f \in \mathcal{BS}$ and $w \in \mathbf{C}$;
(b) $2\hbar \frac{d}{dz} e_w = \bar{w} e_w$ for all $w \in \mathbf{C}$; and
(c) $f = \int \langle f, e_w \rangle e_w \, d\mu(w)$ for all $f \in \mathcal{BS}$, where the integral is taken in the weak sense.

In the $L^2(\mathbf{R})$ picture, the eigenfunctions of $\mathcal{Q} + i\mathcal{P}$ are the functions $e^{-(x-\lambda)^2/2\hbar}$ for $\lambda \in \mathbf{C}$. They are called coherent states. A short computation shows that

$$(\mathcal{Q} + i\mathcal{P}) e^{-(x-\lambda)^2/2\hbar} = \lambda e^{-(x-\lambda)^2/2\hbar},$$

so the eigenvalue is λ. Recalling that $\mathcal{B}(\mathcal{Q} + i\mathcal{P})\mathcal{B}^{-1} = 2\hbar \frac{d}{dx}$, this shows that up to normalization the Bargmann transform takes the coherent states to the reproducing kernel functions e_w with $w = \bar{\lambda}$.

4.4 Quantum complex analysis

If bounded self-adjoint operators on $L^2(\mathbf{R})$ correspond to bounded measurable real-valued functions on the plane, then general bounded operators on $L^2(\mathbf{R})$ should correspond to bounded measurable complex-valued functions on the plane, because any bounded operator A can be written as $A = \mathrm{Re}\, A + i \mathrm{Im}\, A$.

Recall that the unitary groups $e^{-is\mathcal{P}/\hbar}$ and $e^{it\mathcal{Q}/\hbar}$ play the role of coordinate translations on the quantum plane. We can use these groups to take derivatives of operators on $L^2(\mathbf{R})$ in the following way. For any bounded operator A on $L^2(\mathbf{R})$, the operators $e^{-is\mathcal{P}/\hbar} A e^{is\mathcal{P}/\hbar}$ and $e^{it\mathcal{Q}/\hbar} A e^{-it\mathcal{Q}/\hbar}$ are thought of as its shifts by s and t units in the horizontal and vertical directions, respectively. Therefore the "derivative of A in the horizontal direction" should be the operator

$$\lim_{s \to 0} \frac{e^{-is\mathcal{P}/\hbar} A e^{is\mathcal{P}/\hbar} - A}{s}$$

and its "derivative in the vertical direction" should be

$$\lim_{t\to 0} \frac{e^{it\mathcal{Q}/\hbar} A e^{-it\mathcal{Q}/\hbar} - A}{t}.$$

Heuristically, since $e^B = I + B + \cdots$, we expect that

$$\begin{aligned}\lim_{s\to 0} \frac{e^{-is\mathcal{P}/\hbar} A e^{is\mathcal{P}/\hbar} - A}{s} &= \lim_{s\to 0} \frac{(I - \frac{i}{\hbar}s\mathcal{P})A(I + \frac{i}{\hbar}s\mathcal{P}) - A}{s} \\ &= -\frac{i}{\hbar}[\mathcal{P}, A]\end{aligned}$$

where $[\mathcal{P}, A] = \mathcal{P}A - A\mathcal{P}$ is the commutator of $\mathcal{P}$ and A, and similarly

$$\lim_{t\to 0} \frac{e^{it\mathcal{Q}/\hbar} A e^{-it\mathcal{Q}/\hbar} - A}{t} = \frac{i}{\hbar}[\mathcal{Q}, A].$$

(As usual, there are interpretational issues here due to the fact that $\mathcal{P}$ and $\mathcal{Q}$ are not defined on all of $L^2(\mathbf{R})$.)

We now want to determine which operators on $L^2(\mathbf{R})$ correspond to *analytic* functions. Classically, a C^1 function $f : \mathbf{C} \to \mathbf{C}$ is analytic if

$$\frac{\partial f}{\partial y} = i\frac{\partial f}{\partial x};$$

this is a form of the Cauchy-Riemann equations. According to the last paragraph, the quantum analog of this condition is

$$\frac{i}{\hbar}[\mathcal{Q}, A] = \frac{1}{\hbar}[\mathcal{P}, A],$$

or equivalently $[\mathcal{Q} + i\mathcal{P}, A] = 0$. Thus, operators on $L^2(\mathbf{R})$ which commute with the annihilation operator can be viewed as satisfying a quantum version of the Cauchy-Riemann equations.

The first thing we need to do is to make this condition more rigorous. Since we are going to allow A to be unbounded, we must say what we mean by the commutator of two unbounded operators. The appropriate definition here is the following.

DEFINITION 4.4.1 *Let D be a dense subspace of a Hilbert space $\mathcal{H}$ and let A and B be unbounded operators on $\mathcal{H}$. Suppose D is contained in the domains of A, A^*, B, and B^*. Then we say that A and B commute relative to D*

$$\langle Av, B^*w\rangle = \langle Bv, A^*w\rangle$$

for all $v, w \in D$.

If A and B are both bounded, commuting relative to D is equivalent to the condition $\langle (BA - AB)v, w\rangle = 0$; since we assume D is dense, it follows that A and B commute in the usual sense, and conversely.

Also, note that A and B commute relative to D if and only if A^* and B^* commute relative to D. Thus, the quantum analyticity condition is morally equivalent to the condition $[\mathcal{Q} - i\mathcal{P}, A^*] = 0$, which we interpret as saying that A^* is antianalytic.

Let us now pass to the Bargmann-Segal space $\mathcal{BS}$ and replace $\mathcal{Q} + i\mathcal{P}$ with $2\hbar\frac{d}{dz}$. Now $\frac{d}{dz}$ certainly ought to commute with itself, or moreover with any polynomial in $\frac{d}{dz}$, or indeed any function of $\frac{d}{dz}$ to the extent that such an operator can be defined. In fact, it turns out that the *only* operators which commute with $\frac{d}{dz}$ in a reasonable sense are operators which arise from it by a kind of functional calculus. The next result describes the relevant class of functions. Recall the reproducing kernels $e_w(z) = e^{z\bar{w}/2\hbar}$ introduced at the end of the last section.

PROPOSITION 4.4.2

Let ϕ be an entire analytic function on $\mathbf{C}$. Then $\phi e_w \in \mathcal{BS}$ for all $w \in \mathbf{C}$ if and only if $|\phi(z)| = O(e^{|z|^2/4\hbar - N|z|})$ for all $N > 0$.

PROOF Suppose $|\phi(z)| = O(e^{|z|^2/4\hbar - N|z|})$. Then

$$\begin{aligned}||\phi e_w||^2 &= \frac{1}{2\pi\hbar}\int |\phi(z)e_w|^2 e^{-|z|^2/2\hbar}\,dz\\ &\le C\int (e^{|z|^2/4\hbar - N|z|}e^{|w||z|/2\hbar})^2 e^{-|z|^2/2\hbar}\,dz\\ &= C\int e^{(|w|/\hbar - 2N)|z|}\,dz.\end{aligned}$$

We can ensure that this last integral is finite by taking $N > |w|/2\hbar$. This shows that $\phi e_w \in \mathcal{BS}$.

Conversely, suppose $\phi e_w \in \mathcal{BS}$ for all $w \in \mathbf{C}$. Fix $N > 0$ and let $w_k = 2\hbar\sqrt{2}Ne^{2\pi ik/4}$ for $k = 0, 1, 2, 3$. Then the Cauchy-Schwarz inequality yields

$$|\phi(z)e^{z\bar{w}_k/2\hbar}| = |\langle \phi e_{w_k}, e_z\rangle| \le ||\phi e_{w_k}||e^{|z|^2/4\hbar}$$

since $||e_z|| = \langle e_z, e_z\rangle^{1/2} = e^{|z|^2/4\hbar}$. Also, for any $z \in \mathbf{C}$ one of the w_k satisfies $|\mathrm{Arg}(w_k) - \mathrm{Arg}(z)| \le \pi/4$, and for this w_k we have

$$\mathrm{Re}\, z\bar{w}_k \ge \frac{1}{\sqrt{2}}|z\bar{w}_k| = 2\hbar N|z|.$$

Thus

$$|\phi(z)e^{z\bar{w}_k/2\hbar}| = |\phi(z)|e^{\mathrm{Re}\, z\bar{w}_k/2\hbar} \ge |\phi(z)|e^{N|z|},$$

and combining this with the earlier estimate on $|\phi(z)e^{z\bar{w}_k/2\hbar}|$ yields

$$|\phi(z)| \leq Ce^{|z|^2/4\hbar - N|z|}$$

with $C = \max\{\|\phi e_{w_k}\| : 0 \leq k \leq 3\}$. ∎

Thus, for any ϕ satisfying the stated growth condition, the multiplication operator M_ϕ contains all of the functions e_w in its domain. (Naturally, we take $D(M_\phi)$ to be $\{f \in \mathcal{BS} : \phi f \in \mathcal{BS}\}$.) We call such ϕ Bargmann-Segal multipliers. We can regard M_ϕ as the operator obtained by applying ϕ to M_z by a sort of functional calculus, and its adjoint M_ϕ^* as obtained by applying ϕ to $2\hbar\frac{d}{dz}$.

Let $\mathcal{R} = \text{span}\{e_w : w \in \mathbf{C}\}$.

LEMMA 4.4.3
Let ϕ be an analytic function on $\mathbf{C}$ such that $|\phi(z)| = O(e^{|z|^2/4\hbar - N|z|})$. Then $\mathcal{R}$ is a core for M_ϕ^* and $M_\phi^* e_w = \overline{\phi(w)} e_w$ for all $w \in \mathbf{C}$.

PROOF Let $w \in \mathbf{C}$. Then for any $f \in D(M_\phi)$ we have

$$\langle M_\phi f, e_w \rangle = \phi(w) f(w) = \langle f, \overline{\phi(w)} e_w \rangle.$$

This shows that $e_w \in D(M_\phi^*)$ and $M_\phi^* e_w = \overline{\phi(w)} e_w$.

To show that $\mathcal{R}$ is a core for M_ϕ^*, let A be the closure of $M_\phi^*|_\mathcal{R}$. For any $f \in D(A^*)$, if $g = A^* f$ then

$$g(w) = \langle g, e_w \rangle = \langle f, Ae_w \rangle = \phi(w)\langle f, e_w \rangle = \phi(w) f(w)$$

for all $w \in \mathbf{C}$. This shows that $f \in D(M_\phi)$ and $A^* f = M_\phi f$. So $\Gamma(A^*) \subset \Gamma(M_\phi)$, and it follows from the definition of adjoints that $\Gamma(M_\phi^*) \subset \Gamma(A^{**}) = \Gamma(A)$. But it is clear that $\Gamma(A) \subset \Gamma(M_\phi^*)$, so $A = M_\phi^*$, and this shows that $\mathcal{R}$ is a core for M_ϕ^*. ∎

THEOREM 4.4.4
Let A be a closed, unbounded operator on $\mathcal{BS}$ and suppose $\mathcal{R}$ is a core for A and $\mathcal{R} \subset D(A^*)$. Then the following are equivalent:

(a) A and $\frac{d}{dz}$ commute relative to $\mathcal{R}$;
(b) e_w is an eigenvector of A for all $w \in \mathbf{C}$; and
(c) $A = M_\phi^*$ for some entire analytic function ϕ on $\mathbf{C}$ satisfying $|\phi(z)| = O(e^{|z|^2/4\hbar - N|z|})$.

PROOF

(a) $\Rightarrow$ (b): Suppose A and $\frac{d}{dz}$ commute relative to $\mathcal{R}$. Then for any $v, w \in \mathbf{C}$ we have

$$\langle Ae_v, ze_w\rangle = \langle 2\hbar\frac{d}{dz}e_v, A^*e_w\rangle = \langle \bar{v}e_v, A^*e_w\rangle = \langle \bar{v}Ae_v, e_w\rangle.$$

This shows that $Ae_v \in D(M_z^*) = D(2\hbar\frac{d}{dz})$ and $2\hbar\frac{d}{dz}Ae_v = \bar{v}Ae_v$. That is, Ae_v is an eigenvector of $2\hbar\frac{d}{dz}$ with eigenvalue $\bar{v}$, and it follows that Ae_v is a constant multiple of e_v. So e_v is an eigenvector of A.

(b) $\Rightarrow$ (c): Let $\phi = A^*e_0$ and say $Ae_w = \lambda e_w$. Then

$$\begin{aligned}\langle e_v, Ae_w\rangle &= \langle M_{e_v}e_0, \lambda e_w\rangle = \bar{\lambda}\langle e_0, M_{e_v}^*e_w\rangle \\ &= \bar{\lambda}\langle e_0, \overline{e_v(w)}e_w\rangle = e_v(w)\langle e_0, Ae_w\rangle \\ &= \phi(w)e_v(w) = \langle \phi e_v, e_w\rangle.\end{aligned}$$

This shows that $A^*e_v = M_\phi e_v$, and so $|\phi(z)| = O(e^{(|z|^2/2 - N|z|)/2\hbar})$ by Proposition 4.4.2. It also follows that $M_\phi^* = A$ on $\mathcal{R}$, and since $\mathcal{R}$ is a core for both operators (by assumption and the lemma), we conclude that $A = M_\phi^*$.

(c) $\Rightarrow$ (a): This follows from the fact that M_ϕ and $M_z = 2\hbar\frac{d}{dz}$ commute relative to $\mathcal{R}$. ∎

It is unknown whether the condition that $\mathcal{R}$ be a core for A can be replaced by the weaker requirement that $\mathcal{R} \subset D(A)$. In any case, subject to reasonable domain conditions, the unbounded operators which satisfy the quantum version of the Cauchy-Riemann equations are precisely those of the form M_ϕ^* where ϕ is an analytic function satisfying the stated growth condition. We note again that the slightly unnatural appearance of an adjoint in this conclusion could be remedied by modifying the definition of $\mathcal{BS}$; see the comment following Definition 4.3.4.

The next result, which is a quantum version of Liouville's theorem, follows from Theorem 4.4.4 and the classical Liouville theorem.

COROLLARY 4.4.5

Let A be a bounded operator on $L^2(\mathbf{R})$ and suppose $[A^-, A] = 0$ in the sense that $\langle Af, A^+g\rangle = \langle A^-f, A^*g\rangle$ for all $f, g \in D(A^+) = D(A^-)$. Then A is a scalar multiple of the identity.

4.5 Notes

The main ideas in Section 4.1 can be found in any standard book on quantum mechanics; for instance, see [63] and [8] for physical and mathematical perspectives, respectively. Another construction which is also called the "quantum plane" is given in [47]. The tracial representation described in Section 4.2 is based on a more general construction in [60]. See e.g. [28] for a more thorough treatment of the Fourier transform.

Section 4.3 follows the treatment of [29]. For more on coherent states see [1], and for more on the property described in Proposition 4.3.6 (c) see [15].

Section 4.4 is based on [61]. Proposition 4.4.2 is from [54]; see also [55].

Chapter 5

C*-algebras

5.1 The algebras $C(X)$

We have determined that bounded real-valued functions on a set correspond to bounded self-adjoint operators on a Hilbert space. Since any bounded operator can be written in the form $A + iB$ with A and B self-adjoint, there is a sense in which bounded complex-valued functions on a set correspond to general bounded operators. Thus, any structure on a set X which can be described in terms of real- or complex-valued functions is likely to have a Hilbert space analog defined the same way in terms of operators. This is the idea that we will use, in this chapter and the next, to define quantum versions of topologies and measures.

The natural functional analytic object corresponding to a compact Hausdorff topological space X is the space of continuous complex-valued functions $C(X)$. The goal of this section is to relate topological properties of X to algebraic properties of $C(X)$. As we will see, there is a very complete correspondence between the two.

We begin with a definition of the relevant algebraic concepts, given here in a general form which will also accomodate the noncommutative setting that we discuss later.

***DEFINITION* 5.1.1** *Let $\mathcal{A}$ and $\mathcal{B}$ be Banach spaces equipped with a product and an involution. We say they are* unital *if they possess multiplicative units, which we then denote $I_{\mathcal{A}}$ and $I_{\mathcal{B}}$.*

$\mathcal{B}$ is a C*-subalgebra *of $\mathcal{A}$ if it is a closed subspace of $\mathcal{A}$ and $\alpha, \beta \in \mathcal{B}$ implies $\alpha^*, \alpha\beta \in \mathcal{B}$. If $\mathcal{A}$ is unital then $\mathcal{B}$ is a* unital C*-subalgebra *if it is a C*-subalgebra and contains $I_{\mathcal{A}}$.*

$\mathcal{B}$ is a C*-ideal *of $\mathcal{A}$ if it is a C*-subalgebra of $\mathcal{A}$ with the property that $\alpha \in \mathcal{A}$ and $\beta \in \mathcal{B}$ imply $\alpha\beta, \beta\alpha \in \mathcal{B}$.*

A bounded linear map $\pi : \mathcal{A} \to \mathcal{B}$ is a $*$-homomorphism *if $\pi(\alpha\beta) = \pi(\alpha)\pi(\beta)$ and $\pi(\alpha^*) = \pi(\alpha)^*$ for all $\alpha, \beta \in \mathcal{A}$. It is* unital *if $\pi(I_{\mathcal{A}}) = I_{\mathcal{B}}$.*

(This only makes sense if $\mathcal{A}$ and $\mathcal{B}$ are both unital.) A <u>$*$-isomorphism</u> *is an isometric $*$-homomorphism.*

If $\mathcal{A}$ is unital, we define its <u>spectrum</u> *to be the set* $\mathrm{sp}(\mathcal{A})$ *of all unital $*$-homomorphisms from $\mathcal{A}$ into* $\mathbf{C}$*, endowed with the weak* topology it inherits from $\mathcal{A}^*$.*

More generally, we will use the term "(unital) $*$-subalgebra" for a possibly unclosed subspace which (contains $I_{\mathcal{A}}$ and) is stable under formation of products and adjoints. We will also use the term "$*$-homomorphism" in this more general context.

In the present section $\mathcal{A}$ and $\mathcal{B}$ will be $C(X)$ spaces, with product being the pointwise product of functions and involution being the pointwise complex conjugate. Every $C(X)$ has a unit $I_{C(X)} = 1_X$, the function which is constantly 1 on X.

In the case of $C(X)$ spaces, the qualification that $*$-homomorphisms be bounded is superfluous. In fact, every unital $*$-homomorphism $\pi : C(X) \to C(Y)$ is automatically a contraction. To see this, let $f \in C(X)$ and choose $\lambda \in \mathbf{C}$ such that $|\lambda| > \|f\|_\infty$; then $f - \lambda \cdot 1_X$ is invertible in $C(X)$, so $\pi(f - \lambda \cdot 1_X) = \pi(f) - \lambda \cdot 1_Y$ is invertible in $C(Y)$. Thus $\pi(f)(y) \neq \lambda$ for all $y \in Y$, and we conclude that $\|\pi(f)\|_\infty \leq \|f\|_\infty$.

C*-subalgebras of $C(X)$ are related to quotients of X in the following way.

Example 5.1.2

Let X and Y be compact Hausdorff spaces and let $\phi : X \to Y$ be a continuous surjection. Then $\mathcal{A} = \{f \circ \phi : f \in C(Y)\}$ is a unital C*-subalgebra of $C(X)$ and the <u>composition map</u> $C_\phi : f \mapsto f \circ \phi$ is a $*$-isomorphism of $C(Y)$ onto $\mathcal{A}$.

This construction is general, as the following proposition shows.

PROPOSITION 5.1.3

Let X be a compact Hausdorff space and let $\mathcal{A}$ be a unital C-subalgebra of $C(X)$. Then there exists a compact Hausdorff space Y and a continuous surjection $\phi : X \to Y$ such that $\mathcal{A} = C_\phi(C(Y))$.*

PROOF For $x, y \in X$ set $x \sim y$ if $f(x) = f(y)$ for all $f \in \mathcal{A}$. Let Y be the quotient space with the quotient topology, and let $\phi : X \to Y$ be the quotient map. Then Y is compact because it is a continuous image of a compact space. To see that it is Hausdorff let $[x], [y] \in Y$ be distinct; then there exists $f \in \mathcal{A}$ such that $f(x) \neq f(y)$. Hence there exist disjoint open sets $\mathcal{O}, \mathcal{O}' \subset \mathbf{C}$ such that $f(x) \in \mathcal{O}$ and $f(y) \in \mathcal{O}'$, and $f^{-1}(\mathcal{O})$ and

$f^{-1}(\mathcal{O}')$ give rise to disjoint open sets in Y which respectively contain $[x]$ and $[y]$.

Define $C_\phi : C(Y) \to C(X)$ by $C_\phi f = f \circ \phi$. Then C_ϕ is a $*$-isomorphism, and its range contains $\mathcal{A}$. (Any function in $\mathcal{A}$ lifts to a well-defined function on Y, which is continuous by the definition of the quotient topology.) Thus ${C_\phi}^{-1}$ maps $\mathcal{A}$ isometrically into a C*-subalgebra of $C(Y)$ which separates points. By the Stone-Weierstrass theorem, ${C_\phi}^{-1}$ takes $\mathcal{A}$ onto $C(Y)$, and hence C_ϕ maps $C(Y)$ onto $\mathcal{A}$. ∎

The requirement that $\mathcal{A} \subset C(X)$ be unital is not essential. If $1_X \notin \mathcal{A}$, we can work instead with the unital algebra

$$\mathcal{A}^+ = \{f + a \cdot 1_X : f \in A \text{ and } a \in \mathbf{C}\}.$$

Then Proposition 5.1.3 implies that $\mathcal{A}^+ \cong C(Y)$ for some quotient Y of X. Since $\mathcal{A} \subset \mathcal{A}^+$ separates points in Y, the Stone-Weierstrass theorem implies that it corresponds to the set of continuous functions on Y which vanish at some distinguished point y_0.

Thus, $\mathcal{A} \cong C_0(Y - \{y_0\})$, the continuous functions which vanish at infinity on the locally compact space $Y - \{y_0\}$. Indeed, removing a point from a compact space always leaves a locally compact space, and conversely, every locally compact space has a one-point compactification. Nonunital C*-subalgebras correspond to locally compact spaces in the same way that unital C*-subalgebras correspond to compact spaces. We regard the nonunital case as ancilliary to the unital case in this way.

Next, we relate C*-ideals of $C(X)$ to closed subsets of X.

Example 5.1.4

Let K be a closed subset of a compact Hausdorff space X and let $\mathcal{I} = \{f \in C(X) : f|_K = 0\}$. Then $\mathcal{I}$ is a C*-ideal of $C(X)$.

PROPOSITION 5.1.5

Let $\mathcal{I}$ be a C-ideal of $C(X)$. Then there exists a closed subset K of X such that $\mathcal{I} = \{f \in C(X) : f|_K = 0\}$.*

PROOF For $x, y \in X$ set $x \sim y$ if $f(x) = f(y)$ for all $f \in \mathcal{I}$. Observe that any point x at which some $f \in \mathcal{I}$ is nonzero cannot be equivalent to any other point. For if $y \neq x$ then there exists $g \in C(X)$ such that $g(x) = 1$ and $g(y) = 0$, so $fg(x) \neq fg(y)$, and thus x and y are not equivalent.

Let $\mathcal{I}^+ = \{f + a \cdot 1_X : f \in \mathcal{I} \text{ and } a \in \mathbf{C}\}$. This is a unital C*-subalgebra of $C(X)$. Define Y to be the quotient space as in Proposition 5.1.3 and let $C_\phi : C(Y) \to \mathcal{I}^+$ be the corresponding $*$-isomorphism. Also let $K = \{x \in X : f(x) = 0 \text{ for all } f \in \mathcal{I}\}$. It is clear that every $f \in \mathcal{I}$ satisfies $f|_K = 0$. Conversely, if $f|_K = 0$ then f lifts to a continuous function on Y, and hence $f \in C_\phi(C(Y)) = \mathcal{I}^+$. But since $f|_K = 0$ we must have $f \in \mathcal{I}$, so we conclude that $\mathcal{I} = \{f \in C(X) : f|_K = 0\}$. ∎

The preceding result allows us to chacterize C*-quotients of $C(X)$.

PROPOSITION 5.1.6

Let K be a closed subset of X and let $\mathcal{I} = \{f \in C(Y) : f|_K = 0\}$. Then $C(X)/\mathcal{I}$ is $*$-isomorphic to $C(K)$.

PROOF Let $\pi : C(X) \to C(K)$ be the restriction map. It is clear that π is a $*$-homomorphism and has kernel equal to $\mathcal{I}$. For any $f \in C(K)$, by the Tietze extension theorem there exists $g \in C(X)$ with $\|g\|_\infty = \|f\|_\infty$ and $\pi g = g|_K = f$. This shows that π is isometric and onto. ∎

Now we turn to the spectrum. This is the tool that is used to show that the space X can actually be recovered from the algebra $C(X)$, and thus to establish that $C(X)$ contains exactly as much information as X. For $x \in X$ let $\hat{x} : f \mapsto f(x)$ be the evaluation map on $C(X)$.

PROPOSITION 5.1.7

Every compact Hausdorff space X is homeomorphic to sp$(C(X))$ *via the correspondence $x \leftrightarrow \hat{x}$.*

PROOF Let X be a compact Hausdorff space. For any $x \in X$ the map $\hat{x} : f \mapsto f(x)$ is a unital $*$-homomorphism from $C(X)$ to $\mathbf{C}$. Conversely, if $\omega : C(X) \to \mathbf{C}$ is any unital $*$-homomorphism then its kernel is a codimension one C*-ideal, and hence by Proposition 5.1.5 there exists $x \in X$ such that $\ker(\omega) = \{f \in C(X) : f(x) = 0\}$. (The set K in Proposition 5.1.5 must consist of exactly one point in order for $\ker(\omega)$ to have codimension one. This follows from Proposition 5.1.6.) Then for any $f \in C(X)$, letting $a = f(x)$ we have

$$\omega(f) = \omega(f - a \cdot 1_X + a \cdot 1_X) = \omega(f - a \cdot 1_X) + a = a = f(x)$$

since $f - a \cdot 1_X$ vanishes at x. Thus $\omega = \hat{x}$.

We have established a bijection between X and sp$(C(X))$. Suppose x_κ is a net in X which converges to $x \in X$. Then $f(x_\kappa) \to f(x)$ for all $f \in C(X)$, and hence $\hat{x}_\kappa \to \hat{x}$ weak*. This shows that the natural map from X onto $\mathrm{sp}(C(X))$ is continuous. Since X is compact and $\mathrm{sp}(C(X))$ is Hausdorff, it follows that this map is a homeomorphism. ∎

Finally, we establish a correspondence between continuous functions and $*$-homomorphisms.

Example 5.1.8

Let X and Y be compact Hausdorff spaces and let $\phi : Y \to X$ be continuous. Then C_ϕ is a unital $*$-homomorphism from $C(X)$ into $C(Y)$.

PROPOSITION 5.1.9
Let $\pi : C(X) \to C(Y)$ be a unital $*$-homomorphism. Then there is a continuous map $\phi : Y \to X$ such that $\pi = C_\phi$.

PROOF For any $y \in Y$ the map $\hat{y}$ is a unital $*$-homomorphism from $C(Y)$ into $\mathbf{C}$, so $\hat{y} \circ \pi$ is a unital $*$-homomorphism from $C(X)$ into $\mathbf{C}$. By Proposition 5.1.7 there exists a unique point $x \in X$ such that $\hat{y} \circ \pi = \hat{x}$. Define $\phi(y) = x$. For any $f \in C(X)$ and any $y \in Y$ we then have

$$C_\phi f(y) = f(\phi(y)) = \phi(y)\hat{}(f) = \hat{y}(\pi f) = \pi f(y).$$

Thus $\pi = C_\phi$.

To see that ϕ is continuous, let (y_κ) be a net in Y which converges to $y \in Y$. Suppose $\phi(y_\kappa) \not\to \phi(y)$; then there exists an open set $\mathcal{O}$ about $\phi(y)$ such that $\phi(y_\kappa)$ is not eventually in $\mathcal{O}$. Find $f \in C(X)$ such that $f(\phi(y)) = 1$ and $f|_{X-\mathcal{O}} = 0$. Then $g = \pi f$ has the property that $g(y) = 1$ and $g(y_\kappa) = 0$ frequently. But this contradicts the fact that $y_\kappa \to y$. ∎

5.2 *Topologies from functions*

Given any topological space X, we can form the space $C_b(X)$ of bounded continuous functions from X into $\mathbf{C}$. This is a unital C*-subalgebra of $l^\infty(X)$. Conversely, one might hope that every unital C*-subalgebra of $l^\infty(X)$ arises in this way from a topology on X, but that is not exactly true: for example, consider the set of functions $f \in l^\infty(0,1)$ which extend to continuous functions $\tilde{f}$ on $[0,1]$ such that $\tilde{f}(0) = \tilde{f}(1)$. There is no topology on $(0,1)$ for which these are precisely the bounded continuous functions. Still, we have the following result.

PROPOSITION 5.2.1
Let X be a set and let $\mathcal{A}$ be a unital C-subalgebra of $l^\infty(X)$ which*

separates points of X. Then there is a compact Hausdorff space Y such that X is a dense subset of Y and $\mathcal{A} = \{f|_X : f \in C(Y)\}$.

PROOF Let $Y = \mathrm{sp}(\mathcal{A})$. Given $\omega \in Y$ and $f \in \mathcal{A}$, we have $|\omega(f)| \leq \|f\|_\infty$ for the same reason that this is true when $\mathcal{A} = C(X)$; namely, for any $\lambda \in \mathbf{C}$ satisfying $|\lambda| > \|f\|_\infty$ we have

$$(f - \lambda \cdot 1_X)^{-1} = -\frac{1}{\lambda}\left(1_X + \frac{1}{\lambda}f + \left(\frac{1}{\lambda}f\right)^2 + \cdots\right) \in \mathcal{A},$$

so $\omega(f) - \lambda = \omega(f - \lambda \cdot 1_X) \neq 0$.

This shows that Y is contained in the unit ball of $\mathcal{A}^*$. We claim that Y is a compact Hausdorff space. Since dual unit balls are always weak* compact and Hausdorff, we need only show that Y is closed. But if (ω_κ) is a net in Y which weak* converges to $\omega \in \mathcal{A}^*$, it is easy to see that ω is also a unital $*$-homomorphism and hence belongs to Y. So Y is compact Hausdorff.

The embedding $x \mapsto \hat{x}$ allows us to identify X with a subset of Y. Moreover, any $f \in \mathcal{A}$ extends to a continuous function $\tilde{f}$ on Y defined by $\tilde{f}(\omega) = \omega(f)$. (In particular, $\tilde{f}(\hat{x}) = \hat{x}(f) = f(x)$, so $\tilde{f}$ really is an extension of f.) The set $\{\tilde{f} : f \in \mathcal{A}\}$ is then a unital C*-subalgebra of $C(Y)$, and it separates points of Y because $\tilde{f}(\omega_1) = \tilde{f}(\omega_2)$ for all $f \in \mathcal{A}$ implies $\omega_1(f) = \omega_2(f)$ for all f, which implies $\omega_1 = \omega_2$. Thus the extension map $f \mapsto \tilde{f}$ takes $\mathcal{A}$ onto $C(Y)$, and we conclude that $\mathcal{A} = \{f|_X : f \in C(Y)\}$.

Finally, X is dense in Y because any function in $C(Y)$ which vanishes on X restricts to the zero function on X, and hence must be zero on Y by the above isomorphism of $\mathcal{A}$ with $C(Y)$. ∎

Thus, unital C*-subalgebras of $l^\infty(X)$ correspond to compactifications of X, i.e., compact Hausdorff spaces which contain X as a dense subset (up to homeomorphisms fixing X). This is our justification for thinking of C*-subalgebras as corresponding to topologies.

In a way, measure spaces provide a better setting for this kind of result, because we can agree that the extra points that have to be added to X constitute a set of measure zero, and hence do not essentially alter X as a measure space. Here we must assume σ-finiteness, however.

We proceed to formulate a result of this type. It uses the following notion of equivalence of measure spaces.

DEFINITION 5.2.2 *Let $X = (X, \mu)$ and $Y = (Y, \nu)$ be σ-finite* *measure spaces. Then a* <u>measure equivalence</u> *between* X *and* Y *is a bijection* Φ *between the measurable sets in* X *and the measurable sets in* Y*, both modulo null sets, such that*

(a) $\Phi(\emptyset) = \emptyset$ *and* $\Phi(X) = Y$*;*
(b) $\Phi(S^c) = \Phi(S)^c$*;*
(c) $\Phi(\bigcup S_n) = \bigcup \Phi(S_n)$ *and* $\Phi(\bigcap S_n) = \bigcap \Phi(S_n)$*; and*
(d) $\mu(S) = \nu(\Phi(S))$

for all measurable $S, S_n \subset X$*. If a measure equivalence exists then we say that* X *and* Y *are* <u>measurably equivalent</u>.

Suppose X and Y are measurably equivalent and let Φ be a measure equivalence. If $f : X \to \mathbf{C}$ has countable range, then we can partition X into a countable family of positive measure sets on each of which f is constant. Then it is clear how to define a function $\Phi(f) : Y \to \mathbf{C}$ with the property that $\Phi(f^{-1}(S)) = \Phi(f)^{-1}(S)$ (up to null sets) for every positive measure subset S of $\mathbf{C}$. Now for any measurable function $f : X \to \mathbf{C}$ we can find a sequence of measurable functions $f_n : X \to \mathbf{C}$, each with countable range, such that $\|f - f_n\|_\infty \to 0$. Setting $\Phi(f) = \lim \Phi(f_n)$ then yields a correspondence between the measurable functions on X and the measurable functions on Y, both up to null sets, which respects the measure equivalence Φ. In particular, any measure equivalence between X and Y implements an isometric isomorphism between $L^\infty(X)$ and $L^\infty(Y)$.

We also need a notion of "separating points" which is suitable for measurable sets.

DEFINITION 5.2.3 *Let* X *be a* σ*-finite measure space and let* $\mathcal{A} \subset L^\infty(X)$*. Let* Ω *be the smallest* σ*-algebra for which every function in* $\mathcal{A}$ *is measurable. Then we say that* $\mathcal{A}$ <u>measurably separates points</u> *if every measurable subset of* X *has null symmetric difference with a set in* Ω*.*

Note that if $\mathcal{A}$ is closed under complex conjugation then Ω can also be described as the smallest σ-algebra for which every real function in $\mathcal{A}$ is measurable.

PROPOSITION 5.2.4
Let (X, μ) *be a* σ*-finite measure space, let* $\mathcal{A}$ *be a unital C*-subalgebra of* $L^\infty(X)$*, and suppose* $\mathcal{A}$ *measurably separates points. Then there is a compact Hausdorff space* Y*, a regular Borel measure* ν *on* Y*, and a*

measure equivalence between X and Y that takes $\mathcal{A}$ to $C(Y)$.

PROOF Let Y be the spectrum of $\mathcal{A}$. As in Proposition 5.2.1, Y is a compact Hausdorff space and the map $f \mapsto \tilde{f}$, where $\tilde{f}(\omega) = \omega(f)$, is a unital $*$-homomorphism, and hence a contraction, from $\mathcal{A}$ into $C(Y)$. We claim that this map is isometric. To see this, let $f \in \mathcal{A}$ and $\epsilon > 0$ and fix a subset $S_0 \subset X$ such that $|f| \geq \|f\|_\infty - \epsilon$ on S_0. Then let Λ be a maximal family of positive measure subsets of X, each contained in S_0, with the property that $S, T \in \Lambda$ implies $S \cap T \in \Lambda$. Direct Λ by reverse inclusion. For each $g \in \mathcal{A}$ and $S \in \Lambda$ the essential range of $g|_S$ is compact, so the intersection of these essential ranges is nonempty. Moreover, there cannot be more than one intersection point by maximality of Λ. For each $g \in \mathcal{A}$ define $\omega(g)$ to be this intersection point; then ω is a $*$-homomorphism from $\mathcal{A}$ into $\mathbf{C}$ such that $|\omega(f)| \geq \|f\|_\infty - \epsilon$, which establishes the claim. So $\mathcal{A}$ embeds in $C(Y)$, and since its image clearly separates points, the embedding is onto. So $\mathcal{A} \cong C(Y)$. Note that this shows in particular that the real functions in $\mathcal{A}$ are closed under finite lattice operations.

Let μ' be a finite measure on X such that μ and μ' are mutually absolutely continuous. We claim that $\mathcal{A}$ is dense in $L^2(X, \mu')$. To see this first let $S \subset X$ be a set of the form $S = f^{-1}((-\infty, a])$ for some real function $f \in \mathcal{A}$ and some $a \in \mathbf{R}$. Without loss of generality $a > 0$. Then, using the lattice operations in $\mathcal{A}$, we can find a real function $g = a/(f \vee a) \in \mathcal{A}$ such that $0 \leq g \leq 1$ and $S = g^{-1}(1)$, and consequently the powers g^n converge to χ_S in $L^2(X, \mu')$. Now it is straightforward to show that the family of $S \subset X$ such that χ_S is in the L^2 closure of $\mathcal{A}$ constitutes a σ-algebra, and it follows from the above and the definition of measurable separation that every measurable $S \subset X$ belongs to this family. So the L^2 closure of $\mathcal{A}$ contains all characteristic functions, and hence equals $L^2(X, \mu')$.

Now the map $f \mapsto \int f\, d\mu'$ is a bounded linear functional on $\mathcal{A}$, so it corresponds to a bounded linear functional on $C(Y)$, and thus there exists a regular Borel measure ν' on Y such that $\int f\, d\mu' = \int \tilde{f}\, d\nu'$ for all $f \in \mathcal{A}$. Now $\mathcal{A}$ is dense in $L^2(X, \mu')$ by the last paragraph, and it is standard that $C(Y)$ is dense in $L^2(Y, \nu')$. Moreover, the map $f \mapsto \tilde{f}$ is an isometry in L^2 norm by the definition of ν'. So this map extends to a surjective isometric isomorphism $U : L^2(X, \mu') \to L^2(Y, \nu')$. It is easy to see that U takes characteristic functions to characteristic functions, and therefore it defines a measure isomorphism between (X, μ') and (Y, ν'). It is also clear that this measure isomorphism takes $\mathcal{A}$ to $C(Y)$.

Finally, for each Borel set $S \subset Y$ define $\nu(S) = \mu(S')$ where S' is the corresponding set in X. Then ν and ν' are mutually absolutely continuous, and the result for μ' and ν' transfers to μ and ν. ∎

5.3 *Abelian C*-algebras*

We saw in the last section that C*-subalgebras $\mathcal{A}$ of $l^\infty(X)$ correspond to compactifications Y of X, and by the results of Section 5.1 algebraic properties of $\mathcal{A}$ are exactly mirrored in topological properties of Y. C*-algebras are the Hilbert space version of this construction.

DEFINITION 5.3.1 *A* <u>(concrete) C*-algebra</u> *is a C*-subalgebra $\mathcal{A}$ of $B(\mathcal{H})$, for some Hilbert space $\mathcal{H}$. If $AB = BA$ for all $A, B \in \mathcal{A}$ then $\mathcal{A}$ is* <u>abelian</u>.

If $I \notin \mathcal{A}$, then we define the <u>unitization</u> *of $\mathcal{A}$ to be the C*-algebra $\mathcal{A}^+ = \{A + aI : A \in \mathcal{A}$ and $a \in \mathbf{C}\}$.*

We will discuss abstract C*-algebras in Section 5.6.

Just as for $C(X)$, every unital $*$-homomorphism $\pi : \mathcal{A} \to \mathcal{B}$ between C*-algebras is automatically contractive. For if $A \in \mathcal{A}$ is self-adjoint and $|\lambda| > \|A\|$ then $A - \lambda I = -\lambda(I - \frac{1}{\lambda}A)$ is invertible in $\mathcal{A}$ by Lemma 3.2.3. (The series that defines the inverse converges in $\mathcal{A}$.) It follows that $\pi(A - \lambda I) = \pi(A) - \lambda$ is invertible in $\mathcal{B}$, so that $\lambda \notin \mathrm{sp}(\pi(A))$; as this is true whenever $|\lambda| > \|A\|$, we have $\|\pi(A)\| \leq \|A\|$ by Proposition 3.2.4. Now for any $A \in \mathcal{A}$ the preceding implies that $\|\pi(A)\|^2 = \|\pi(A^*A)\| \leq \|A^*A\| = \|A\|^2$. So π is a contraction.

Moreover, if π has null kernel then $\mathrm{sp}(A) = \mathrm{sp}(\pi(A))$ for all self-adjoint $A \in \mathcal{A}$; otherwise, let $f \in C(\mathrm{sp}(A))$ vanish on $\mathrm{sp}(\pi(A)) \subset \mathrm{sp}(A)$ but not on all of $\mathrm{sp}(A)$, and observe that $f(A) \neq 0$ but $\pi(f(A)) = f(\pi(A)) = 0$, contradicting injectivity. It follows that $\|A\| = \|\pi(A)\|$ for all self-adjoint $A \in \mathcal{A}$, and hence $\|A\|^2 = \|A^*A\| = \|\pi(A^*A)\| = \|\pi(A)\|^2$ for all $A \in \mathcal{A}$. This shows that a $*$-homomorphism with null kernel must be isometric, i.e., a $*$-isomorphism.

Example 5.3.2

Let $A \in B(\mathcal{H})$ be self-adjoint and let $C^*(A)$ be the norm closure of the polynomials in A. This is the unital C*-algebra generated by A.

But by the continuous functional calculus, the norm closure of the polynomials in A is precisely the set $\{f(A) : f \in C(\mathrm{sp}(A))\}$. Therefore, we have $C^*(A) \cong C(\mathrm{sp}(A))$, and so Proposition 5.1.7 implies that $\mathrm{sp}(C^*(A))$ (in the sense of Definition 5.1.1) is naturally identified with $\mathrm{sp}(A)$ (in the sense of Definition 3.2.1).

Example 5.3.3

Let X be a compact Hausdorff space, let μ be a regular Borel measure on X, and let $\mathcal{X}$ be a measurable Hilbert bundle over X. Then the set of multiplication operators $\{M_f : f \in C(X)\}$ on $L^2(X; \mathcal{X})$ is a unital C*-algebra.

The map $f \mapsto M_f$ is a unital $*$-homomorphism, and it is a $*$-isomorphism if neither μ nor $\mathcal{X}$ is constantly 0 on any open set.

Both of the above examples are abelian, and the second one is general over separable Hilbert spaces. We now proceed to prove this.

LEMMA 5.3.4

Let $\{A_\kappa : \kappa \in J\}$ be a pairwise commuting set of self-adjoint operators in $B(\mathcal{H})$. Assume $\|A_\kappa\| \leq 1$ for all κ. Then there is a spectral measure E on the set $X = [-1, 1]^J$ which on each coordinate gives rise to the spectral measure of A_κ in the manner of Lemma 3.5.2.

PROOF For any indices $\kappa_1, \ldots, \kappa_n$ and any function f on X of the form

$$f(x) = \sum_{j=1}^{m} a_j \chi_{S_1^j}(x_{\kappa_1}) \cdots \chi_{S_n^j}(x_{\kappa_n})$$

with $x = (x_\kappa)$ and $S_1^j, \ldots, S_n^j \subset [0, 1]$, we define $\pi(f)$ to be the operator

$$\pi(f) = \sum_{j=1}^{m} a_j \chi_{S_1^j}(A_{\kappa_1}) \cdots \chi_{S_n^j}(A_{\kappa_n}).$$

Since the A_κ commute, so do polynomials in the A_κ, and by continuity so do bounded Borel functions of the A_κ obtained by functional calculus. It follows that π is a $*$-homomorphism.

If f has a bounded inverse then its inverse is also of the above form, and this allows us to prove that π is contractive by the argument that showed this for $*$-homomorphisms between C*-algebras given at the beginning of this section. Then π extends by continuity to any continuous function on X — every coordinate function is clearly in the norm closure of the set of functions considered above, and by the Stone-Weierstrass theorem this implies that all of $C(X)$ is in the norm closure — and it extends further to the bounded Borel functions on X precisely as in Theorem 3.3.3. We can then define, for any Borel set $S \subset X$, a projection $E(S)$ by $P_{E(S)} = \pi(\chi_S)$. Verification that this is a spectral measure is routine. By construction, E agrees on each coordinate κ with the spectral measure of A_κ. ∎

THEOREM 5.3.5

Let $\mathcal{A} \subset B(\mathcal{H})$ be a unital abelian C-algebra and suppose $\mathcal{H}$ is separable. Then there is a probability measure μ on* $\mathrm{sp}(\mathcal{A})$*, a measurable Hilbert bundle $\mathcal{X}$ over* $\mathrm{sp}(\mathcal{A})$*, and an isometric isomorphism U :*

$L^2(\mathrm{sp}(\mathcal{A}); \mathcal{X}) \cong \mathcal{H}$ such that $\mathcal{A} = UC(\mathrm{sp}(\mathcal{A}))U^{-1}$ where $C(\mathrm{sp}(\mathcal{A}))$ acts on $L^2(\mathrm{sp}(\mathcal{A}); \mathcal{X})$ by multiplication.

PROOF Let $\{A_\kappa : \kappa \in J\}$ be the set of all self-adjoint operators in $\mathcal{A}$ of norm at most 1, let E be the spectral measure on $X = [-1, 1]^J$ provided by the lemma, and let μ, $\mathcal{X}$, and U be as in Corollary 3.4.3. Note that μ is regular because E is regular, in the same sense.

Define a $*$-homomorphism $\pi : C(X) \to B(\mathcal{H})$ by $\pi(f) = UM_fU^{-1}$. This map takes the coordinate functions on X to the operators A_κ, so it takes polynomials in the coordinates into $\mathcal{A}$, and by continuity its image is contained in $\mathcal{A}$. Conversely, every operator in $\mathcal{A}$ is a linear combination of self-adjoint operators of norm at most 1, so we conclude that the image of π equals $\mathcal{A}$.

Let $\mathcal{I} = \ker(\pi)$. By Proposition 5.1.5 there is a closed subset K of X such that $\mathcal{I} = \{f : f|_K = 0\}$, and π descends to a map $\tilde{\pi} : C(K) \cong C(X)/\mathcal{I} \to \mathcal{A}$ which is a bijective $*$-homomorphism, and therefore a $*$-isomorphism.

K can be identified with $\mathrm{sp}(A)$ by Proposition 5.1.7, and it is clear that μ is supported on K, so we obtain the desired result. ∎

In the nonseparable case, we can still say that there is a spectral measure E on $\mathrm{sp}(\mathcal{A})$ which gives rise to a $*$-isomorphism from $C(\mathrm{sp}(\mathcal{A}))$ onto $\mathcal{A}$ via approximation of continuous functions by simple functions, as in the proof of Lemma 5.3.4.

The comment about invariants that we made after Theorem 3.5.1 also applies here: $\mathrm{sp}(\mathcal{A})$, μ (up to measure class), and (X_n) (up to null sets) constitute a complete set of invariants for the unital C*-algebra $\mathcal{A}$, up to unitary equivalence.

5.4 The quantum plane

In this section we will define a quantum analog of the C*-algebra $C_0(\mathbf{R}^2)$ and determine its structure. We view this C*-algebra as endowing the quantum plane with topological structure.

The idea is to develop a functional calculus, so that we can define operators of the type $f(\mathcal{Q}, \mathcal{P})$ where f is a continuous function on the plane which goes to zero at infinity and $\mathcal{Q}$ and $\mathcal{P}$ are the position and momentum operators of Section 4.1. Even though $\mathcal{Q}$ and $\mathcal{P}$ are unbounded and do not commute, we can still do this for sufficiently smooth functions. The result is called the Weyl functional calculus. We will use functions in the Schwartz class $\mathcal{S}(\mathbf{R}^2)$ consisting of all C^∞ functions which, together with their partial derivatives of every order, are $o((|x| + |y|)^{-n})$ as $x, y \to \infty$. The key facts about $\mathcal{S}(\mathbf{R}^2)$ are that it is closed under sums and products and that it is taken onto itself by the Fourier transform. Also note that $C_c^\infty(\mathbf{R}^2) \subset \mathcal{S}(\mathbf{R}^2)$, so $\mathcal{S}(\mathbf{R}^2)$ is dense in $C_0(\mathbf{R}^2)$. After we have defined the operators $f(\mathcal{Q}, \mathcal{P})$ for $f \in \mathcal{S}(\mathbf{R}^2)$, the desired C*-algebra will simply be their norm closure.

It is convenient to work in the tracial representation of Section 4.2.

DEFINITION 5.4.1 *Let $\hbar \geq 0$. For $t_1, t_2 \in \mathbf{R}$ define a unitary operator L_{t_1,t_2} on $L^2(\mathbf{R}^2)$ by*

$$(L_{t_1,t_2}g)(x_1, x_2) = e^{i(x_1t_1+x_2t_2)}g(x_1 + \frac{1}{2}\hbar t_2, x_2 - \frac{1}{2}\hbar t_1);$$

for $f, g \in \mathcal{S}(\mathbf{R}^2)$ define the <u>*twisted product*</u> *$f \times_\hbar g \in \mathcal{S}(\mathbf{R}^2)$ by*

$$\begin{aligned}(f \times_\hbar g)(x_1, x_2) &= \frac{1}{2\pi}\int\int \hat{f}(t_1, t_2)L_{t_1,t_2}g\, dt_1dt_2\\ &= \frac{1}{2\pi}\int\int \hat{f}(t_1, t_2)g(x_1 + \frac{1}{2}\hbar t_2, x_2 - \frac{1}{2}\hbar t_1)e^{i(x_1t_1+x_2t_2)}\, dt_1dt_2;\end{aligned}$$

and for $f \in \mathcal{S}(\mathbf{R}^2)$ define the <u>*twisted multiplication operator*</u> *L_f on $L^2(\mathbf{R}^2)$ by $L_f(g) = f \times_\hbar g$.*

We are using the Fourier transform in two variables defined by

$$\hat{f}(t_1, t_2) = \frac{1}{2\pi}\int\int f(x_1, x_2)e^{-i(x_1t_1+x_2t_2)}\, dx_1dx_2.$$

One can prove that $\|f \times_\hbar g\|_2 \leq \|\hat{f}\|_1\|g\|_2$ using Minkowski's inequality for integrals, just as one proves the same inequality for the convolution of $\hat{f}$ and g. Thus L_f is a bounded operator with norm at most $\|\hat{f}\|_1$.

The operators L_{t_1,t_2} generalize the unitaries U_s and V_t introduced in Section 4.2; indeed, $L_{s,0} = U_s$ and $L_{0,t} = V_t$. They obey the commutation relation

$$L_{s_1,s_2}L_{t_1,t_2} = e^{i\hbar(s_2t_1-s_1t_2)}L_{t_1,t_2}L_{s_1,s_2},$$

which is a general version of the Weyl commutation relations.

The operator L_f plays the role of $f(\mathcal{Q}, \mathcal{P})$ in the tracial representation. The intuition is that L_{t_1,t_2} is $e^{i(t_1\mathcal{Q}+t_2\mathcal{P})}$, so we obtain $f(\mathcal{Q}, \mathcal{P})$ by Fourier expansion:

$$f(\mathcal{Q}, \mathcal{P}) = \frac{1}{2\pi}\int\int \hat{f}(t_1, t_2)e^{i(t_1\mathcal{Q}+t_2\mathcal{P})}\, dt_1dt_2.$$

Applying the right side of this formula to g yields the formula for $L_f g$.

The following result is a fairly straightforward computation.

PROPOSITION 5.4.2
(a) Let $f, g \in \mathcal{S}(\mathbf{R}^2)$ and $a, b \in \mathbf{C}$. Then $aL_f + bL_g = L_{af+bg}$.
(b) Let $f, g, h \in \mathcal{S}(\mathbf{R}^2)$. Then $(f \times_\hbar g) \times_\hbar h = f \times_\hbar (g \times_\hbar h)$. Thus $L_f L_g = L_{f \times_\hbar g}$.
(c) $L^*_{t_1,t_2} = L_{-t_1,-t_2}$ for all $t_1, t_2 \in \mathbf{R}$. For $f \in \mathcal{S}(\mathbf{R}^2)$ we have $L^*_f = L_g$ where $g(x_1, x_2) = f(-x_1, -x_2)$.

DEFINITION 5.4.3 *$C_0^\hbar(\mathbf{R}^2)$ is the norm closure of the operators L_f for $f \in \mathcal{S}(\mathbf{R}^2)$.*

It follows from Proposition 5.4.2 that $C_0^\hbar(\mathbf{R}^2)$ is a C*-algebra.

The next result is trivial.

PROPOSITION 5.4.4
Suppose $\hbar = 0$. Then $L_f = M_f$ for all $f \in \mathcal{S}(\mathbf{R}^2)$, and $C_0^\hbar(\mathbf{R}^2) \cong C_0(\mathbf{R}^2)$.

Next we consider the action of translation operators on $C_0^\hbar(\mathbf{R}^2)$. Recall the unitary operator T_{t_1,t_2} on $L^2(\mathbf{R}^2)$ introduced following Definition 4.2.1. For $A \in B(L^2(\mathbf{R}^2))$ define $\theta_{t_1,t_2}(A) = T_{t_1,t_2} A T^{-1}_{t_1,t_2}$.

PROPOSITION 5.4.5
For any $t_1, t_2 \in \mathbf{R}$ and $f \in \mathcal{S}(\mathbf{R}^2)$ we have

$$\theta_{t_1,t_2}(L_f) = L_{T_{t_1,t_2} f}.$$

The map θ_{t_1,t_2} restricts to a $*$-isomorphism from $C_0^\hbar(\mathbf{R}^2)$ onto itself. This yields an action of $\mathbf{R}^2$ by automorphisms of $C_0^\hbar(\mathbf{R}^2)$. Moreover, for any $A \in C_0^\hbar(\mathbf{R}^2)$ the map $(t_1, t_2) \mapsto \theta_{t_1,t_2}(A)$ is continuous in norm.

PROOF The first equality is a straightforward calculation, and it implies that θ_{t_1,t_2} takes $C_0^\hbar(\mathbf{R}^2)$ into itself. The inverse map $\theta_{-t_1,-t_2}$, has the same property, so θ_{t_1,t_2} restricts to a $*$-isomorphism from $C_0^\hbar(\mathbf{R}^2)$ onto itself. It is clear that the map $(t_1, t_2) \to \theta_{t_1,t_2}$ is a group homomorphism.

For $A = L_f$ ($f \in \mathcal{S}(\mathbf{R}^2)$) we have

$$\|\theta_{t_1,t_2}(A) - A\| = \|L_{T_{t_1,t_2} f} - L_f\| = \|L_{T_{t_1,t_2} f - f}\|.$$

This converges to 0 as $t_1, t_2 \to 0$ since $\|(T_{t_1,t_2} f - f)\hat{}\|_1 \to 0$. But for any $A \in C_0^\hbar(\mathbf{R}^2)$ and any $\epsilon > 0$ we can find L_f such that $\|L_f - A\| \leq \epsilon/3$ and $\delta > 0$ such that $|t_1|, |t_2| \leq \delta$ implies $\|\theta_{t_1,t_2}(L_f) - L_f\| \leq \epsilon/3$. Then

$|t_1|, |t_2| \leq \delta$ also implies

$$\|\theta_{t_1,t_2}(A) - A\| \leq \|\theta_{t_1,t_2}(A - L_f)\| + \|\theta_{t_1,t_2}(L_f) - L_f\| + \|L_f - A\| \leq \epsilon.$$

This shows that the map $(t_1, t_2) \mapsto \theta_{t_1,t_2}(A)$ is norm continuous. ∎

In particular, we have

$$T_{s,0} L_f T_{s,0}^{-1} = L_{T_{-s,0}f} = V_{-s/\hbar} L_f V_{-s/\hbar}^{-1}$$

and

$$T_{0,t} L_f T_{0,t}^{-1} = L_{T_{0,-t}f} = U_{t/\hbar} L_f U_{t/\hbar}^{-1}.$$

This is the sense in which $V_{-s/\hbar}$ and $U_{t/\hbar}$ are "equivalent" to $T_{s,0}$ and $T_{0,t}$, as we indicated at the end of Section 4.2.

The remainder of this section is devoted to characterizing the structure of $C_0^\hbar(\mathbf{R}^2)$. We observe first that $C_0^\hbar(\mathbf{R}^2) \cong C_0^{a\hbar}(\mathbf{R}^2)$ for all $a > 0$, i.e., these algebras are mutually $*$-isomorphic for all nonzero values of $\hbar$. Indeed, the unitary operator U defined on $L^2(\mathbf{R}^2)$ by $Uf(x_1, x_2) = af(\sqrt{a}x_1, \sqrt{a}x_2)$ satisfies

$$U^{-1} L_{t_1,t_2} U g(x_1, x_2) = e^{i(x_1 t_1 + x_2 t_2)/\sqrt{a}} g(x_1 + \frac{1}{2}\hbar\sqrt{a}t_2, x_2 - \frac{1}{2}\hbar\sqrt{a}t_1),$$

so $U^{-1}L_{t_1,t_2}U$ is the operator $L_{t_1/\sqrt{a}, t_2/\sqrt{a}}$ for $\hbar' = a\hbar$; and integration against $\hat{f}$ yields that conjugation by U takes L_f (for $\hbar$) to $L_{U^{-1}f}$ (for $\hbar' = a\hbar$). Thus conjugation by U defines a $*$-isomorphism from $C_0^\hbar(\mathbf{R}^2)$ onto $C_0^{a\hbar}(\mathbf{R}^2)$.

To determine the structure of $C_0^\hbar(\mathbf{R}^2)$ more explicitly we must pass to the $L^2(\mathbf{R})$ representation of Section 4.1. In this model the operator $L_f \in B(L^2(\mathbf{R}^2))$ is replaced by the operator $\tilde{L}_f \in B(L^2(\mathbf{R}))$ defined by

$$\tilde{L}_f g(x) = \frac{1}{2\pi} \int\int \hat{f}(t_1, t_2) e^{it_1(x + \frac{1}{2}\hbar t_2)} g(x + \hbar t_2)\, dt_1 dt_2.$$

We have $L_f = W^{-1}(\tilde{L}_f \otimes I)W$ where W is the unitary operator of Proposition 4.2.2. This can be proven by a direct computation using the definition of W, or more easily by using the formula $L_{t_1,t_2} = e^{i\hbar t_1 t_2/2} U_{t_1} V_{t_2}$ and the fact that $L_f g = \frac{1}{2\pi} \int\int \hat{f}(t) L_t g\, dt$.

Let $\tilde{C}_0^\hbar(\mathbf{R}^2)$ be the norm closure of the operators $\tilde{L}_f$ on $L^2(\mathbf{R})$. Then the map $L_f \mapsto \tilde{L}_f$ extends to a $*$-isomorphism from $C_0^\hbar(\mathbf{R}^2)$ onto $\tilde{C}_0^\hbar(\mathbf{R}^2)$. We will prove that $\tilde{C}_0^\hbar(\mathbf{R}^2) = K(L^2(\mathbf{R}))$, the C*-algebra of compact operators on $L^2(\mathbf{R})$. First we must develop some general information about compact operators.

DEFINITION 5.4.6 *An operator $A \in B(\mathcal{H})$ is* compact *if the image under A of the unit ball of $\mathcal{H}$ is precompact in norm, or equivalently, for any bounded sequence (v_n) in $\mathcal{H}$, the sequence (Av_n) has a cluster point. The set of all compact operators is denoted $K(\mathcal{H})$.*

For example, a projection is compact if and only if its range is finite dimensional.

For $A \in B(\mathcal{H})$ let $|A|$ be the operator $|A| = (A^*A)^{1/2}$ defined using continuous functional calculus on the self-adjoint operator A^*A. The next result is known as polar decomposition. (To see why, apply it to an operator in $B(\mathbf{C}) \cong \mathbf{C}$.)

LEMMA 5.4.7
Let $A \in B(\mathcal{H})$. Then there is a unique $U \in B(\mathcal{H})$ such that $A = U|A|$, $U = 0$ on $\operatorname{ran}(|A|)^\perp$, and U is an isometry on $\operatorname{ran}(|A|)$.

PROOF For any $v \in \mathcal{H}$ we have

$$\||A|v\|^2 = \langle |A|v, |A|v\rangle = \langle |A|^2 v, v\rangle = \langle A^*Av, v\rangle = \langle Av, Av\rangle = \|Av\|^2.$$

Thus the map $U : |A|v \mapsto Av$ is well-defined and isometric on $\operatorname{ran}(|A|)$, and we can extend it to $\mathcal{H}$ by setting $U = 0$ on $\operatorname{ran}(|A|)^\perp$. Uniqueness is clear. ∎

PROPOSITION 5.4.8
$K(\mathcal{H})$ is a C-ideal of $B(\mathcal{H})$.*

PROOF It is straightforward to verify that $K(\mathcal{H})$ is a linear subspace of $B(\mathcal{H})$, and that $A \in K(\mathcal{H})$ and $B \in B(\mathcal{H})$ implies $AB, BA \in K(\mathcal{H})$. Next we show that $K(\mathcal{H})$ is closed.

Let (A_n) be a sequence in $K(\mathcal{H})$ which converges in norm to $A \in B(\mathcal{H})$. We may assume that $\|A_j - A_k\| \leq \frac{1}{n}$ for $j, k \geq n$. We must show that A is compact. To see this, let (v_k) be a sequence of vectors of norm at most 1. Define $v_k^0 = v_k$, and for $n > 0$ inductively let (v_k^n) be a subsequence of (v_k^{n-1}) such that $\|A_n v_j^n - A_n v_k^n\| \leq \frac{1}{n}$ for all j, k. This is possible because A_n is compact. Finally, define $w_k = v_k^k$; this is a subsequence of (v_k) such that

$$\begin{aligned}\|Aw_j - Aw_k\| &= \|Aw_j - A_n w_j + A_n w_j - A_n w_k + A_n w_k - Aw_k\| \\ &\leq \|A - A_n\|(\|w_j\| + \|w_k\|) + \|A_n w_j - A_n w_k\| \\ &\leq 3/n\end{aligned}$$

for $j, k \geq n$. Thus (Aw_k) converges, showing that A is compact.

Finally, we must show that $A \in K(\mathcal{H})$ implies $A^* \in K(\mathcal{H})$. Let $A \in K(\mathcal{H})$ and write $A = U|A|$ as in the lemma. Then $|A|$ is compact because every polynomial in A^*A is compact and $K(\mathcal{H})$ is closed in norm, and therefore $A^* = |A|U^*$ is compact by the ideal property of $K(\mathcal{H})$. ∎

Self-adjoint compact operators have a particularly simple structure.

PROPOSITION 5.4.9
Let $A \in B(\mathcal{H})$ be self-adjoint. Then A is compact if and only if $A = \sum a_n P_n$ where the P_n are orthogonal projections with finite dimensional ranges and $a_n \to 0$.

PROOF The reverse direction follows from the fact that $K(\mathcal{H})$ is closed. For the converse, suppose A is compact and write $A = UM_xU^{-1}$ as in Theorem 3.5.1. It will suffice to show that for each $\epsilon > 0$ the set $S = \{x \in \mathrm{sp}(A) : |x| \geq \epsilon\}$ is finite and the corresponding spectral projection $P = \chi_S(A)$ has finite dimensional range; for then we can write $A = \sum a_n P_n$ where the a_n enumerate $\mathrm{sp}(A) - \{0\}$ and $P_n = \chi_{a_n}(A)$.

Fix ϵ, S, and P. Then for any $v \in \mathcal{H}$ we have $\|Av\| \geq \epsilon\|Pv\|$. So compactness of A implies compactness of P, and it follows that P has finite dimensional range, and hence S is finite, as desired. ∎

Next we introduce another concept which will lead to the key theorem that we need in order to identify the C*-algebra $C_0^\hbar(\mathbf{R}^2)$.

DEFINITION 5.4.10 *Let $\mathcal{A} \subset B(\mathcal{H})$ be a C*-algebra. Then $\mathcal{A}$ is* irreducible *if for every $v \in \mathcal{H}$ the set $\mathcal{A}v = \{Av : A \in \mathcal{A}\}$ is dense in $\mathcal{H}$.*

By considering the compact operators $v \mapsto \langle v, w\rangle w$ (for $w \in \mathcal{H}$) one easily sees that $K(\mathcal{H})$ is irreducible.

THEOREM 5.4.11
Let $\mathcal{A} \subset B(\mathcal{H})$ be an irreducible C-algebra and suppose $\mathcal{A} \subset K(\mathcal{H})$. Then $\mathcal{A} = K(\mathcal{H})$.*

PROOF It follows from Proposition 5.4.9 and the continuous functional calculus that $\mathcal{A}$ contains a projection with finite dimensional range. Let $P \in \mathcal{A}$ be such a projection whose range has minimal (nonzero) dimension. We claim this dimension is one. Otherwise, let $v, w \in \mathrm{ran}(P)$ be orthogonal. For any self-adjoint operator $A \in \mathcal{A}$, the operator PAP is self-adjoint and by Proposition 5.4.9 it can be decomposed into projections. But $\mathrm{ran}(PAP) \subset \mathrm{ran}(P)$, so minimality of P implies $PAP = aP$ for some $a \in \mathbf{C}$. Thus

$$\langle Av, w\rangle = \langle PAPv, w\rangle = a\langle v, w\rangle = 0,$$

so that w is orthogonal to Av. Since any $A \in \mathcal{A}$ is a linear combination of self-adjoint operators in $\mathcal{A}$, we have $w \perp \mathcal{A}v$, contradicting irreducibility. This proves the claim.

Now say the range of P is spanned by the unit vector v, so that $Pu = \langle u, v\rangle v$ for all $u \in \mathcal{H}$, and let w be any other unit vector. Find a sequence $(A_n) \subset \mathcal{A}$ such that $A_n v \to w$ and $\|A_n v\| = 1$ for all n. Let P' be the projection onto the span of w. Then for every $u \in \mathcal{H}$

$$\begin{aligned}\|A_n P A_n^* u - P'u\| &= \|\langle u, A_n v\rangle A_n v - \langle u, w\rangle w\| \\ &= \|\langle u, A_n v\rangle(A_n v - w) + \langle u, A_n v - w\rangle w\| \\ &\leq 2\|A_n v - w\|\|u\|.\end{aligned}$$

We conclude that $A_n P A_n^* \to P'$ in norm, and hence $P' \in \mathcal{A}$, so $\mathcal{A}$ contains the projections onto all one-dimensional subspaces of $\mathcal{H}$. By Proposition 5.4.9 and decomposition into real and imaginary parts, it follows that $\mathcal{A}$ contains every compact operator. ∎

We now proceed to prove that $C_0^\hbar(\mathbf{R}^2) \cong K(L^2(\mathbf{R}))$.

LEMMA 5.4.12
Let $k = (k_{mn}) \in l^2(\mathbf{N} \times \mathbf{N})$. Then the operator A on $l^2(\mathbf{N})$ defined by

$$Av = \sum_{m,n} k_{mn}\langle v, e_m\rangle e_n$$

satisfies $\|A\| \leq \|k\|$ and is compact.

PROOF Let $v = (a_n), w = (b_n) \in l^2(\mathbf{N})$. Then $\bar{v} \otimes w = (\bar{a}_m b_n) \in l^2(\mathbf{N} \times \mathbf{N})$, so

$$|\langle Av, w\rangle| = \left|\sum_{m,n} k_{mn} a_m \bar{b}_n\right| = |\langle k, \bar{v} \otimes w\rangle| \leq \|k\|\|v\|\|w\|.$$

Thus $\|A\| \leq \|k\|$.

If only finitely many k_{mn} are nonzero then A has finite dimensional range, and hence is compact. In general the norm estimate shows that every A is approximated by operators of this type, so compactness of A follows from the fact that $K(\mathcal{H})$ is closed. ∎

THEOREM 5.4.13
If $\hbar \neq 0$ *then* $C_0^\hbar(\mathbf{R}^2) \cong K(L^2(\mathbf{R}))$.

PROOF Let $\mathcal{A} \cong \tilde{C}_0^\hbar(\mathbf{R}^2) \subset B(L^2(\mathbf{R}))$ be the norm closure of the operators $\tilde{L}_f$ for $f \in \mathcal{S}(\mathbf{R}^2)$. As we noted earlier, $\mathcal{A}$ is naturally $*$-isomorphic to $C_0^\hbar(\mathbf{R}^2)$. We will show that $\mathcal{A} = K(L^2(\mathbf{R}))$. By Theorem 5.4.11, it will suffice to show that $\mathcal{A} \subset K(L^2(\mathbf{R}))$ and $\mathcal{A}$ is irreducible.

Since $K(L^2(\mathbf{R}))$ is closed, to prove the first assertion we need only show that each $\tilde{L}_f$ is in $K(L^2(\mathbf{R}))$. In fact, since $\|\tilde{L}_f\| \leq \|\hat{f}\|_1$, passing to the Fourier transform picture shows that every $\tilde{L}_f$ is approximated by linear combinations of operators $\tilde{L}_f$ with f of the form $f(x_1, x_2) = f_1(x_1)f_2(x_2)$ for $f_1, f_2 \in \mathcal{S}(\mathbf{R})$. So it will be enough to show that every operator of this form is compact.

Fix such a function $f(x_1, x_2) = f_1(x_1)f_2(x_2)$. Then for $g \in \mathcal{S}(\mathbf{R})$ we have

$$\begin{aligned}\tilde{L}_f g(x) &= \frac{1}{2\pi}\int\int \hat{f}_1(t_1)\hat{f}_2(t_2)e^{it_1(x+\frac{1}{2}\hbar t_2)}g(x+\hbar t_2)\,dt_1dt_2\\ &= \frac{1}{\sqrt{2\pi}}\int f_1(x+\frac{1}{2}\hbar t_2)\hat{f}_2(t_2)g(x+\hbar t_2)\,dt_2\\ &= \frac{1}{\sqrt{2\pi}}\int k(x,s)g(s)\,ds\end{aligned}$$

where $k(x, s) = \frac{1}{\hbar}f_1((s+x)/2)\hat{f}_2((s-x)/\hbar)$ and we have made the substitution $s = x + \hbar t_2$. We have $k \in L^2(\mathbf{R}^2)$.

Write $k = \sum k_{mn}\overline{h_m(s)}h_n(x)$ where (h_n) is the Hermite basis of $L^2(\mathbf{R})$ and $(k_{mn}) \in l^2(\mathbf{N} \times \mathbf{N})$. Then

$$\begin{aligned}\int k(x,s)h_i(s)\,ds &= \int \sum_{m,n} k_{mn}\overline{h_m(s)}h_n(x)h_i(s)\,ds\\ &= \sum_n k_{in}h_n(x).\end{aligned}$$

By linearity, this shows that $\tilde{L}_f$ is an operator of the form given in the lemma, and therefore $\tilde{L}_f$ is compact. Thus $\mathcal{A} \subset K(L^2(\mathbf{R}))$.

To verify irreducibility let $a \in \mathbf{R}$ and for $\epsilon > 0$ find $f_2^\epsilon \in \mathcal{S}(\mathbf{R})$ such that $\hat{f}_2^\epsilon \geq 0$, $\int \hat{f}_2^\epsilon = 1$, and $\hat{f}_2^\epsilon$ is supported on the interval $[a-\epsilon, a+\epsilon]$. This can be done because the inverse Fourier transform takes $C_c^\infty(\mathbf{R})$ into $\mathcal{S}(\mathbf{R})$.

Now for any $f_1, g \in \mathcal{S}(\mathbf{R})$, defining $f(x_1, x_2) = f_1(x_1)f_2(x_2)$ yields

$$\tilde{L}_f g(x) = \frac{1}{\sqrt{2\pi}}\int f_1(x+\frac{1}{2}\hbar t_2)\hat{f}_2(t_2)g(x+\hbar t_2)\,dt_2$$

$$\to \frac{1}{\sqrt{2\pi}} f_1(x + \frac{1}{2}\hbar a) g(x + \hbar a)$$

as $\epsilon \to 0$. That is, $\tilde{L}_f g \to \frac{1}{\sqrt{2\pi}} M_f T_{-\hbar a} g$ where $f(x) = f_1(x + \frac{1}{2}\hbar a)$. Moreover, the convergence is uniform in g, so we conclude that the same limit obtains, and hence $\frac{1}{\sqrt{2\pi}} M_f T_{-\hbar a} g \in \overline{\mathcal{A}g}$, for all $g \in L^2(\mathbf{R})$. Since a and f_1 were arbitrary, this shows that $\overline{\mathcal{A}g}$ contains all L^2 functions whose support is contained in a translation of the support of g. Thus $\overline{\mathcal{A}g} = L^2(\mathbf{R})$ for any $g \in L^2(R)$, and we conclude that $\mathcal{A}$ is irreducible. ∎

Together with Proposition 5.4.4, this determines the structure of every $C_0^\hbar(\mathbf{R}^2)$.

5.5 *Quantum tori*

Quantum tori are related to the quantum plane, but they are technically easier to deal with owing to the fact that the Fourier transform takes the torus $\mathbf{T}^2 = \mathbf{R}^2/2\pi\mathbf{Z}^2$ to the discrete space $\mathbf{Z}^2$.

First we describe the untransformed model. The underlying Hilbert space is $L^2(\mathbf{T}^2)$. By analogy with the operators U_s and V_t of Section 4.2, we define

$$Uf(x_1, x_2) = e^{ix_1} f(x_1, x_2 - \frac{1}{2}\hbar)$$

and

$$Vf(x_1, x_2) = e^{ix_2} f(x_1 + \frac{1}{2}\hbar, x_2)$$

for $f \in L^2(\mathbf{T}^2)$. These are unitary operators, and they obey the commutation relation $UV = e^{-i\hbar} VU$.

We regard the operators U and V as "exponential functions" on a quantum torus, so the C*-algebra they generate should be the algebra of "continuous functions."

DEFINITION 5.5.1 *Let $C^\hbar(\mathbf{T}^2)$ be the closed span in $B(L^2(\mathbf{T}^2))$ of the operators $U^m V^n$ for $m, n \in \mathbf{Z}$.*

Equivalently, $C^\hbar(\mathbf{T}^2)$ is the C*-algebra generated by U and V.

The next result is trivial.

PROPOSITION 5.5.2
If $\hbar = 0$ then $C^\hbar(\mathbf{T}^2) \cong C(\mathbf{T}^2)$.

Now we pass to the Fourier transform picture. Let $\tilde{\mathcal{F}}_2 : L^2(\mathbf{T}^2) \mapsto l^2(\mathbf{Z}^2)$ be the Fourier transform given by

$$(\tilde{\mathcal{F}}_2 f)_{mn} = \frac{1}{2\pi}\int_0^{2\pi}\int_0^{2\pi} f(x_1, x_2)e^{-i(mx_1+nx_2)}\,dx_1 dx_2$$

(cf. Example 2.3.6). Then conjugation by $\tilde{\mathcal{F}}_2$ gives rise to operators $\hat{U} = \tilde{\mathcal{F}}_2 U \tilde{\mathcal{F}}_2^{-1}$ and $\hat{V} = \tilde{\mathcal{F}}_2 V \tilde{\mathcal{F}}_2^{-1}$ on $l^2(\mathbf{Z}^2)$ which satisfy

$$\hat{U}e_{m,n} = e^{-i\hbar n/2}e_{m+1,n} \qquad \text{and} \qquad \hat{V}e_{m,n} = e^{i\hbar m/2}e_{m,n+1}$$

where $\{e_{m,n}\}$ is the standard orthonormal basis of $l^2(\mathbf{Z}^2)$. More generally, for any $k, l \in \mathbf{Z}$ define

$$L_{k,l}e_{m,n} = e^{i\hbar(ml-nk)/2}e_{m+k,n+l} = e^{i\hbar kl/2}\hat{U}^k\hat{V}^l e_{m,n}.$$

These are analogous to (a Fourier transform version of) the operators L_{t_1,t_2} of the last section.

Let $\hat{C}^\hbar(\mathbf{T}^2)$ be the C*-algebra generated by the operators $\hat{U}$ and $\hat{V}$. It is $*$-isomorphic to $C^\hbar(\mathbf{T}^2)$ via conjugation by $\tilde{\mathcal{F}}_2$.

Any polynomial in $\hat{U}$ and $\hat{V}$ can be written as a finite sum $A = \sum a_{k,l}L_{k,l}$. We have the following basic facts.

PROPOSITION 5.5.3

Let $A = \sum a_{k,l}L_{k,l}$ be a polynomial in the operators $L_{k,l}$. Then

$$Ae_{m,n} = \sum_{k,l} e^{i\hbar(ml-nk)/2}a_{k,l}e_{m+k,n+l}$$

and

$$\langle Ae_{m,n}, e_{m+k,n+l}\rangle = e^{i\hbar(ml-nk)/2}a_{k,l}.$$

For any $A \in \hat{C}^\hbar(\mathbf{T}^2)$ we also have

$$\langle Ae_{m,n}, e_{m+k,n+l}\rangle = e^{i\hbar(ml-nk)/2}\langle Ae_{0,0}, e_{k,l}\rangle.$$

The last part of this proposition holds for polynomials in the $L_{k,l}$ by direct computation, and for any $A \in \hat{C}^\hbar(\mathbf{T}^2)$ by continuity.

We regard the $a_{k,l}$ as the "Fourier coefficients" of $\sum a_{k,l}L_{k,l}$. The preceding suggests that for any $A \in \hat{C}^\hbar(\mathbf{T}^2)$, not just polynomials, we should define the Fourier coefficients of A to be

$$a_{k,l} = \langle Ae_{0,0}, e_{k,l}\rangle.$$

We must then ask in what sense $A = \sum a_{k,l}L_{k,l}$. This is not true in the sense of norm convergence; that already fails in the case $\hbar = 0$.

(There exists a continuous function on the circle, and hence also one on the torus, such that the partial sums of its Fourier series do not converge uniformly.) Rather, we must follow the $\hbar = 0$ case and consider convergence of Cesàro means. We now proceed to do this.

DEFINITION 5.5.4 *For $A \in \hat{C}^{\hbar}(\mathbf{T}^2)$ and $m, n \in \mathbf{Z}$ define*

(a) $a_{k,l}(A) = \langle Ae_{0,0}, e_{k,l}\rangle$ (the <u>*Fourier coefficients*</u> *of A);*
(b) $s_N(A) = \sum_{|k|,|l|\leq N} a_{k,l}(A)L_{k,l}$ (the <u>*partial sums*</u> *of the Fourier series); and*
(c) $\sigma_N(A) = \frac{1}{(N+1)^2}\sum_{|k|,|l|\leq N}(N+1-|k|)(N+1-|l|)a_{k,l}(A)$ (the <u>*Cesàro means*</u> *of the Fourier series).*

In order to prove convergence in the Cesàro sense of the Fourier series, we need an alternative formula for the Fourier coefficients which involves the unitary operators $M_{e^{-i(sm+tn)}}$ on $B(l^2(\mathbf{Z}^2))$ $(s, t \in \mathbf{R})$. In the untransformed picture, these are coordinate translations. For $A \in B(l^2(\mathbf{Z}))$ define

$$\hat{\theta}_{s,t}(A) = M_{e^{-i(sm+tn)}}AM^{-1}_{e^{-i(sm+tn)}}.$$

We have the following result.

PROPOSITION 5.5.5
For any $s, t \in \mathbf{R}$ and $k, l \in \mathbf{Z}$ we have $\hat{\theta}_{s,t}(L_{k,l}) = e^{-i(sk+tl)}L_{k,l}$. The map $\hat{\theta}_{s,t}$ restricts to a $*$-isomorphism from $\hat{C}^{\hbar}(\mathbf{T}^2)$ onto itself. This defines an action of $\mathbf{R}^2$ by automorphisms of $\hat{C}^{\hbar}(\mathbf{T}^2)$. Moreover, for any $A \in \hat{C}^{\hbar}(\mathbf{T}^2)$ the map $(s,t) \mapsto \hat{\theta}_{s,t}(A)$ is continuous in norm.

PROOF It is clear that $\hat{\theta}_{s,t}$ is a $*$-isomorphism from $B(l^2(\mathbf{Z}^2))$ onto itself. Since $\hat{\theta}_{s,t}(L_{k,l}) = e^{-i(sk+tl)}L_{k,l}$ (an easy computation) it follows that $\hat{\theta}_{s,t}$ takes polynomials in $\hat{U}$ and $\hat{V}$ to polynomials in $\hat{U}$ and $\hat{V}$, and by continuity it takes $\hat{C}^{\hbar}(\mathbf{T}^2)$ into itself. As $\hat{\theta}^{-1}_{s,t} = \hat{\theta}_{-s,-t}$ has the same property, it follows that $\hat{\theta}_{s,t}$ maps $\hat{C}^{\hbar}(\mathbf{T}^2)$ onto itself. It is clear that the map $(s,t) \mapsto \hat{\theta}_{s,t}$ is a group homomorphism.

The equations $\hat{\theta}^{-1}_{s,t} = \hat{\theta}_{-s,-t}$ and $\hat{\theta}_{s,t} \circ \hat{\theta}_{s',t'} = \hat{\theta}_{s+s',t+t'}$ are trivial, so $\hat{\theta}$ is an action of $\mathbf{R}^2$ by automorphisms of $\hat{C}^{\hbar}(\mathbf{T}^2)$.

If A is a polynomial in the $L_{k,l}$, the above expression for $\hat{\theta}_{s,t}(L_{k,l})$ shows that the map $(s,t) \mapsto \hat{\theta}_{s,t}(A)$ is continuous in norm. For any $A \in \hat{C}^{\hbar}(\mathbf{T}^2)$ and any $\epsilon > 0$, find a polynomial B in the $L_{k,l}$ such that $||A-B|| \leq \epsilon/3$ and find $\delta > 0$ such that $|s|, |t| \leq \delta$ implies $||\hat{\theta}_{s,t}(B)-B|| \leq$

$\epsilon/3$. Then $|s|, |t| \leq \delta$ implies

$$\|\hat{\theta}_{s,t}(A) - A\| \leq \|\hat{\theta}_{s,t}(A - B)\| + \|\hat{\theta}_{s,t}(B) - B\| + \|B - A\| \leq \epsilon.$$

This shows that the map $(s,t) \mapsto \hat{\theta}_{s,t}(A)$ is norm continuous. ∎

Now we can characterize the Fourier coefficients of A differently.

PROPOSITION 5.5.6
Let $A \in \hat{C}^\hbar(\mathbf{T}^2)$. Then

$$a_{k,l}(A)L_{k,l} = \frac{1}{(2\pi)^2}\int_0^{2\pi}\int_0^{2\pi} e^{i(sk+tl)}\hat{\theta}_{s,t}(A)\,dsdt.$$

PROOF The integral is taken in the sense of Definition 3.8.3. Since finite linear combinations of basis vectors $e_{m,n}$ are dense in $l^2(\mathbf{Z}^2)$, it suffices to check equality when both sides are paired against these basis vectors. But

$$\begin{aligned}
&\frac{1}{(2\pi)^2}\int_0^{2\pi}\int_0^{2\pi} e^{i(sk+tl)}\langle\hat{\theta}_{s,t}(A)e_{m,n}, e_{m',n'}\rangle\,dsdt\\
&= \frac{1}{(2\pi)^2}\int_0^{2\pi}\int_0^{2\pi} e^{i(sk+tl)}e^{-i((m'-m)s+(n'-n)t)}\langle Ae_{m,n}, e_{m',n'}\rangle\,dsdt\\
&= \begin{cases}\langle Ae_{m,n}, e_{m',n'}\rangle & \text{if } m' = m+k \text{ and } n' = n+l\\ 0 & \text{otherwise}\end{cases}\\
&= \langle a_{k,l}(A)L_{k,l}e_{m,n}, e_{m',n'}\rangle.
\end{aligned}$$

So the desired equality holds. ∎

Using this result, the proof that the Cesàro means of A converge is an easy adaptation of its proof in the classical case.

THEOREM 5.5.7
Let $A \in \hat{C}^\hbar(\mathbf{T}^2)$. Then $\sigma_N(A) \to A$ in norm.

PROOF Let K_N be the Fejér kernel,

$$K_N(t) = \sum_{n=-N}^{N}\left(1 - \frac{|n|}{N+1}\right)e^{int} = \frac{1}{N+1}\left(\frac{\sin((N+1)t/2)}{\sin(t/2)}\right)^2.$$

Then

$$A = \frac{1}{(2\pi)^2}\int_{-\pi}^{\pi}\int_{-\pi}^{\pi} AK_N(s)K_N(t)\,dsdt$$

since $\int_{-\pi}^{\pi} K_N(s)ds = 2\pi$. Also

$$\sigma_N(A) = \frac{1}{(2\pi)^2} \int_{-\pi}^{\pi} \int_{-\pi}^{\pi} \hat{\theta}_{s,t}(A) K_N(s) K_N(t)\, dsdt$$

by Proposition 5.5.6. Therefore

$$\begin{aligned} A - \sigma_N(A) &= \frac{1}{(2\pi)^2} \int\int (A - \hat{\theta}_{s,t}(A)) K_N(s) K_N(t)\, dsdt \\ &= \frac{1}{(2\pi)^2} \int\int (A - \hat{\theta}_{s,0}(A)) K_N(s) K_N(t)\, dsdt \\ &\quad + \frac{1}{(2\pi)^2} \int\int \hat{\theta}_{s,0}(A - \hat{\theta}_{0,t}(A)) K_N(s) K_N(t)\, dsdt \\ &= \frac{1}{2\pi} \int (A - \hat{\theta}_{s,0}(A)) K_N(s)\, ds \\ &\quad + \frac{1}{(2\pi)^2} \int \hat{\theta}_{s,0} \left(\int (A - \hat{\theta}_{0,t}(A)) K_N(t)\, dt \right) K_N(s)\, ds. \end{aligned}$$

This reduces the problem to showing that both of the last two integrals go to zero as $N \to \infty$.

The first integral is small for large N since the map $s \mapsto (A - \hat{\theta}_{s,0}(A))$ is norm continuous and is zero for $s = 0$, while $\frac{1}{2\pi} \int |K_N(s)| ds = 1$ and for any $\epsilon > 0$ we have $\int_{|s| \geq \epsilon} |K_N(s)| ds \to 0$ as $N \to \infty$. The same argument shows that the inner part of the second integral is small, from which it follows that the whole second integral is small because $\hat{\theta}_{s,0}$ is an isometry and $\frac{1}{2\pi} \int |K_N(t)| dt = 1$. ∎

Unlike the quantum plane, quantum tori cannot be scaled onto each other for different values of $\hbar$. In fact, the C*-algebras $C^\hbar(\mathbf{T}^2)$ are generally not $*$-isomorphic for different values of $\hbar$. We will not prove the full uniqueness result, but only the fact that $C^\hbar(\mathbf{T}^2) \not\cong C^{\hbar'}(\mathbf{T}^2)$ if $\hbar$ is a rational multiple of π and $\hbar'$ is an irrational multiple of π. We need the following interesting fact.

THEOREM 5.5.8

Let $\mathcal{H}$ be a Hilbert space, suppose $\hbar$ is an irrational multiple of π, and let $\tilde{U}, \tilde{V} \in B(\mathcal{H})$ be unitaries which satisfy $\tilde{U}\tilde{V} = e^{-i\hbar}\tilde{V}\tilde{U}$. Then the C-algebra $C^*(\tilde{U}, \tilde{V})$ generated by $\tilde{U}$ and $\tilde{V}$ has no proper C*-ideals. In particular, this is true of $C^\hbar(\mathbf{T}^2)$.*

PROOF For each $m, n \in \mathbf{Z}$ and $A \in C^*(\tilde{U}, \tilde{V})$ define

$$\tilde{\theta}_{s,t}(A) = (\tilde{U}^m \tilde{V}^n) A (\tilde{U}^m \tilde{V}^n)^{-1}$$

where $s = -\hbar n + 2\pi k$, $t = \hbar m + 2\pi l$, and k and l are arbitrary. Note that since $\hbar/\pi$ is irrational, distinct values of k and n give rise to distinct values of s and distinct values of l and m give rise to distinct values of t, so $\tilde{\theta}_{s,t}$ is well-defined. We have $\tilde{\theta}_{s,t}(\tilde{U}) = e^{-is}\tilde{U}$ and $\tilde{\theta}_{s,t}(\tilde{V}) = e^{-it}\tilde{V}$ for all s, t (so it agrees with the automorphism $\hat{\theta}_{s,t}$ on $\hat{C}^{\hbar}(\mathbf{T}^2)$).

The values of s and t for which $\tilde{\theta}_{s,t}$ is defined are dense in $\mathbf{R}$. Moreover, it is clear that the map $(s,t) \mapsto \tilde{\theta}_{s,t}(A)$ extends to a norm continuous map from $\mathbf{R}^2$ into $C^*(\tilde{U}, \tilde{V})$ if $A = \tilde{U}$ or $\tilde{V}$. Norm continuity of the sum and product in $B(\mathcal{H})$ implies that the same is true of any polynomial in $\tilde{U}$ and $\tilde{V}$, and it then holds for every $A \in C^*(\tilde{U}, \tilde{V})$ by an $\epsilon/3$ argument. Thus, $\tilde{\theta}_{s,t}$ extends to an action of $\mathbf{R}^2$ on $C^*(\tilde{U}, \tilde{V})$ by automorphisms, which we also denote by $\tilde{\theta}_{s,t}$.

Consider the map $\tau : C^*(\tilde{U}, \tilde{V}) \to C^*(\tilde{U}, \tilde{V})$ defined by

$$\tau(A) = \frac{1}{(2\pi)^2} \int_0^{2\pi} \int_0^{2\pi} \tilde{\theta}_{s,t}(A)\, dsdt.$$

We have $\tau(I) = I$ and $\tau(\tilde{U}^m \tilde{V}^n) = 0$ when m and n are not both zero. By continuity, $\tau(A)$ is a complex multiple of I for every $A \in C^*(\tilde{U}, \tilde{V})$.

Now let $\mathcal{I}$ be a nonzero C*-ideal of $C^*(\tilde{U}, \tilde{V})$ and choose $A \in \mathcal{I}$, $A \neq 0$. Let $v \in \mathcal{H}$ satisfy $Av \neq 0$. Then

$$\langle \tilde{\theta}_{0,0}(A^*A)v, v\rangle = \langle A^*Av, v\rangle = \|Av\|^2 > 0,$$

and

$$\langle \tilde{\theta}_{s,t}(A^*A)v, v\rangle = \langle \tilde{\theta}_{s,t}(A)^*\tilde{\theta}_{s,t}(A)v, v\rangle = \|\tilde{\theta}_{s,t}(A)v\|^2 \geq 0$$

for all s, t. Since $\langle \tilde{\theta}_{s,t}(A^*A)v, v\rangle$ is a continuous function of s and t, it follows that $\langle \tau(A^*A)v, v\rangle > 0$. In particular, $\tau(A^*A) \neq 0$.

But the integral which defines $\tau(A^*A)$ is approximated in norm by finite sums of the form

$$\frac{1}{N} \sum \tilde{\theta}_{s,t}(A^*A)$$

where (s,t) ranges over a discrete N-element subset of $\mathbf{T}^2$; as each term of such a sum belongs to $\mathcal{I}$, we must have $\tau(A^*A) \in \mathcal{I}$. We conclude that $\mathcal{I}$ contains a nonzero multiple of the identity operator, and hence that $\mathcal{I} = C^*(\tilde{U}, \tilde{V})$. ∎

COROLLARY 5.5.9

Under the same hypotheses as the theorem, $C^*(\tilde{U}, \tilde{V}) \cong C^{\hbar}(\mathbf{T}^2)$.

PROOF For $k = 1, 2$ suppose $\tilde{U}_k, \tilde{V}_k \in B(\mathcal{H}_k)$ are unitaries which satisfy $\tilde{U}_k\tilde{V}_k = e^{-i\hbar}\tilde{V}_k\tilde{U}_k$. Then let $\mathcal{A} \subset B(\mathcal{H}_1 \oplus \mathcal{H}_2)$ be the C*-algebra generated by the unitary operators $\tilde{U}_1 \oplus \tilde{U}_2$ and $\tilde{V}_1 \oplus \tilde{V}_2$. These operators satisfy the commutation relations required by the theorem, so we conclude that $\mathcal{A}$ has no proper C*-ideals.

Consider the restriction maps $\pi_k : \mathcal{A} \to C^*(\tilde{U}_k, \tilde{V}_k)$. These are unital $*$-homomorphisms, and $\pi_k(\mathcal{A})$ contains $\tilde{U}_k$ and $\tilde{V}_k$, so the kernel of π_k is not all of $\mathcal{A}$. Since $\mathcal{A}$ has no proper C*-ideals, it follows that π_k has null kernel, and hence is a $*$-isomorphism. Therefore

$$C^*(\tilde{U}_1, \tilde{V}_1) \cong \mathcal{A} \cong C^*(\tilde{U}_2, \tilde{V}_2).$$

In particular, this is true if we take $\tilde{U}_1 = U$ and $\tilde{V}_1 = V$ to be the generators of $C^\hbar(\mathbf{T}^2)$. ∎

The theorem and corollary both fail when $\hbar$ is a rational multiple of π. Suppose $\hbar = 2\pi p/q$ with $p, q \in \mathbf{N}$ and consider the operators $\tilde{U}$ and $\tilde{V}$ acting on $l^2((\mathbf{Z}/q)^2)$ by the some formulas as $\hat{U}$ and $\hat{V}$, namely

$$\tilde{U} e_{m,n} = e^{-i\hbar n/2} e_{m+1,n} \qquad \text{and} \qquad \tilde{V} e_{m,n} = e^{i\hbar m/2} e_{m,n+1},$$

but now with $m, n \in \mathbf{Z}/q$. It is easy to check that $\tilde{U}\tilde{V} = e^{-i\hbar}\tilde{V}\tilde{U}$ still holds, so Corollary 5.5.9 clearly fails, as $C^*(\tilde{U}, \tilde{V})$ is finite dimensional.

We claim that there is a $*$-homomorphism from $\hat{C}^\hbar(\mathbf{T}^2)$ to $C^*(\tilde{U}, \tilde{V})$ which takes $\hat{U}$ to $\tilde{U}$ and $\hat{V}$ to $\tilde{V}$; this will imply that $\hat{C}^\hbar(\mathbf{T}^2)$ has a proper C*-ideal, and hence is not $*$-isomorphic to $\hat{C}^{\hbar'}(\mathbf{T}^2)$ for $\hbar'$ any irrational multiple of π.

Since $\tilde{U}$ and $\tilde{V}$ obey the same commutation relation as $\hat{U}$ and $\hat{V}$, the map $\hat{U} \mapsto \tilde{U}$, $\hat{V} \mapsto \tilde{V}$ extends $*$-homomorphically to the $*$-algebra generated by $\hat{U}$ and $\hat{V}$. To prove the claim we must show that this map is contractive, i.e., $\|f(\tilde{U}, \tilde{V})\| \leq \|f(\hat{U}, \hat{V})\|$ for any function f of the form $f(x, y) = \sum_{|k|,|l| \leq N} a_{k,l} x^k y^l$. Now given $w = (b_{m,n}) \in l^2((\mathbf{Z}/q)^2)$ $(0 \leq m, n < q)$ and $M > 0$, define $w^M \in l^2(\mathbf{Z}^2)$ by

$$w^M_{m+kq,n+lq} = \begin{cases} b_{m,n} & \text{for } |k|, |l| \leq M \\ 0 & \text{otherwise.} \end{cases}$$

Then

$$\|f(\hat{U}, \hat{V}) w^M\| / \|w^M\| \to \|f(\tilde{U}, \tilde{V}) w\| / \|w\|$$

as $M \to \infty$. This implies that $\|f(\tilde{U}, \tilde{V})\| \leq \|f(\hat{U}, \hat{V})\|$, as desired. We have proven:

COROLLARY 5.5.10
If $\hbar$ is a rational multiple of π then $C^\hbar(\mathbf{T}^2)$ has a proper C-ideal. If also $\hbar'$ is an irrational multiple of π then $C^\hbar(\mathbf{T}^2) \not\cong C^{\hbar'}(\mathbf{T}^2)$.*

5.6 The GNS construction

We have seen several instances of apparently different C*-algebras which turned out to be $*$-isomorphic. In such cases it can be helpful to draw a distinction between the abstract C*-algebra, viewed as a Banach space equipped with a product and an involution, and its realization(s) as a C*-subalgebra of $B(\mathcal{H})$.

In keeping with this point of view, we now present a general technique for finding $*$-homomorphisms from a given C*-algebra into some $B(\mathcal{H})$. Here the order structure in C*-algebras becomes important.

DEFINITION 5.6.1 *For $A, B \in B(\mathcal{H})$ we define $A \geq B$ if $\langle Av, v\rangle \geq \langle Bv, v\rangle$ for all $v \in \mathcal{H}$. We say A is positive if $A \geq 0$.*

A multiplication operator M_f is positive if and only if $f \geq 0$ almost everwhere, and it follows from the spectral theorem that $A \leq \|A\| I$ for all self-adjoint $A \in B(\mathcal{H})$. It is also easy to see that $A \leq B$ implies $C^*AC \leq C^*BC$ for all $A, B, C \in B(\mathcal{H})$.

PROPOSITION 5.6.2
Let $A \in B(\mathcal{H})$. The following are equivalent:

(a) A is positive;
(b) A is self-adjoint and $\mathrm{sp}(A) \subset [0, \infty)$; and
(c) $A = B^*B$ for some $B \in B(\mathcal{H})$.

PROOF
(a) $\Rightarrow$ (b): Suppose $\langle Av, v\rangle \geq 0$ for all $v \in \mathcal{H}$. Then by polarization

$$\begin{aligned}\overline{\langle Av, w\rangle} &= \sum_{k=0}^{3} (-i)^k \left\langle A(v + i^k w), (v + i^k w)\right\rangle \\ &= \sum_{k=0}^{3} (-i)^k \left\langle A(w + (-i)^k v), (w + (-i)^k v)\right\rangle \\ &= \langle Aw, v\rangle,\end{aligned}$$

so A is self-adjoint. By the spectral theorem we may suppose A is multiplication by some bounded real-valued function f on some $L^2(X)$; then positivity easily implies that $f \geq 0$ almost everywhere, so that $\mathrm{sp}(A)$, which is the essential range of f (Example 3.2.2), is contained in $[0, \infty)$.

(b) $\Rightarrow$ (c): Suppose A is self-adjoint and $\mathrm{sp}(A) \subset [0, \infty)$. Then without loss of generality $A = M_f$, and the spectrum condition implies that $f \geq 0$ almost everywhere. Thus $A = M_g^2 = M_g^* M_g$ where $g = \sqrt{f}$.

(c) $\Rightarrow$ (a): Suppose $A = B^*B$. Then for any $v \in \mathcal{H}$ we have

$$\langle Av, v\rangle = \langle B^*Bv, v\rangle = \langle Bv, Bv\rangle = \|Bv\|^2 \geq 0,$$

so $A \geq 0$. ∎

By functional calculus, if A belongs to a unital C*-algebra $\mathcal{A}$ and $A - \lambda I$ is invertible in $B(\mathcal{H})$, then $(A - \lambda I)^{-1} \in \mathcal{A}$. Thus, part (b) of Proposition 5.6.2 shows that positivity of an element of an abstract C*-algebra $\mathcal{A}$ is independent of the realization of $\mathcal{A}$ in $B(\mathcal{H})$. That is, positivity is well-defined in abstract C*-algebras.

Next we describe the objects that are used to construct representations.

DEFINITION 5.6.3 *Let $\mathcal{A}$ be a C*-algebra. A state on $\mathcal{A}$ is a bounded linear functional $\omega \in \mathcal{A}^*$ such that $\|\omega\| = 1$ and $A \geq B$ implies $\omega(A) \geq \omega(B)$.*

*More generally, a weight on $\mathcal{A}$ is a linear map ω from a dense *-subalgebra of $\mathcal{A}$ into $\mathbf{C}$ with the property that $A \geq B$ implies $\omega(A) \geq \omega(B)$.*

Observe that by functional calculus any self-adjoint operator A can be written as a difference of positive operators, so $\omega(A)$ must be real for any state or weight ω.

If $\mathcal{A}$ is unital, states can equivalently be defined as linear functionals ω which preserve positivity and satisfy $\omega(I_{\mathcal{A}}) = 1$. This can be shown as follows. If ω preserves positivity then for any self-adjoint $A \in \mathcal{A}$ we have

$$\omega(A) \leq \omega(\|A\| I_{\mathcal{A}}) = \|A\|\omega(I_{\mathcal{A}});$$

then for any $A \in \mathcal{A}$, letting $a = \overline{\omega(A)}$ we have

$$\begin{aligned} |\omega(A)|^2 &= \omega(aA) = \omega(\mathrm{Re}\, aA + i\mathrm{Im}\, aA) \\ &= \omega(\mathrm{Re}\, aA) \leq \|\mathrm{Re}\, aA\|\omega(I_{\mathcal{A}}) \leq |a|\omega(I_{\mathcal{A}})\|A\|. \end{aligned}$$

Thus $\|\omega\| \leq \omega(I_{\mathcal{A}})$. The reverse inequality is automatic, so we conclude that in the presence of positivity and a unit we have $\|\omega\| = \omega(I_{\mathcal{A}})$.

Example 5.6.4

Let X be a compact Hausdorff space and let μ be a regular Borel measure on X with the property that any $x \in X$ has a neighborhood $\mathcal{O}$ such that $\mu(\mathcal{O}) < \infty$. Then integration against μ defines a weight on $C(X)$ whose domain is $L^1(\mu) \cap C(X)$. It is a state if and only if μ is a probability measure.

Example 5.6.5

Let $\mathcal{A} \subset B(\mathcal{H})$ be a C*-algebra and let $v \in \mathcal{H}$ be a unit vector. Then $A \mapsto \langle Av, v \rangle$ is a state on $\mathcal{A}$.

As the second example indicates, the concept of a state on a C*-algebra is related to the physical notion of a state. The next theorem, which describes the so-called GNS construction, is a kind of converse.

Let $\mathcal{A}$ be a C*-algebra and $\omega : \mathcal{A}_0 \to \mathbf{C}$ a weight. Define a pseudo inner product on $\mathcal{A}_0$ by setting $\langle A, B \rangle = \omega(B^*A)$. Let $\mathcal{H}_\omega$ denote the Hilbert space formed by factoring out null vectors and completing, and write $\bar{A}$ for the element of $\mathcal{H}_\omega$ corresponding to $A \in \mathcal{A}_0$.

THEOREM 5.6.6

Let $\mathcal{A}$ be a C-algebra and let $\omega : \mathcal{A}_0 \to \mathbf{C}$ be a weight on $\mathcal{A}$. Then there is a unique $*$-homomorphism $\pi : \mathcal{A} \to B(\mathcal{H}_\omega)$ such that $\pi(A)\bar{B} = \overline{AB}$ for all $A, B \in \mathcal{A}_0$. If $\mathcal{A}$ is unital and ω is a state, then $\bar{I}_\mathcal{A}$ is a unit vector in $\mathcal{H}_\omega$ and we have $\omega(A) = \langle \pi(A)\bar{I}_\mathcal{A}, \bar{I}_\mathcal{A} \rangle$.*

PROOF Let $A \in \mathcal{A}_0$. We show first that the map $\bar{B} \mapsto \overline{AB}$ defines a bounded operator $\pi(A)$ on $\mathcal{H}_\omega$ of norm at most $\|A\|$. This is so because $B^*A^*AB \leq \|A^*A\|B^*B$, and hence

$$\|\overline{AB}\|^2 = \omega(B^*A^*AB) \leq \|A^*A\|\omega(B^*B) = \|A\|^2\|\bar{B}\|^2.$$

Thus we have a contraction $\pi : \mathcal{A}_0 \to B(\mathcal{H}_\omega)$. It is clearly linear, and it preserves products and adjoints because

$$\pi(AB)\bar{C} = \overline{ABC} = \pi(A)\pi(B)\bar{C}$$

and

$$\langle \pi(A)\bar{B}, \bar{C} \rangle = \omega(C^*AB) = \omega((A^*C)^*B) = \langle \bar{B}, \pi(A^*)\bar{C} \rangle$$

for all $A, B, C \in \mathcal{A}_0$. Since π is contractive on $\mathcal{A}_0$ it extends to a $*$-homomorphism from $\mathcal{A}$ into $B(\mathcal{H}_\omega)$. Uniqueness of π follows from density of $\mathcal{A}_0$ in $\mathcal{A}$.

If $\mathcal{A}$ is unital and ω is a state, the remark made after Definition 5.6.3 implies that $\|\bar{I}_\mathcal{A}\|^2 = \omega(I_\mathcal{A}) = 1$. The final assertion is trivial. ∎

It follows from this theorem that if ω is a state on a unital C*-algebra $\mathcal{A}$ then there is a representation of $\mathcal{A}$ on a Hilbert space $\mathcal{H}$ (i.e., a $*$-homomorphism from $\mathcal{A}$ into $B(\mathcal{H})$) which contains a unit vector that gives rise to the state in the manner of Example 5.6.5.

Next we indicate how some of the representations we have been using can be obtained from the GNS construction.

Example 5.6.7

(a) Consider the quantum torus algebra $\hat{C}^{\hbar}(\mathbf{T}^2) \subset B(l^2(\mathbf{Z}^2))$. The map $\omega(A) = \langle Ae_{0,0}, e_{0,0}\rangle$ is a state on $\hat{C}^{\hbar}(\mathbf{T}^2)$, and the GNS construction gives rise to a representation of $\hat{C}^{\hbar}(\mathbf{T}^2)$ which is equivalent to the original representation on $l^2(\mathbf{Z}^2)$. The identification of $\mathcal{H}_\omega$ with $l^2(\mathbf{Z}^2)$ takes $\bar{L}_{k,l}$ to $e_{k,l}$.

The state ω can also be defined on $C^{\hbar}(\mathbf{T}^2)$ in terms of the automorphisms $\tilde{\theta}$ used in the proof of Theorem 5.5.8: we have $\tau(A) = \omega(A)I$. This definition is less circular than the one above because it does not depend on already having the $l^2(\mathbf{Z}^2)$ representation.

(b) For the noncommutative plane algebra $C_0^{\hbar}(\mathbf{R}^2)$, we have a weight defined by $\omega(L_f) = 2\pi\hat{f}(0,0) = \int f$ for $f \in \mathcal{S}(\mathbf{R}^2)$. Then the GNS construction gives rise to our representation of $C_0^{\hbar}(\mathbf{R}^2)$ on $L^2(\mathbf{R}^2)$. In particular, for $f, g \in \mathcal{S}(\mathbf{R}^2)$ we have

$$\omega(L_g^* L_f) = \omega(L_{\bar{g}\times_\hbar f}) = \int f\bar{g},$$

so the correspondence of $\bar{L}_f \in \mathcal{H}_\omega$ with $f \in \mathcal{S}(\mathbf{R}^2) \subset L^2(\mathbf{R}^2)$ is isometric.

We now indicate why we call the $L^2(\mathbf{R}^2)$ model of the quantum plane the "tracial" representation. For certain Hilbert space operators it is possible to define a trace in a way that generalizes the trace of finite dimensional matrices (see Section 6.3). The operators $\tilde{L}_f \in B(L^2(\mathbf{R}))$ fall into this category, and we have $\mathrm{tr}(\tilde{L}_f) = \frac{1}{2\pi\hbar}\int f$ for $f \in \mathcal{S}(\mathbf{R}^2)$. Thus, $2\pi\hbar$ times the trace on $B(L^2(\mathbf{R}))$ restricts to a weight on $C_0^{\hbar}(\mathbf{R}^2)$ such that the resulting GNS representation is the $L^2(\mathbf{R}^2)$ model.

C*-algebras can be characterized abstractly. They are just those Banach algebras equipped with an antilinear involution * such that $(\alpha\beta)^* = \beta^*\alpha^*$ and $\|\alpha^*\alpha\| = \|\alpha\|^2$ for all α, β. This result, known as the Gelfand-Neumark theorem, is proven by using the GNS construction to find a Hilbert space representation of such an abstractly given algebra. It is presented in nearly every standard exposition of C*-algebras; we will not prove it here, because it requires significant Banach algebra preliminaries that we prefer to avoid.

Actually, one often does not need to use the Gelfand-Neumark theorem, because C*-algebras are usually defined in terms of their representations, or can at least be given a representation without much effort. The most prominent exception is the quotient construction, which is not evidently represented on any Hilbert space but is in fact always an abstract C*-algebra. One can prove this using the Gelfand-Neumark theorem, but it is not much more difficult to prove it directly from Theorem 5.6.6; we do this now.

For any C*-ideal $\mathcal{I}$ of a C*-algebra $\mathcal{A}$, write

$$\|A+\mathcal{I}\| = \inf\{\|A+B\| : B \in \mathcal{I}\}$$

for the quotient norm in $\mathcal{A}/\mathcal{I}$.

LEMMA 5.6.8

Let $\mathcal{I}$ be a C-ideal of a unital C*-algebra $\mathcal{A}$ and let $A \in \mathcal{A}$. Then $\|A^*A+\mathcal{I}\| = \|A+\mathcal{I}\|^2$.*

PROOF One inequality is easy: for any $B \in \mathcal{I}$ we have

$$\|A^*A+\mathcal{I}\| \leq \|(A^*+B^*)(A+B)\| = \|A+B\|^2,$$

and taking the infimum over $B \in \mathcal{I}$ yields $\|A^*A+\mathcal{I}\| \leq \|A+\mathcal{I}\|^2$.

For the converse, observe first that if A is self-adjoint then

$$\|A+\mathcal{I}\| = \inf\{\|A+B\| : B \in \mathcal{I}, B = B^*\}$$

because for any B we have

$$\|A+\operatorname{Re} B\| \leq \frac{1}{2}(\|A+B\| + \|A+B^*\|) = \|A+B\|.$$

Thus let $A \in \mathcal{A}$ and $B \in \mathcal{I}$ be self-adjoint and let $\epsilon > 0$. For any real-valued $f \in C(\operatorname{sp}(B))$ such that $f(0) = 0$, $0 \leq f(x) \leq 1$ for all x, and $f(x) = 1$ for $|x| \geq \epsilon$ we have

$$\begin{aligned}
\|A^*A+B\| &\geq \|(A^*A+B)(I_{\mathcal{A}}-f(B))\| \\
&\geq \|A^*A(I_{\mathcal{A}}-f(B))\| - \|B(I_{\mathcal{A}}-f(B))\| \\
&\geq \|A^*A(I_{\mathcal{A}}-f(B))\| - \epsilon \\
&\geq \|(I_{\mathcal{A}}-f(B))A^*A(I_{\mathcal{A}}-f(B))\| - \epsilon \\
&= \|A(I_{\mathcal{A}}-f(B))\|^2 - \epsilon \\
&\geq \|A+\mathcal{I}\|^2 - \epsilon.
\end{aligned}$$

Finally, taking the infimum over all self-adjoint operators $B \in \mathcal{I}$ yields $\|A^*A+\mathcal{I}\| \geq \|A+\mathcal{I}\|^2 - \epsilon$, which is enough. ∎

We need one other, slightly more technical lemma.

LEMMA 5.6.9

Let $\mathcal{I}$ be a C-ideal of a unital C*-algebra $\mathcal{A}$ and let $A \in \mathcal{A}$ be positive. Then*

$$\|A+bI_{\mathcal{A}}+\mathcal{I}\| \geq \big|\|A+\mathcal{I}\|+b\big|$$

for all $b \in \mathbf{R}$.

PROOF Say $\mathcal{A} \subset B(H)$. First consider the case $b \geq -\|A+\mathcal{I}\|$. Let $B \in \mathcal{I}$; as in Lemma 5.6.8, we may assume $B = B^*$. Fix $\epsilon > 0$ and a real-valued function $f \in C(\mathrm{sp}(B))$ such that $f(0) = 0$, $0 \leq f(x) \leq 1$ for all x, and $f(x) = 1$ for $|x| \geq \epsilon$, just as in the proof of Lemma 5.6.8. Let $U = f(B)$. Find a unit vector $v_0 \in \mathcal{H}$ such that

$$\langle (I-U)A(I-U)v_0, v_0\rangle \geq \|(I-U)A(I-U)\| - \epsilon$$

and let $w_0 = (I-U)v_0/\|(I-U)v_0\|$. Then choose $g \in C(\mathrm{sp}(B))$ such that $g(x) = 0$ for $|x| \leq \epsilon$, $0 \leq g(x) \leq 1$ for all x, and $g(x) = 1$ for $|x| \geq 2\epsilon$. Let $V = g(B)$. Then $(1-g)(1-f) = 1-f$, so

$$\begin{aligned}\|(A+bI)(I-V)\| &\geq \langle (A+bI)(I-V)w_0, (I-V)w_0\rangle \\ &= \langle (A+bI)(I-U)v_0, (I-U)v_0\rangle/\|(I-U)v_0\|^2 \\ &\geq \|(I-U)A(I-U)\| + b - \epsilon \\ &\geq \|A+\mathcal{I}\| + b - \epsilon.\end{aligned}$$

But

$$\|A+bI+B\| \geq \|(A+bI+B)(I-V)\| \geq \|(A+bI)(I-V)\| - 2\epsilon$$

as in Lemma 5.6.8, so we conclude that $\|A+bI+B\| \geq \|A+\mathcal{I}\|+b-3\epsilon$. Taking $\epsilon \to 0$ completes the proof in the case $b \geq -\|A+\mathcal{I}\|$.

If $b < -\|A+\mathcal{I}\|$ then let $B_0 \in \mathcal{I}$ satisfy $B_0^* = B_0$ and $\|A+B_0\| < -b$. Then $A' = \|A+B_0\|I - (A+B_0)$ is positive, so the $b \geq -\|A+\mathcal{I}\|$ case implies

$$\begin{aligned}\|A+bI+B_0-B\| &= \|A' - (\|A+B_0\|+b)I + B\| \\ &\geq \|A'+\mathcal{I}\| - (\|A+B_0\|+b) \\ &\geq \big|\|A+B_0\|+b\big|\end{aligned}$$

for all $B \in \mathcal{I}$. Thus

$$\|A+bI+\mathcal{I}\| \geq \big|\|A+B_0\|+b\big|,$$

and taking the infimum over B_0 yields the desired inequality. ∎

THEOREM 5.6.10

Let $\mathcal{I}$ be a C-ideal of a C*-algebra $\mathcal{A}$. Then there is a Hilbert space $\mathcal{H}$ and a $*$-isomorphism from $\mathcal{A}/\mathcal{I}$ into $B(\mathcal{H})$.*

PROOF It is clear that the product and $*$-operation on $\mathcal{A}$ descend to $\mathcal{A}/\mathcal{I}$. We will find a Hilbert space $\mathcal{H}$ and a $*$-homomorphism $\pi : \mathcal{A} \to \mathcal{H}$

such that $\ker(\pi) = \mathcal{I}$ and $\|\pi(A)\| = \|A + \mathcal{I}\|$ for all $A \in \mathcal{A}$. If $\mathcal{A}$ has no unit we can adjoin one without loss of generality, because $\mathcal{I}$ will still be an ideal of $\mathcal{A}^+$ and $\mathcal{A}/\mathcal{I}$ will be a C*-subalgebra of $\mathcal{A}^+/\mathcal{I}$. Thus, we assume $\mathcal{A}$ is unital.

For any positive element $A \in \mathcal{A}$, let

$$E = \{aA + bI_{\mathcal{A}} + B : a, b \in \mathbf{R},\, B \in \mathcal{I},\, B = B^*\}$$

be the self-adjoint part of the span of the ideal $\mathcal{I}$ and the elements A and $I_{\mathcal{A}}$. Then define $\omega_1 : E \to \mathbf{R}$ by

$$\omega_1(aA + bI_{\mathcal{A}} + B) = a \cdot \|A + \mathcal{I}\| + b.$$

(This map is not obviously well-defined if $A = b'I_{\mathcal{A}} + B'$ for some $b' \in \mathbf{C}$ and $B' \in \mathcal{I}$. One can either prove that it indeed is well-defined even in this case using Lemma 5.6.9 with $A = 0$, or one can observe that this particular case is not actually needed in the sequel.) Lemma 5.6.9 implies that $\|\omega_1\| \le 1$, and since $\omega_1(I_{\mathcal{A}}) = 1$ we have $\|\omega_1\| = 1$. By the Hahn-Banach theorem we can extend ω_1 to a real linear functional ω_2 on the self-adjoint part of $\mathcal{A}$ with $\|\omega_2\| = 1$. Moreover, if $B \in \mathcal{A}$ is positive then $\|B\|I_{\mathcal{A}} - B$ has norm at most $\|B\|$, so

$$\|B\| - \omega_2(B) = \omega_2(\|B\|I_{\mathcal{A}} - B) \le \|B\|,$$

and hence $\omega_2(B) \ge 0$. Thus the complex linear extension ω of ω_2 to $\mathcal{A}$ is a state.

For each positive $A \in \mathcal{A}$ let ω be such a state and let $\pi_\omega : \mathcal{A} \to B(\mathcal{H}_\omega)$ be the $*$-homomorphism given by Theorem 5.6.6; then let

$$\pi = \bigoplus \pi_\omega : \mathcal{A} \to B\left(\bigoplus \mathcal{H}_\omega\right)$$

be the direct sum of these $*$-homomorphisms. It is clearly itself a $*$-homomorphism and hence a contraction, and its kernel contains $\mathcal{I}$, so it descends to a contraction from $\mathcal{A}/\mathcal{I}$ into $B(\bigoplus \mathcal{H}_\omega)$. Conversely, for each $A \in \mathcal{A}$ we have

$$\langle \pi_\omega(A^*A)\bar{I}_{\mathcal{A}}, \bar{I}_{\mathcal{A}}\rangle = \omega(A^*A) = \|A^*A + \mathcal{I}\|$$

for some ω, so

$$\|\pi(A)\|^2 = \|\pi(A^*A)\| \ge \|A^*A + \mathcal{I}\| = \|A + \mathcal{I}\|^2$$

(using Lemma 5.6.8). Thus π is isometric for the quotient norm. This completes the proof. ∎

Thus, quotients of C*-algebras are themselves C*-algebras.

Since separable Hilbert spaces are better-behaved than nonseparable ones, it is worth pointing out that we can take $\mathcal{H}$ to be separable in Theorem 5.6.10 provided $\mathcal{A}$ is a separable C*-algebra. For it is sufficient to ensure that π be isometric on a countable dense subset of $\mathcal{A}$, so that we only need to use countably many states ω, and it is also easy to see that each $\mathcal{H}_\omega$ is separable if $\mathcal{A}$ is. So $\mathcal{H} = \bigoplus \mathcal{H}_\omega$ will also be separable.

5.7 Notes

Basic material on C*-algebras is covered in many texts. Besides the books cited in earlier chapters, we also mention [40] on this topic.

For more on the quantum plane see [60], and for more on quantum tori see [60] and [16]. The latter are also called irrational rotation algebras when $\hbar/\pi$ is irrational, the case of greatest interest. Our treatment of $K(\mathcal{H})$ in Section 5.4 follows [3]. The discussion of Fourier series on noncommutative tori in Section 5.5 follows [71].

Chapter 6

Von Neumann algebras

6.1 The algebras $l^\infty(X)$

In this chapter we discuss quantum measure theory. This is formally rather similar to quantum topology; in both cases we treat the given structure (topological or measure theoretic) in terms of the algebra of complex-valued functions which are compatible with that structure. The algebras which arise in the latter case are actually a special case of the former: every von Neumann algebra is a C*-algebra (although this is not always the best way to think about them).

We begin with the functional analytic objects which reflect pure set theoretic structure, the algebras $l^\infty(X)$. The theory of these algebras closely parallels that of the algebras $C(X)$ presented in Section 5.1. The key difference is that $l^\infty(X) \cong l^1(X)^*$ is always a dual space, and the weak* topology is favored over the norm topology.

Two observations about the weak* topology on $l^\infty(X)$ will be used repeatedly. First, if (f_κ) is a bounded net in $l^\infty(X)$ then $f_\kappa \to f$ weak* if and only if $f_\kappa \to f$ pointwise. The forward direction follows by pairing with the characteristic functions $\chi_x \in l^1(X)$: we have

$$f_\kappa(x) = \sum f_\kappa \chi_x \to \sum f\chi_x = f(x)$$

for all $x \in X$. Conversely, if $f_\kappa \to f$ pointwise then $\sum f_\kappa g \to \sum fg$ for every finitely supported function $g \in l^1(X)$; since (f_κ) is bounded, an $\epsilon/3$ argument then shows that $\sum f_\kappa g \to \sum fg$ for all $g \in l^1(X)$.

Second, we require the Krein-Smulian theorem, which states that a linear subspace E of a dual Banach space $\mathcal{V}$ is weak* closed if and only if the intersection of E with the closed unit ball of $\mathcal{V}$ is weak* closed. That is, E is weak* closed if it is stable under weak* convergence of *bounded* nets. This means that we usually only need to consider weak* topologies on bounded sets, where they are typically easier to describe, as we just saw for $l^\infty(X)$.

The following are the basic concepts surrounding l^∞ spaces. As in Definition 5.1.1, we immediately formulate the fundamental definitions at a level of generality that encompasses the noncommutative (quantum) setting.

DEFINITION 6.1.1 *Let $\mathcal{M}$ be a dual Banach space equipped with a product and an involution.*

A W-subalgebra of $\mathcal{M}$ is a weak* closed C*-subalgebra of $\mathcal{M}$.*

A W-ideal of $\mathcal{M}$ is a weak* closed C*-ideal of $\mathcal{M}$.*

If $\mathcal{M}$ is unital, the weak spectrum of $\mathcal{M}$ is the set $\mathrm{sp}^*(\mathcal{M})$ of all weak* continuous, unital $*$-homomorphisms from $\mathcal{M}$ to $\mathbf{C}$.*

The main tool we need in this section is a version of the Stone-Weierstrass theorem for $l^\infty(X)$. It follows easily from the next lemma.

LEMMA 6.1.2

Let $\mathcal{M}$ be a unital W-subalgebra of $l^\infty(X)$, let $f \in \mathcal{M}$ be real-valued, and let $S \subset \mathbf{R}$. Then $\chi_{f^{-1}(S)} \in \mathcal{M}$.*

PROOF Let $a = \|f\|_\infty$ and let $(g_\kappa) \subset C[-a,a]$ be a bounded net of continuous functions which converges pointwise to χ_S on $[-a,a]$. By the Stone-Weierstrass theorem, we may assume every g_κ is a polynomial. Then the net $(g_\kappa \circ f)$ is bounded in $\mathcal{M}$ and converges pointwise to $\chi_S \circ f = \chi_{f^{-1}(S)}$, so this limit also belongs to $\mathcal{M}$ by weak* closure of $\mathcal{M}$ in $l^\infty(X)$. ∎

PROPOSITION 6.1.3

Let X be a set and let $\mathcal{M}$ be a unital W-subalgebra of $l^\infty(X)$ which separates points. Then $\mathcal{M} = l^\infty(X)$.*

PROOF Let $g \in l^\infty(X)$ and let $x, y \in X$ be distinct. Then there exists $f \in \mathcal{M}$ such that $f(x) \neq f(y)$, and by the lemma the characteristic function of some set which contains x but not y will belong to $\mathcal{M}$. By taking products of such functions, for any finite set $S \subset X$ not containing x we can find in $\mathcal{M}$ a characteristic function which vanishes on S and takes the value 1 on x. But χ_x is a weak* limit of such functions, so $\chi_x \in \mathcal{M}$. As this is true for every $x \in X$, we conclude that $\mathcal{M}$ contains every finitely supported function, and hence, by weak* closure, it equals $l^\infty(X)$. ∎

Now we proceed to the relation between algebraic structure in $l^\infty(X)$

and set theoretic structure in X. Quotients and subalgebras are related in the $l^\infty(X)$ setting in the same way they are related in the $C(X)$ setting.

Example 6.1.4

Let $\phi : X \to Y$ be a surjection. Then the map $C_\phi : f \mapsto f \circ \phi$ weak* continuously and $*$-isomorphically embeds $l^\infty(Y)$ in $l^\infty(X)$, and $C_\phi(l^\infty(Y))$ is a unital W*-subalgebra of $l^\infty(X)$.

In this example the fact that $C_\phi(l^\infty(Y))$ is weak* closed in $l^\infty(X)$ is a consequence of the Krein-Smulian theorem. If $(C_\phi f_\kappa)$ is a bounded net in $C_\phi(l^\infty(Y))$ and $C_\phi f_\kappa \to g$ pointwise, then (f_κ) is bounded in $l^\infty(Y)$ so we can pass to a subnet which converges pointwise to $f \in l^\infty(Y)$, and then weak* continuity implies that $C_\phi f = g$. Thus $C_\phi(l^\infty(Y))$ is weak* closed.

PROPOSITION 6.1.5
Let X be a set and let $\mathcal{M}$ be a unital W-subalgebra of $l^\infty(X)$. Then there is a set Y and a surjection $\phi : X \to Y$ such that $\mathcal{M} = C_\phi(l^\infty(Y))$.*

PROOF For $x, y \in X$ let $x \sim y$ if $f(x) = f(y)$ for all $f \in \mathcal{M}$. Take $Y = X/\sim$ and let $\phi : X \to Y$ be the projection map. Then the map $C_\phi : f \mapsto f \circ \phi$ takes $l^\infty(Y)$ $*$-isomorphically into $l^\infty(X)$, and as every function in $\mathcal{M}$ respects $\sim$ it follows that $\mathcal{M}$ is contained in the image of $l^\infty(Y)$ under C_ϕ. Also, $C_\phi{}^{-1}$ takes $\mathcal{M}$ onto a unital W*-subalgebra of $l^\infty(Y)$ which separates points, and hence onto $l^\infty(Y)$ by Proposition 6.1.3. Thus $\mathcal{M} = C_\phi(l^\infty(Y))$. ∎

Next, we turn to ideals and subsets.

Example 6.1.6

Let X be a set and let $K \subset X$. Then $\mathcal{I} = \{f \in l^\infty(X) : f|_K = 0\}$ is a W*-ideal of $l^\infty(X)$.

PROPOSITION 6.1.7
Let X be a set and let $\mathcal{I}$ be a W-ideal of $l^\infty(X)$. Then there is a subset $K \subset X$ such that $\mathcal{I} = \{f \in l^\infty(X) : f|_K = 0\}$.*

PROOF Let $K = \{x \in X : f(x) = 0 \text{ for all } x \in \mathcal{I}\}$ and let $g \in l^\infty(X)$ satisfy $g|_K = 0$; we must show that $g \in \mathcal{I}$. Let $\mathcal{I}^+ = \mathcal{I} + \mathbf{C} \cdot 1_X$; then $\mathcal{I}^+$ is a unital W*-subalgebra of $l^\infty(X)$. Observe that if $x, y \notin K$ then

x and y are not identified by the equivalence relation in the proof of Proposition 6.1.5. Thus, that result implies $g \in \mathcal{I}^+$. If $K \neq \emptyset$ then this immediately yields $g \in \mathcal{I}$.

However, if $K = \emptyset$ then the ideal property implies that for every $x \in X$ the characteristic function χ_x belongs to $\mathcal{I}$. Thus every finitely supported function on X belongs to $\mathcal{I}$, and weak* closure then implies $\mathcal{I} = l^\infty(X)$. ∎

The following three corollaries are all routine; their proofs are similar to, but easier than, those of the corresponding results in Section 5.1.

COROLLARY 6.1.8
If $\mathcal{I}$ is a W-ideal of $l^\infty(X)$ then $l^\infty(X)/\mathcal{I} \cong l^\infty(K)$, where $K \subset X$ is the set of points on which every function in $\mathcal{I}$ vanishes.*

For each $x \in X$ let $\hat{x} : f \mapsto f(x)$ be the corresponding evaluation functional on $l^\infty(X)$.

COROLLARY 6.1.9
The correspondence $x \leftrightarrow \hat{x}$ is a bijection between X and $\mathrm{sp}^*(l^\infty(X))$.

COROLLARY 6.1.10
The weak continuous, unital $*$-homomorphisms from $l^\infty(X)$ to $l^\infty(Y)$ are in a natural bijection with the functions from Y to X.*

6.2 The algebras $L^\infty(X)$

There are results for $L^\infty(X)$ (for X a σ-finite measure space) which are analogous to some of the results of the last section for $l^\infty(X)$. The main difference is that the role of a "quotient" of X is played here by a coarser σ-algebra on X.

If X is σ-finite then $L^\infty(X) \cong L^1(X)^*$, and for a bounded net (f_κ) in $L^\infty(X)$ we have $f_\kappa \to f$ weak* if and only if $\int_S f_\kappa \to \int_S f$ for all finite measure subsets S of X. The forward implication is simply a consequence of the fact that $\chi_S \in L^1(X)$, because $\int_S f = \int f\chi_S$ is the pairing of $f \in L^\infty(X)$ with χ_S. For the reverse implication, observe that if $\int_S f_\kappa \to \int_S f$ for all $S \subset X$ of finite measure, then $\int f_\kappa g \to \int fg$ for all simple functions $g \in L^1(X)$, by linearity; then density of the simple functions in $L^1(X)$ and the fact that (f_κ) is bounded imply that $\int f_\kappa g \to \int fg$ for all $g \in L^1(X)$, i.e., $f_\kappa \to f$ weak*.

Again, before proceeding we need a Stone-Weierstrass theorem for $L^\infty(X)$.

LEMMA 6.2.1

Let X be a σ-finite measure space, let $\mathcal{M}$ be a unital W-subalgebra of $L^\infty(X)$, let $f \in \mathcal{M}$ be real-valued, and let $S \subset \mathbf{R}$ be Borel measurable. Then $\chi_{f^{-1}(S)} \in \mathcal{M}$.*

PROOF Let $a = \|f\|_\infty$. The map $\mu \mapsto \mu(S)$ is a bounded linear functional of norm one on $M[-a,a] \cong C[-a,a]^*$, so there is a bounded net $(g_\kappa) \subset C[-a,a]$ which converges to it weak*. That is, we have $\int g_\kappa d\mu \to \mu(S)$ for all $\mu \in M[-a,a]$. By the Stone-Weierstrass theorem we may assume each g_κ is a polynomial.

Let $h \in L^1(X)$. Then the map $g \mapsto \int_X (g \circ f)h$ is a bounded linear functional on $C[-a,a]$, so there is a measure $\mu \in M[-a,a]$ such that $\int_X (g \circ f)h = \int g d\mu$ for all g. Then $\mu(K) = \int_{f^{-1}(K)} h$ holds for any closed set $K \subset [-a,a]$ and hence for every Borel set. Therefore

$$\int_X (g_\kappa \circ f)h = \int g_\kappa d\mu \to \mu(S) = \int_X \chi_{f^{-1}(S)} h.$$

This shows that $g_\kappa \circ f \to \chi_{f^{-1}(S)}$ weak*, and so $\chi_{f^{-1}(S)} \in \mathcal{M}$. ∎

PROPOSITION 6.2.2

Let X be a σ-finite measure space and suppose $\mathcal{M}$ is a unital W-subalgebra of $L^\infty(X)$ which measurably separates points. Then $\mathcal{M} = L^\infty(X)$.*

PROOF Let Ω be the smallest σ-algebra with respect to which every function in $\mathcal{M}$ is measurable and let Ω' be the collection of sets $S \subset X$ such that $\chi_S \in \mathcal{M}$. The lemma implies that Ω' generates Ω as a σ-algebra. But Ω' is a σ-algebra itself, so $\Omega' = \Omega$. By the definition of measurable separation of points (Definition 5.2.3), it follows that $\mathcal{M}$ contains the characteristic function of every measurable set, and hence $\mathcal{M} = L^\infty(X)$. ∎

Now we show how W*-subalgebras correspond to sub σ-algebras.

Example 6.2.3

Let X and Y be σ-finite measure spaces and let $\phi : X \to Y$ be a measurable surjection such that $\phi^{-1}(S)$ is null in X if and only if S is null in Y. Define $\mathcal{M} = \{f \circ \phi : f \in L^\infty(Y)\}$; this is a unital W*-subalgebra of $L^\infty(X)$, and the map $C_\phi : f \mapsto f \circ \phi$ is a weak* continuous $*$-isomorphism from $L^\infty(Y)$ onto $\mathcal{M}$.

In this example the map C_ϕ is well-defined because the inverse image under ϕ of a null set is null, and it is an isometry because the inverse image of a positive measure set has positive measure.

To see that $\mathcal{M}$ is weak* closed in the example, let Ω be the σ-algebra

$$\Omega = \{\phi^{-1}(S) : S \subset Y \text{ is measurable}\}.$$

It is contained, perhaps properly, in the σ-algebra of measurable subsets of X. Let μ and ν be the original measures on X and Y, and let μ' be the inverse image of ν, defined on Ω. Possibly replacing μ and ν with equivalent measures, we may assume that μ is finite and $\nu(S) = \mu(\phi^{-1}(S))$ for all measurable $S \subset Y$, so that $\mu' = \mu|_\Omega$. Then $\mathcal{M} = L^\infty(X, \mu')$. (It is clear that every function in $\mathcal{M}$ is μ'-measurable, and conversely $\mathcal{M}$ contains the characteristic function of every μ'-measurable set, so it contains $L^\infty(X, \mu')$.)

Now let (f_κ) be a bounded net in $\mathcal{M}$ and suppose $f_\kappa \to f$ weak* in $L^\infty(X)$. Since μ is assumed to be finite, we have $L^2(X, \mu) \subset L^1(X, \mu)$. Then $\langle f, g\rangle = 0$ for every $g \in L^2(X, \mu)$ which is orthogonal to $\mathcal{M}$ because $\langle f_\kappa, g\rangle = 0$ for every such g, for all κ. Thus f belongs to the closure of $\mathcal{M}$ in $L^2(X, \mu)$, which is clearly $L^2(X, \mu')$, and since f is essentially bounded this implies $f \in \mathcal{M}$. Thus, $\mathcal{M}$ is weak* closed.

PROPOSITION 6.2.4

Let X be a σ-finite measure space and let $\mathcal{M}$ be a W-subalgebra of $L^\infty(X)$. Then there is a σ-finite measure space Y and a measurable surjection $\phi : X \to Y$ such that $\phi^{-1}(S)$ is null in X if and only if S is null in Y, and $\mathcal{M} = C_\phi(L^\infty(Y))$.*

PROOF Let μ be the given measure on X; possibly replacing it with an equivalent measure, we may assume it is finite. Let Ω be the σ-algebra generated by the sets $f^{-1}(S)$ for $S \subset \mathbf{R}$ Borel measurable and $f \in \mathcal{M}$; let $\nu = \mu|_\Omega$; and let Y be the set X equipped with the measure ν. The identity map $\phi : X \to Y$ is then a measurable surjection for which $\phi^{-1}(S)$ is null in X if and only if S is null in Y.

It is clear that $\mathcal{M} \subset L^\infty(Y)$, and the reverse containment follows from Proposition 6.2.2. ∎

Next we characterize W*-ideals.

Example 6.2.5

Let X be a σ-finite measure space and let $K \subset X$ be a measurable subset. Then

$$\mathcal{I} = \{f \in L^\infty(X) : f|_K = 0\}$$

is a W*-ideal of $L^\infty(X)$.

PROPOSITION 6.2.6
Let $\mathcal{I}$ be a W-ideal of $L^\infty(X)$. Then there is a measurable subset K of X such that $\mathcal{I} = \{f \in L^\infty(X) : f|_K = 0\}$.*

PROOF Let Λ be the collection of subsets $S \subset X$ with the property that $\chi_S \in \mathcal{I}$. Ordered by inclusion, it is easy to see that Λ is directed upwards. Then the net (χ_S) for $S \in \Lambda$ converges weak* because every function in $L^1(X)$ can be written as a linear combination of positive functions in $L^1(X)$, and $\int \chi_S h$ converges for every positive $h \in L^1(X)$. So there must exist $f \in \mathcal{I}$ such that $\chi_S \to f$ weak*.

Now f is the characteristic function of a set $K' \subset X$ with the property that every $S \in \Lambda$ is almost everywhere contained in K'. Letting $K = X - K'$, the fact that $\chi_{K'} \in \mathcal{I}$ and the ideal property of $\mathcal{I}$ imply that every function which vanishes almost everywhere on K is in $\mathcal{I}$. Conversely, if $f \in \mathcal{I}$ and f does not vanish almost everywhere on K then (again using the ideal property of $\mathcal{I}$) it follows that $\mathcal{I}$ contains the characteristic function of some positive measure subset of K. But this contradicts the definition of K, so every function in $\mathcal{I}$ must vanish on K. ∎

PROPOSITION 6.2.7
Let $\mathcal{I} = \{f \in L^\infty(X) : f|_K = 0\}$ be a weak closed ideal of $L^\infty(X)$. Then $L^\infty(X)/\mathcal{I} \cong L^\infty(K)$.*

The last proposition is trivial.

6.3 Trace class operators

Like $l^\infty(X)$ and $L^\infty(X)$, $B(\mathcal{H})$ is a dual space. The goal of the present section is to prove this result. Then we will discuss its W*-subalgebras and ideals in the following sections.

It is not too hard to prove that $B(\mathcal{H})$ is a dual space using abstract methods. However, we prefer to spend a little time describing its predual concretely. The predual of $B(\mathcal{H})$ is naturally realized as an ideal (but not weak* closed, or even norm closed) in $B(\mathcal{H})$, much as $l^1(X)$ is an ideal in $l^\infty(X)$. It is called the "ideal of trace class operators on $\mathcal{H}$."

We develop the properties of trace class operators through a series of lemmas.

LEMMA 6.3.1

Let $A \in B(\mathcal{H})$ be positive. If the sum $\sum \langle Ae_\kappa, e_\kappa\rangle$ is finite for some orthonormal basis (e_κ) of $\mathcal{H}$ then it is finite, and converges to the same value, for every orthonormal basis. If this happens then A is compact.

PROOF Let (e_κ) and $(\tilde{e}_{\tilde{\kappa}})$ be orthonormal bases of $\mathcal{H}$. Then

$$\sum_\kappa \langle Ae_\kappa, e_\kappa\rangle = \sum_\kappa \|A^{1/2}e_\kappa\|^2 = \sum_{\kappa,\tilde{\kappa}} \langle A^{1/2}e_\kappa, \tilde{e}_{\tilde{\kappa}}\rangle^2$$

where $A^{1/2}$ is defined by functional calculus, and similarly we have $\sum_{\tilde{\kappa}} \langle A\tilde{e}_{\tilde{\kappa}}, \tilde{e}_{\tilde{\kappa}}\rangle = \sum_{\kappa,\tilde{\kappa}} \langle A^{1/2}\tilde{e}_{\tilde{\kappa}}, e_\kappa\rangle^2$. Thus the two sums are equal; in particular, if one is finite so is the other.

Now suppose A is not compact. By Proposition 5.4.9 it follows that there exists $\epsilon > 0$ such that $E(S)$ is infinite dimensional, where E is the spectral measure of A and $S = \{x \in \mathrm{sp}(A) : |x| \geq \epsilon\}$. Letting (e_κ) be an orthonormal basis of $\mathcal{H}$ which contains an orthonormal basis of $E(S)$, we then have $\langle Ae_\kappa, e_\kappa\rangle \geq \epsilon$ for every $e_\kappa \in E(S)$. This shows that $\sum \langle Ae_\kappa, e_\kappa\rangle$ does not converge. ∎

For any $A \in B(\mathcal{H})$ we define $\mathrm{tr}(A)$ to be the sum $\sum \langle Ae_\kappa, e_\kappa\rangle$, provided that it converges absolutely, and to the same value, for every orthonormal basis (e_κ) of $\mathcal{H}$. Otherwise we say that $\mathrm{tr}(A)$ is not well-defined.

The rank of an operator is the dimension of its range, and an operator has finite rank if its range is finite dimensional. Equivalently, $F \in B(\mathcal{H})$ has finite rank if its kernel has finite codimension. Since $\mathrm{ran}(F)^\perp = \ker(F^*)$, it follows that F^* has finite rank if F does.

LEMMA 6.3.2

Suppose $F \in B(\mathcal{H})$ has finite rank. Then $\mathrm{tr}(F)$ is well-defined.

PROOF Let $\mathcal{K} = \ker(F)^\perp \vee \mathrm{ran}(F)$. We can write F as $F = F|_\mathcal{K} \oplus 0$ on $\mathcal{K} \oplus \mathcal{K}^\perp$. Then $F|_\mathcal{K} \in B(\mathcal{K})$ can be decomposed into its real and imaginary parts, each of which can then be written as a difference of two positive operators. So F can be expressed as a linear combination of four positive finite rank operators. Each of these has a well-defined trace by Lemma 6.3.1, so the same must be true of F. ∎

LEMMA 6.3.3

Let $A, F \in B(\mathcal{H})$ and suppose F has finite rank. Then $\mathrm{tr}(AF) = \mathrm{tr}(FA)$ and $|\mathrm{tr}(AF)| \leq \mathrm{tr}(|A|)\|F\|$.

PROOF It is clear that AF and FA both have finite rank, so $\mathrm{tr}(AF)$ and $\mathrm{tr}(FA)$ are well-defined. For the first statement it suffices to take F self-adjoint. In that case let (e_κ) be an orthonormal basis of $\mathcal{H}$ consisting of eigenvectors of F and let λ_κ be the corresponding eigenvalues (all but finitely many of which are zero). Then

$$\sum \langle AFe_\kappa, e_\kappa\rangle = \sum \lambda_\kappa \langle Ae_\kappa, e_\kappa\rangle = \sum \langle FAe_\kappa, e_\kappa\rangle,$$

so $\mathrm{tr}(AF) = \mathrm{tr}(FA)$ as claimed.

For the second statement, if $\mathrm{tr}(|A|) = \infty$ we are done. Otherwise $|A|$ is compact and we can find an orthonormal basis $(\tilde{e}_{\tilde{\kappa}})$ of $\mathcal{H}$ consisting of eigenvectors of $|A|$, with corresponding eigenvalues $\tilde{\lambda}_{\tilde{\kappa}}$. Write $A = U|A|$ as in Lemma 5.4.7; then

$$|\mathrm{tr}(FA)| = \left|\sum \langle FU|A|\tilde{e}_{\tilde{\kappa}}, \tilde{e}_{\tilde{\kappa}}\rangle\right| \leq \sum \tilde{\lambda}_{\tilde{\kappa}} \|F\| = \mathrm{tr}(|A|)\|F\|,$$

as desired. ∎

LEMMA 6.3.4
Let $A, F \in B(\mathcal{H})$ and suppose F has finite rank. Then

$$\begin{aligned} \mathrm{tr}(|A|) &= \sup\{|\mathrm{tr}(AF)| : F \textit{ has finite rank and } \|F\| \leq 1\} \\ &= \sup\{|\mathrm{tr}(FA)| : F \textit{ has finite rank and } \|F\| \leq 1\}. \end{aligned}$$

PROOF We will show that $\mathrm{tr}(|A|) \leq \sup\{|\mathrm{tr}(FA)|\}$; Lemma 6.3.3 implies the rest. Let (e_κ) be an orthonormal basis of $\mathcal{H}$, write $A = U|A|$ as in Lemma 5.4.7, and let P be the orthogonal projection onto the span of some finite set of basis vectors. Then, taking $F = PU^*$, we have $\mathrm{tr}(FA) = \mathrm{tr}(P|A|)$; and as P tends to I the right side tends to $\mathrm{tr}(|A|)$, which is enough. ∎

We are now ready for the following definition.

DEFINITION 6.3.5
Let $A \in B(\mathcal{H})$. Then A is trace class if its trace norm $\|A\|_{TC} = \mathrm{tr}(|A|)$ is finite. The set of all trace class operators is denoted $TC(\mathcal{H})$.

LEMMA 6.3.6
Let $A \in TC(\mathcal{H})$ be self-adjoint. Then we can write $A = A_+ - A_-$ where $0 \leq A_+, A_- \leq |A|$ and $\mathrm{tr}(A_+), \mathrm{tr}(A_-) \leq \mathrm{tr}(|A|)$.

PROOF Define $A_+ = \frac{1}{2}(|A| + A)$ and $A_- = \frac{1}{2}(|A| - A)$; regarding A as a multiplication operator, it is clear that $0 \leq A_+, A_- \leq |A|$, and this immediately implies that $\mathrm{tr}(A_+), \mathrm{tr}(A_-) \leq \mathrm{tr}(|A|)$. ∎

PROPOSITION 6.3.7

$TC(\mathcal{H})$ is a vector space and $\|\cdot\|_{TC}$ is a complete norm. If $A \in TC(\mathcal{H})$ and $B \in B(\mathcal{H})$ then A^*, AB, and BA all belong to $TC(\mathcal{H})$.

PROOF Everything but completeness is a straightforward consequence of Lemma 6.3.4. To prove completeness, let $(A_n) \subset TC(\mathcal{H})$ and suppose $\sum \|A_n\|_{TC} < \infty$; by Lemma 2.1.7, it will suffice to show that $\sum A_n$ converges in $TC(\mathcal{H})$. Observe first that $\|\mathrm{Re}\, A_n\|_{TC}, \|\mathrm{Im}\, A_n\|_{TC} \leq \|A_n\|_{TC}$. Thus by decomposing into real and imaginary parts we can reduce to the case where each A_n is self-adjoint. Then by Lemma 6.3.6 we can reduce to the case where each A_n is positive.

Now $\|A_n\| \leq \mathrm{tr}(A_n)$ in this case, so $\sum A_n$ converges in $B(H)$. Let A be its limit. Then we have $\langle Av, v\rangle = \sum \langle A_n v, v\rangle$ for all $v \in H$, and summing over an orthonormal basis yields $\mathrm{tr}(A) = \sum \mathrm{tr}(A_n)$ and

$$\mathrm{tr}(A - \sum_{n=1}^{N} A_n) = \mathrm{tr}(\sum_{n=N+1}^{\infty} A_n) \to 0.$$

So $A \in TC(\mathcal{H})$ and $\sum A_n = A$, and we conclude that $TC(\mathcal{H})$ is complete. ∎

LEMMA 6.3.8

For any $A \in TC(\mathcal{H})$, $\mathrm{tr}(A)$ is well-defined. For any $A \in TC(\mathcal{H})$ and $B \in B(\mathcal{H})$ we have $\mathrm{tr}(AB) = \mathrm{tr}(BA)$ and $|\mathrm{tr}(AB)| \leq \|A\|_{TC}\|B\|$. The finite rank operators are dense in $TC(\mathcal{H})$, and $TC(\mathcal{H}) \subset K(\mathcal{H})$.

PROOF It follows from Lemmas 6.3.6 and 6.3.7 that every trace class operator can be expressed as a linear combination of positive trace class operators. This implies the first assertion; together with the easy fact that positive trace class operators are approximated by finite rank operators, it implies density of the finite rank operators in $TC(\mathcal{H})$; and together with Lemma 6.3.1 it implies that every trace class operator is compact. The remainder is proven exactly as in Lemma 6.3.3. ∎

The preceding results give a fairly complete picture of $TC(\mathcal{H})$, and we can now use them to identify the dual of $TC(\mathcal{H})$ with $B(\mathcal{H})$.

THEOREM 6.3.9
For all $B \in B(\mathcal{H})$ the map $\omega_B : A \mapsto \mathrm{tr}(AB)$ defines a bounded linear functional on $TC(\mathcal{H})$. The map $B \mapsto \omega_B$ isometrically identifies $B(\mathcal{H})$ with $TC(\mathcal{H})^*$. On bounded subsets of $B(\mathcal{H})$ the weak topology agrees with the weak operator topology.*

PROOF Boundedness of ω_B follows from Lemma 6.3.8; in fact, this shows $\|\omega_B\| \leq \|B\|$. Conversely, for any vectors $v, w \in \mathcal{H}$ let $A_{v,w}$ be the operator $A_{v,w}u = \langle u, w\rangle v$; then $\|A_{v,w}\|_{TC} = \|v\|\|w\|$ and $|\mathrm{tr}(BA_{v,w})| = |\langle Bv, w\rangle|$. Taking the supremum over all unit vectors v and w yields $\|\omega_B\| \geq \|B\|$.

Next, given $\omega \in TC(\mathcal{H})^*$, the map $(v, w) \mapsto \omega(A_{v,w})$ is a bounded sesquilinear form, and hence there exists $B \in B(\mathcal{H})$ such that $\langle Bv, w\rangle = \omega(A_{v,w})$ for all $v, w \in \mathcal{H}$. Thus $\omega_B(A_{v,w}) = \mathrm{tr}(BA_{v,w}) = \omega(A_{v,w})$. But the operators $A_{v,w}$ span the finite rank operators, and by density of the latter in $TC(\mathcal{H})$ we have $\omega = \omega_B$. So every bounded linear functional on $TC(\mathcal{H})$ is of the form ω_B for some $B \in B(\mathcal{H})$.

If (B_κ) is a bounded net in $B(\mathcal{H})$, then $B_\kappa \to B$ weak* if and only if $\mathrm{tr}(B_\kappa A) \to \mathrm{tr}(BA)$ for all $A \in TC(\mathcal{H})$, which holds if and only if $\mathrm{tr}(B_\kappa A_{v,w}) \to \mathrm{tr}(BA_{v,w})$ for all $v, w \in \mathcal{H}$ by density of the finite rank operators in $TC(\mathcal{H})$. That is, $B_\kappa \to B$ weak* if and only if $\langle B_\kappa v, w\rangle \to \langle Bv, w\rangle$ for all $v, w \in \mathcal{H}$. ∎

The weak* topology on $B(\mathcal{H})$ is also called the <u>ultraweak</u> or <u>σ-weak</u> topology.

6.4 *The algebras $B(\mathcal{H})$*

Now that we have a weak* topology on $B(\mathcal{H})$, we can consider W*-subalgebras and ideals. We will discuss general W*-subalgebras in the next section. W*-ideals, however, are trivial:

PROPOSITION 6.4.1
Let $\mathcal{I}$ be a W-ideal of $B(\mathcal{H})$. Then $\mathcal{I} = 0$ or $\mathcal{I} = B(\mathcal{H})$.*

PROOF Suppose $\mathcal{I} \neq 0$ and let $A \in \mathcal{I}$ be nonzero. Then for a suitable operator B the product $AB \in \mathcal{I}$ has rank one. Multiplying AB on the left and right by appropriate rank one operators shows that $\mathcal{I}$ contains every rank one operator, and hence $\mathcal{I}$ contains all finite rank operators.

Now if $A \in TC(\mathcal{H})$ and the map $B \mapsto \mathrm{tr}(AB)$ annihilates every finite rank operator then we must have $\|A\|_{TC} = 0$ and hence $A = 0$.

This shows that the finite rank operators are weak* dense in $B(\mathcal{H})$, and therefore we conclude that $\mathcal{I} = B(\mathcal{H})$. ∎

Although this result is somewhat disappointing, there is a substitute for the notion of an ideal which yields a better analog of Propositions 6.1.7 and 6.2.6. First we introduce the substitute notion and show that it is the same thing as an ideal in the commutative case.

DEFINITION 6.4.2 *Let $\mathcal{A}$ be a C*-algebra and let $\mathcal{B} \subset \mathcal{A}$ be a C*-subalgebra. $\mathcal{B}$ is* *hereditary* *if $A \in \mathcal{A}$, $B \in \mathcal{B}$, and $0 \leq A \leq B$ imply $A \in \mathcal{B}$.*

PROPOSITION 6.4.3

Let $\mathcal{A}$ be an abelian C-algebra. Then a C*-subalgebra of $\mathcal{A}$ is hereditary if and only if it is an ideal.*

PROOF By Theorem 5.3.5 (and the comment following it) we may assume that $\mathcal{A} = C(X)$ for some compact Hausdorff space X. Since any C*-ideal of $C(X)$ consists of all functions which vanish on some closed subset $K \subset X$, it is easy to see that every C*-ideal is hereditary. Conversely, suppose $\mathcal{B} \subset \mathcal{A}$ is a hereditary C*-subalgebra and let $f \in \mathcal{A}$ and $g \in \mathcal{B}$; we must show that $fg \in \mathcal{B}$. Let $f = \sum_{k=0}^{3} i^k f_k$ and $g = \sum_{k=0}^{3} i^k g_k$ be decompositions of f and g into positive functions. By functional calculus each g_k belongs to $\mathcal{B}$, and it will suffice to show that $f_j g_k \in \mathcal{B}$ for all j, k. But for $\epsilon = 1/\|f_j\|_\infty$ we have $0 \leq \epsilon f_j g_k \leq g_k$, and hence $f_j g_k \in \mathcal{B}$ as desired. ∎

Now we show that the hereditary W*-subalgebras of $B(\mathcal{H})$ correspond to closed subspaces of $\mathcal{H}$.

Example 6.4.4

Let P be a projection in $B(\mathcal{H})$. Then

$$PB(\mathcal{H})P = \{PAP : A \in B(\mathcal{H})\}$$

is a hereditary W*-subalgebra of $B(\mathcal{H})$.

PROPOSITION 6.4.5

Let $\mathcal{M}$ be a hereditary W-subalgebra of $B(\mathcal{H})$. Then $\mathcal{M} = PB(\mathcal{H})P$ for some projection $P \in \mathcal{M}$.*

PROOF Let $A \in \mathcal{M}$ be self-adjoint and define $|A|^{1/n}$ by functional calculus. Then $|A|^{1/n}$ converges weak operator, and hence weak*, to $\chi_{\mathbf{R}-\{0\}}(A)$. So if $A \in \mathcal{M}$ is self-adjoint, then the projection onto its range also belongs to $\mathcal{M}$.

If P_1 and P_2 are projections and $v \perp \mathrm{ran}(P_1 + P_2)$ then

$$\|P_1 v\|^2 = \langle P_1 v, v\rangle \leq \langle (P_1 + P_2)v, v\rangle = 0.$$

This shows that the range of P_1 is contained in the closure of the range of $P_1 + P_2$. Thus, by the first paragraph of the proof the sequence $(P_1 + P_2)^{1/n}$ converges to a projection in $\mathcal{M}$ which is larger than P_1, and likewise it is larger than P_2. So the set of projections in $\mathcal{M}$ is directed upwards, and by weak* closure its limit belongs to $\mathcal{M}$. Let P be this limit.

Now if $A \in B(\mathcal{H})$ is positive, then PAP is also positive and $0 \leq PAP \leq \|A\|P$, so $PAP \in \mathcal{M}$. This shows that $PB(\mathcal{H})P \subset \mathcal{M}$. Conversely, if $A \in \mathcal{M}$ is positive then P dominates the projection onto the range of A, and hence $A = PAP$. By linearity we conclude that $\mathcal{M} \subset PB(\mathcal{H})P$ as well. ∎

We conclude this section by classifying the weak* continuous automorphisms of $B(\mathcal{H})$. With a little more work the assumption of weak* continuity can be dropped, and one can actually classify all $*$-homomorphisms from $B(\mathcal{H})$ into $B(\mathcal{K})$, but we will not do this. Our result resembles Corollary 6.1.10 in the special case that $Y = X$, when the latter implies that the unital $*$-isomorphisms from $l^\infty(X)$ onto itself are in one-to-one correspondence with the automorphisms of X. The following is the corresponding construction for $B(\mathcal{H})$.

Example 6.4.6

Let $U \in B(\mathcal{H})$ be unitary. Then the map $A \mapsto UAU^*$ is a weak* continuous unital $*$-isomorphism from $B(\mathcal{H})$ onto itself.

PROPOSITION 6.4.7

Let $\pi : B(\mathcal{H}) \cong B(\mathcal{H})$ be a weak continuous unital $*$-isomorphism. Then $\pi(A) = UAU^*$ for some unitary $U \in B(\mathcal{H})$.*

PROOF Fix a a rank one projection $P \in B(\mathcal{H})$ and unit vectors $v \in \mathrm{ran}(P)$ and $w \in \mathrm{ran}(\pi(P))$. Observe that for any $A \in B(\mathcal{H})$ we have $PAP = f(A)P$ for some complex number $f(A)$. Thus

$$\begin{aligned}\|\pi(A)w\|^2 &= \|\pi(AP)w\|^2 = \langle \pi(PA^*AP)w, w\rangle \\ &= f(A^*A)\langle \pi(P)w, w\rangle = f(A^*A),\end{aligned}$$

and $\|Av\|^2 = f(A^*A)$ by a similar computation. So if $A, B \in B(\mathcal{H})$ satisfy $Av = Bv$ then $\|\pi(A-B)v\|^2 = \|(A-B)v\|^2 = 0$, and this shows that the map $U : Av \mapsto \pi(A)w$ is well-defined. The preceding also shows that U is an isometry.

For any $A, B \in B(\mathcal{H})$ we have

$$UAU^*(\pi(B)w) = UABv = \pi(AB)w = \pi(A)(\pi(B)w).$$

Thus $\pi(A) = UAU^*$. Finally, if U is not surjective then $\pi(A)v = 0$ for any v orthogonal to the range of U, contradicting surjectivity of π. Thus U is surjective, and hence unitary. ∎

6.5 Von Neumann algebras

A von Neumann algebra (or W*-algebra) is a W*-subalgebra of some $B(\mathcal{H})$. As with C*-algebras, sometimes we will consider them as abstract spaces. In fact, they have a simple abstract characterization, the easy direction of which is the following.

PROPOSITION 6.5.1
Every von Neumann algebra is a dual space.

This proposition follows from standard Banach space facts. If $\mathcal{M} \subset B(\mathcal{H})$ is a von Neumann algebra then $\mathcal{M} \cong (TC(\mathcal{H})/E)^*$ where

$$E = \{A \in TC(\mathcal{H}) : \mathrm{tr}(AB) = 0 \text{ for all } B \in \mathcal{M}\}$$

is the preannihilator of $\mathcal{M}$.

The above property actually characterizes abstract von Neumann algebras: an abstract C*-algebra is an abstract von Neumann algebra if and only if it is a dual space. This result is known as Sakai's theorem; we will not prove it here. Note, however, that the slightly stronger assumption that the C*-algebra $\mathcal{A}$ has an *order* predual (i.e., an ordered Banach space whose dual is isometrically isomorphic to $\mathcal{A}$ as an ordered Banach space) easily implies that $\mathcal{A}$ is a von Neumann algebra. For in this case there trivially exist a family of weak* continuous states on $\mathcal{A}$ sufficient, using the GNS construction, to weak* continuously embed $\mathcal{A}$ as a W*-subalgebra of some $B(\mathcal{H})$.

Next we determine the structure of abelian von Neumann algebras.

Example 6.5.2

Let X be a σ-finite measure space, let $\mathcal{X}$ be a measurable Hilbert bundle over X, and let $\mathcal{H} = L^2(X;\mathcal{X})$. Then $\mathcal{M} = \{M_f : f \in L^\infty(X)\}$ is a von Neumann algebra, and it is $*$-isomorphic to $L^\infty(X)$.

THEOREM 6.5.3

Suppose $\mathcal{H}$ is separable and let $\mathcal{M} \subset B(\mathcal{H})$ be an abelian von Neumann algebra. Then there is a probability measure μ on $\mathrm{sp}(\mathcal{M})$, a measurable Hilbert bundle $\mathcal{X}$ over $\mathrm{sp}(\mathcal{M})$, and an isometric $*$-isomorphism $U : L^2(\mathrm{sp}(\mathcal{M}); \mathcal{X}) \cong \mathcal{H}$ such that $\mathcal{M} = UL^\infty(\mathrm{sp}(\mathcal{M}))U^{-1}$.

PROOF Let Ω be the σ-algebra on $\mathrm{sp}(\mathcal{M})$ generated by the continuous functions, i.e., the Baire σ-algebra. Apply Theorem 5.3.5 to $\mathcal{M}$; this provides a spectral measure on the Borel sets of $\mathrm{sp}(\mathcal{M})$ which restricts to a spectral measure E on Ω. Then by Corollary 3.4.3 there is a probability measure μ on Ω, a measurable Hilbert bundle $\mathcal{X}$, and a surjective isometry $U : L^2(\mathrm{sp}(\mathcal{M}); \mathcal{X}) \to \mathcal{H}$ such that $U\chi_S U^{-1} = E(S)$ for every Baire set S. As every continuous function on $\mathrm{sp}(\mathcal{M})$ can be approximated by Baire measurable simple functions, this implies that $\mathcal{M} = UC(\mathrm{sp}(\mathcal{M}))U^{-1}$. But $C(\mathrm{sp}(\mathcal{M})) \cong \mathcal{M}$ is a W*-subalgebra of $L^\infty(\mathrm{sp}(\mathcal{M}))$ which measurably separates points, so we have $C(\mathrm{sp}(\mathcal{M})) = L^\infty(\mathrm{sp}(\mathcal{M}))$ by Proposition 6.2.2. ∎

In the nonseparable case, as one might expect, $\mathcal{M}$ is still $*$-isomorphic to some $L^\infty(Y)$.

The usual remark about invariants and unitary equivalence also applies here; see the comments following Theorems 3.5.1 and 5.3.5.

Now we turn to W*-ideals and hereditary W*-subalgebras. These have particularly simple structures.

Example 6.5.4

Let $\mathcal{M}$ be a von Neumann algebra and let $P \in \mathcal{M}$ be a projection. Then $P\mathcal{M}P$ is a hereditary W*-subalgebra. If P belongs to the center of $\mathcal{M}$, i.e., $PA = AP$ for all $A \in \mathcal{M}$, then $P\mathcal{M}P = P\mathcal{M}$ is a W*-ideal of $\mathcal{M}$.

PROPOSITION 6.5.5

Let $\mathcal{M}$ be a von Neumann algebra. If $\mathcal{N}$ is a hereditary W-subalgebra of $\mathcal{M}$ then there is a projection $P \in \mathcal{N}$ such that $\mathcal{N} = P\mathcal{M}P$. If $\mathcal{N}$ is a W*-ideal then P is in the center of $\mathcal{M}$.*

PROOF The proof that every hereditary W*-subalgebra has the form $P\mathcal{M}P$ exactly follows the proof of the same result for $B(\mathcal{H})$ (Proposition 6.4.5). The same argument also applies to any W*-ideal $\mathcal{N}$ to show that there is a maximal projection $P \in \mathcal{N}$ and $\mathcal{N} \subset P\mathcal{M}P$. Conversely, $P\mathcal{M}P \subset \mathcal{N}$ follows immediately from the ideal property.

To show that P is central, let $A \in \mathcal{M}$ and let E be the range of P. Since $AP \in \mathcal{N}$ we must have $AP = PBP$ for some $B \in \mathcal{M}$, and therefore $A(E) \subset E$. We have $A^*(E) \subset E$ similarly, which implies that $A(E^\perp) \subset E^\perp$. Thus A and P commute. ∎

In particular, since every von Neumann algebra is a W*-ideal of itself, there always exists a projection $P \in \mathcal{M}$ which satisfies $AP = A$ for all $A \in \mathcal{M}$. That is, every von Neumann algebra has a unit (although it need not be the identity operator in $B(\mathcal{H})$).

Next, we present von Neumann's celebrated "double commutant" theorem, which provides a fundamental algebraic characterization of von Neumann algebras.

DEFINITION 6.5.6 *Let $\mathcal{M}$ be a subset of $B(\mathcal{H})$. Its* <u>*commutant*</u> *$\mathcal{M}^c$ is the set of all operators in $B(\mathcal{H})$ which commute with every operator in $\mathcal{M}$.*

THEOREM 6.5.7
Let $\mathcal{M} \subset B(\mathcal{H})$ be a unital C-algebra. Then $\mathcal{M}$ is a von Neumann algebra if and only if $\mathcal{M} = \mathcal{M}^{cc}$.*

PROOF Suppose (A_κ) is a bounded net in $B(\mathcal{H})$ and $A_\kappa \to A$ weak*. If $B \in B(\mathcal{H})$ commutes with each A_κ then B commutes with A, because

$$\langle ABv, w\rangle = \lim \langle A_\kappa Bv, w\rangle = \lim \langle A_\kappa v, B^*w\rangle = \langle BAv, w\rangle$$

for all $v, w \in \mathcal{H}$. This shows that the commutant of any set is weak* closed. Also, if $\mathcal{M}$ is a $*$-algebra it is easy to check that $\mathcal{M}^c$ is also a $*$-algebra. So if $\mathcal{M} = \mathcal{M}^{cc}$ then $\mathcal{M}$ is a weak* closed $*$-algebra, i.e., a von Neumann algebra.

For the converse, we first prove a seemingly modest density result. Let $A_0 \in \mathcal{M}^{cc}$ and $v \in \mathcal{H}$, and let P be the projection onto $\overline{\mathcal{M}v}$. If $A, B \in \mathcal{M}$ then

$$AP(Bv) = ABv = PA(Bv),$$

so AP and PA agree on vectors in $\overline{\mathcal{M}v}$. But also, if $w \perp \overline{\mathcal{M}v}$ then

$$\langle Aw, Bv\rangle = \langle w, A^*Bv\rangle = 0,$$

so that $Aw \perp \overline{\mathcal{M}v}$; thus AP and PA are both zero on $\overline{\mathcal{M}v}^\perp$. This shows that $P \in \mathcal{M}^c$, so we must have $PA_0 = A_0P$. Applying this equation to v we get $PA_0v = A_0Pv = A_0v$, which means that $A_0v \in \overline{\mathcal{M}v}$. Thus, for any $\epsilon > 0$ we can find $A \in \mathcal{M}$ such that $\|(A - A_0)v\| \leq \epsilon$.

Now it is clear that $\mathcal{M} \subset \mathcal{M}^{cc}$; to prove equality we will show that $\mathcal{M}$ is weak* dense in $\mathcal{M}^{cc}$. Thus let $A_0 \in \mathcal{M}^{cc}$, $B \in TC(\mathcal{H})$, and $\epsilon > 0$; we must find $A \in \mathcal{M}$ such that $|\mathrm{tr}(AB) - \mathrm{tr}(A_0B)| \leq \epsilon$. We may assume $B > 0$.

Let (e_κ) be a basis of $\mathcal{H}$ consisting of eigenvectors of B and say $Be_\kappa = \lambda_\kappa e_\kappa$. Since $\sum \lambda_\kappa < \infty$, it follows that $v = \bigoplus \sqrt{\lambda_\kappa} e_\kappa$ belongs to the direct sum $\mathcal{H}^\infty$ of infinitely many copies of $\mathcal{H}$. (This could even be an uncountable sum.) Let $\mathcal{M}^\infty \cong \mathcal{M}$ be the set of operators on $\mathcal{H}^\infty$ of the form $A^\infty = A \oplus A \oplus \cdots$ for $A \in \mathcal{M}$, and observe that the corresponding operator A_0^∞ belongs to the double commutant of $\mathcal{M}^\infty$. (The commutant of $\mathcal{M}^\infty$ consists of operators C on $\mathcal{H}^\infty$ such that $P_\kappa C P_{\tilde{\kappa}} \in \mathcal{M}^c$ for all $\kappa, \tilde{\kappa}$ where P_κ is the projection of $\mathcal{H}^\infty$ onto the κth copy of $\mathcal{H}$.) By the last paragraph there exists $A \in \mathcal{M}$ such that $\|(A^\infty - A_0^\infty)v\| \leq \epsilon/\|v\|$, and hence

$$|\mathrm{tr}(AB) - \mathrm{tr}(A_0B)| = \langle (A^\infty - A_0^\infty)v, v\rangle \leq \epsilon,$$

for any $\epsilon > 0$. This shows that $\mathcal{M}$ is weak* dense in $\mathcal{M}^{cc}$. ∎

We conclude this section with the factor decomposition of an arbitrary von Neumann algebra. A von Neumann algebra is a <u>factor</u> if it has no proper W*-ideals, or equivalently (Proposition 6.5.5) no proper central projections. The idea of factor decomposition is to express any von Neumann algebra as a "measurable direct sum" of factors.

Given a measurable Hilbert bundle $\mathcal{X} = \bigcup (X_n \times \mathcal{H}_n)$ over X, a <u>field of operators</u> on $\mathcal{X}$ is a function $\tilde{A}$ on X such that $\tilde{A}(x) \in B(\mathcal{H}_x)$ for almost every x, where $\mathcal{H}_x$ is the Hilbert space lying over x. It is <u>weakly measurable</u> if for any $v, w \in \mathcal{H}_n$ the function $x \mapsto \langle \tilde{A}(x)v, w\rangle$ is measurable on X_n.

We require the following fact: if $\mathcal{M} \subset B(\mathcal{H})$ is a von Neumann algebra and $\mathcal{H}$ is separable then $\mathcal{M}$ is both separable and first countable for the weak* topology. This follows from the fact that the predual of $\mathcal{M}$ is a quotient of $TC(\mathcal{H})$ and hence is separable; weak* separability and first countability of $\mathcal{M}$ are then general Banach space facts.

THEOREM 6.5.8

Let $\mathcal{M} \subset B(\mathcal{H})$ be a von Neumann algebra and suppose $\mathcal{H}$ is separable. Then there is a σ-finite measure space X, a measurable Hilbert bundle $\mathcal{X}$ over X, and a family of factor von Neumann algebras $\mathcal{M}_x \subset B(\mathcal{H}_x)$ such that $\mathcal{H}$ can be identified with $L^2(X; \mathcal{X})$ and $\mathcal{M}$ with the bounded weakly measurable fields of operators which lie in $\mathcal{M}_x$ almost everywhere.

PROOF The center of $\mathcal{M}$,

$$Z(\mathcal{M}) = \{A \in \mathcal{M} : AB = BA \text{ for all } B \in \mathcal{M}\},$$

is an abelian W*-subalgebra of $\mathcal{M}$. Find X, $\mathcal{X} = \bigcup(X_n \times \mathcal{H}_n)$, and U as in Theorem 6.5.3 and identify $\mathcal{H}$ with $L^2(X, \mathcal{X})$ and $Z(\mathcal{M})$ with $L^\infty(X) \subset B(L^2(X; \mathcal{X}))$ acting as multiplication operators.

Let $\{A_k\}$ be a countable weak* dense subset of $\mathcal{M}$; we may suppose this set is closed under sums, products, adjoints, and multiplication by scalars in $\mathbf{Q} + i\mathbf{Q}$. For each k, n, and $1 \leq i \leq n$ write $A_k(1_{X_n} e_i) = \sum f^k_{i,j} e_j$; then define a weakly measurable field of operators $\tilde{A}_k$ by $\tilde{A}_k(x) : e_i \mapsto \sum_j f^k_{i,j}(x) e_j$. As each $f^k_{i,j}$ is well-defined almost everwhere, so is $\tilde{A}_k$. Each $\tilde{A}_k$ is essentially bounded because for almost every x and all $v \in \mathcal{H}_x$ we have $\tilde{A}_k(x)v = (A_k v)(x)$. Let $\mathcal{N}_x$ be the weak* closure in $B(\mathcal{H}_x)$ of the set $\{\tilde{A}_k(x)\}$.

By the choice of the family $\{A_k\}$, it is clear that almost every $\mathcal{N}_x$ is a von Neumann algebra. Let $\mathcal{N}$ be the set of bounded measurable fields of operators $\tilde{B}$ such that $\tilde{B}(x) \in \mathcal{N}_x$ for almost every x, and let $\pi : \mathcal{N} \to B(L^2(X; \mathcal{X}))$ be the natural action of $\mathcal{N}$ on $L^2(X; \mathcal{X})$. It is routine to verify that π is a $*$-isomorphism and that $\pi(\mathcal{N})$ is a von Neumann algebra. Next we claim that $\pi(\tilde{A}_k) = A_k$ for all k. By the definition of $\tilde{A}_k$, equality holds when both sides are applied to vectors of the form $1_{X_n} e_i$; since both sides commute with $L^\infty(X) \cong Z(\mathcal{M})$, it follows that

$$\pi(\tilde{A}_k)(f e_i) = M_f \pi(\tilde{A}_k)(1_{X_n} e_i) = M_f A_k(1_{X_n} e_i) = A_k(f e_i)$$

for all $f \in L^\infty(X)$, and taking linear combinations yields that $\pi(\tilde{A}_k) = A_k$ on a dense set of vectors in $L^2(X; \mathcal{X})$. So the claim is proven.

It follows that $\mathcal{M} \subset \pi(\mathcal{N})$. Conversely, if $B \in \mathcal{M}^c$ then we can define $\tilde{B}$ in the same way we defined the $\tilde{A}_k$, and $BA_k = A_k B$ implies that $\tilde{B}(x)$ and $\tilde{A}_k(x)$ commute almost everywhere. Thus $B = \pi(\tilde{B}) \in \pi(\mathcal{N})^c$, and by Theorem 6.5.7 we conclude that $\mathcal{M} = \pi(\mathcal{N})$.

Finally, we must show that almost every $\mathcal{N}_x$ is a factor. The idea is that if $Z(\mathcal{N}_x)$ were nontrivial on a set of positive measure then there would be a nontrivial field of operators in $Z(\mathcal{N}_x)$ and this would contradict the fact that $Z(\mathcal{N}) = Z(\mathcal{M}) = L^\infty(X)$. To prove this rigorously we must ensure that it is possible to find a nontrivial *measurable* field of operators. This can be done as follows.

Suppose $Z(\mathcal{N}_x)$ is nontrivial, i.e., it contains an operator which is not a scalar multiple of the identity, on a set S of positive measure. Without loss of generality we can assume $S \subset X_n$ for some n. Let D be a countable dense subset of $\mathcal{H}_n$. Then for each $x \in S$ there exists $B_x \in Z(\mathcal{N}_x)$, $v, w \in D$, and $\epsilon > 0$ such that $\|B_x\| \leq 1$, $\|B_x v - w\| < \epsilon$, and $||av - w|| \geq 2\epsilon$ for every $a \in \mathbf{C}$. Since D is countable and ϵ can be assumed rational, by restricting S we can ensure that there exists a single triple (v, w, ϵ) such that for every $x \in S$, some $B_x \in Z(\mathcal{N}_x)$ satisfies the preceding. Thus, writing $D = \{v_i : i \in \mathbf{N}\}$, for each x and each $m \in \mathbf{N}$ there is a finite linear combination $\tilde{B}_m(x) = \sum a_k \tilde{A}_k(x)$ with coefficients $a_k \in \mathbf{Q}+i\mathbf{Q}$ such that $||\tilde{B}_m(x)|| \leq 1$, $||\tilde{B}_m(x)v - w|| < \epsilon$, and

$$|\langle(\tilde{B}_m(x)\tilde{A}_k(x) - \tilde{A}_k(x)\tilde{B}_m(x))v_i, v_j\rangle| < \frac{1}{m}||v_i||||v_j||$$

for all $i, j, k \leq m$. Also, for each such coefficient sequence $(a_k) \subset \mathbf{Q}+i\mathbf{Q}$ (with only finitely many terms nonzero) the set of x such that $\sum a_k \tilde{A}_k(x)$ has the above property is measurable; we can therefore find, for each m, a sequence $f_k^m \in L^\infty(S)$ such that $\sum f_k^m \tilde{A}_k = \tilde{B}_m$ has the desired property almost everywhere. Then $B_m = \sum f_k^m A_k$ belongs to $\mathcal{M}$, and $||B_m|| \leq 1$ for all m, so there is a weak* cluster point $B \in \mathcal{M}$ of the sequence (B_m), and $\pi^{-1}(B) = \tilde{B}$ evidently satisfies $\tilde{B}(x) \in Z(\mathcal{N}_x)$ but $\tilde{B}(x)v \neq av$, for almost every x and all $a \in \mathbf{C}$. Thus $B \in Z(\mathcal{M})$ but $B \notin L^\infty(X)$, a contradiction. So almost every $\mathcal{N}_x$ must be a factor. ∎

Measure theoretic complications make it difficult to formulate a meaningful version of this theorem in the nonseparable case, but it seems morally true in general.

6.6 The quantum plane and tori

We return to the quantum plane and tori. The operator analog of the set of bounded measurable functions is the weak* closure of the operator analog of the set of continuous functions (which vanish at infinity). This motivates the following definition.

DEFINITION 6.6.1 *Let $L^\infty_\hbar(\mathbf{R}^2)$ and $L^\infty_\hbar(\mathbf{T}^2)$ respectively be the weak* closures of the C*-algebras $C_0^\hbar(\mathbf{R}^2) \subset B(L^2(\mathbf{R}^2))$ and $C^\hbar(\mathbf{T}^2) \subset B(L^2(\mathbf{T}^2))$ defined in Sections 5.4 and 5.5. Also let $\hat{L}^\infty_\hbar(\mathbf{T}^2)$ be the weak* closure of $\hat{C}^\hbar(\mathbf{T}^2) \subset B(l^2(\mathbf{Z}^2))$.*

Thus $\hat{L}^\infty_\hbar(\mathbf{T}^2)$ and $L^\infty_\hbar(\mathbf{T}^2)$ are unitarily equivalent via the Fourier transform on the torus.

Our first result follows from Propositions 5.4.4 and 5.5.2, together with Proposition 6.2.2.

PROPOSITION 6.6.2
If $\hbar = 0$ then $L^\infty_\hbar(\mathbf{R}^2) \cong L^\infty(\mathbf{R}^2)$ and $L^\infty_\hbar(\mathbf{T}^2) \cong L^\infty(\mathbf{T}^2)$.

Recall the automorphisms $\theta_{s,t}$ of $B(L^2(\mathbf{R}^2))$ and $\hat{\theta}_{s,t}$ of $B(l^2(\mathbf{Z}^2))$ introduced in Sections 5.4 and 5.5. These were shown to restrict to automorphisms of $C_0^\hbar(\mathbf{R}^2)$ and $\hat{C}^\hbar(\mathbf{T}^2)$ in Propositions 5.4.5 and 5.5.5, respectively. We now note the corresponding statement for $L_\hbar^\infty(\mathbf{R}^2)$ and $\hat{L}_\hbar^\infty(\mathbf{T}^2)$.

PROPOSITION 6.6.3

For each $s,t \in \mathbf{R}$ the map $\theta_{s,t}$ restricts to a weak continuous automorphism of $L_\hbar^\infty(\mathbf{R}^2)$ and the map $\hat{\theta}_{s,t}$ restricts to a weak* continuous automorphism of $\hat{L}_\hbar^\infty(\mathbf{T}^2)$. This defines actions of $\mathbf{R}^2$ by automorphisms of $L_\hbar^\infty(\mathbf{R}^2)$ and $\hat{L}_\hbar^\infty(\mathbf{T}^2)$. Moreover, for every $A \in L_\hbar^\infty(\mathbf{R}^2)$ and $B \in \hat{L}_\hbar^\infty(\mathbf{T}^2)$ the maps $(s,t) \mapsto \theta_{s,t}(A)$ and $(s,t) \mapsto \hat{\theta}_{s,t}(B)$ are weak* continuous.*

Most of the proof of this proposition resembles the proofs of Propositions 5.4.5 and 5.5.5. Each $\theta_{s,t}$ and $\hat{\theta}_{s,t}$ is weak* continuous because it is given by conjugation with a unitary. Weak* continuity of the maps $(s,t) \mapsto \theta_{s,t}(A)$ and $(s,t) \mapsto \hat{\theta}_{s,t}(B)$ follows from the fact that the associated unitaries $U_{s,t}$ converge "strongly" to I as $s,t \to 0$ in the sense that $U_{s,t}v \to v$ for all $v \in \mathcal{H}$.

We will now show how $\hat{C}^\hbar(\mathbf{T}^2)$ is distinguished inside of $\hat{L}_\hbar^\infty(\mathbf{T}^2)$. Unfortunately, no such result is available for $C_0^\hbar(\mathbf{R}^2)$, but we will indicate a substitute.

For $A \in \hat{L}^\infty(\mathbf{T}^2)$ define Fourier coefficients $a_{k,l}(A)$, partial sums $s_N(A)$, and Cesàro means just as in Definition 5.5.4. We have the following characterization of $\hat{L}_\hbar^\infty(\mathbf{T}^2)$ (cf. Proposition 5.5.3).

PROPOSITION 6.6.4

An operator $A \in B(l^2(\mathbf{Z}^2))$ belongs to $\hat{L}_\hbar^\infty(\mathbf{T}^2)$ if and only if

$$\langle Ae_{m,n}, e_{m+k,n+l}\rangle = e^{i\hbar(ml-nk)/2}\langle Ae_{0,0}, e_{k,l}\rangle$$

for all $k,l,m,n \in \mathbf{Z}$. The partial Fourier sums $s_N(A)$ converge weak operator to A for all $A \in \hat{L}_\hbar^\infty(\mathbf{T}^2)$.

PROOF It is easy to verify that the stated equality is satisfied for $A = \hat{U}^m\hat{V}^n$ and hence for any polynomial in $\hat{U}$ and $\hat{V}$. Since these polynomials are weak* dense in $\hat{L}_\hbar^\infty(\mathbf{T}^2)$, the forward implication follows.

Conversely, let $A \in B(l^2(\mathbf{Z}^2))$ and suppose A satisfies the stated equality. Then $s_N(A) \to A$ weak operator because

$$\langle Av, w\rangle = \sum_{k,l,m,n} \langle v, e_{k,l}\rangle\langle Ae_{k,l}, e_{m,n}\rangle\langle e_{m,n}, w\rangle$$

and

$$\langle s_N(A)v, w\rangle = \sum_{|k-m|,|l-n|\le N} \langle v, e_{k,l}\rangle\langle Ae_{k,l}, e_{m,n}\rangle\langle e_{m,n}, w\rangle$$

for any $v, w \in l^2(\mathbf{Z}^2)$. Since $s_N(A)$ belongs to $\hat{L}^\infty_\hbar(\mathbf{T}^2)$ for all N, this implies that A belongs to the weak operator closure of $\hat{L}^\infty_\hbar(\mathbf{T}^2)$. But the double commutant theorem (Theorem 6.5.7) implies that $\hat{L}^\infty_\hbar(\mathbf{T}^2)$ is weak operator closed. So $A \in \hat{L}^\infty_\hbar(\mathbf{T}^2)$. ▮

PROPOSITION 6.6.5
Let

$$\mathcal{A}_1 = \{A \in \hat{L}^\infty_\hbar(\mathbf{T}^2) : (s,t) \mapsto \hat{\theta}_{s,t}(A) \text{ is norm continuous}\}$$
$$\mathcal{A}_2 = \{A \in \hat{L}^\infty_\hbar(\mathbf{T}^2) : \|A - \sigma_N(A)\| \to 0\}.$$

Then $\hat{C}^\hbar(\mathbf{T}^2) = \mathcal{A}_1 = \mathcal{A}_2$.

PROOF It is straightforward to check that $\mathcal{A}_1$ is a C*-algebra which contains $\hat{U}$ and $\hat{V}$, and therefore $\hat{C}^\hbar(\mathbf{T}^2) \subset \mathcal{A}_1$. The fact that $\mathcal{A}_2 \subset \hat{C}^\hbar(\mathbf{T}^2)$ is trivial because $\hat{C}^\hbar(\mathbf{T}^2)$ is norm closed and every $\sigma_N(A) \in \hat{C}^\hbar(\mathbf{T}^2)$. It remains to show that $\mathcal{A}_1 \subset \mathcal{A}_2$. But the only property of $\hat{C}^\hbar(\mathbf{T}^2)$ used in the proof of Theorem 5.5.7 was the fact that $\hat{\theta}_{s,t}(A)$ is norm continuous for all $A \in \hat{C}^\hbar(\mathbf{T}^2)$. Thus that argument actually shows $\mathcal{A}_1 \subset \mathcal{A}_2$. ▮

The corresponding statement for $C_0^\hbar(\mathbf{R}^2)$ and $\theta_{s,t}$ already fails in the $\hbar = 0$ case: the functions in $L^\infty(\mathbf{R}^2)$ for which translations are norm continuous are precisely the bounded uniformly continuous functions, not the continuous functions vanishing at infinity. Following this model, we can define the corresponding algebra $C_b^\hbar(\mathbf{R}^2)$ for $\hbar \neq 0$ to be the set of $A \in L^\infty_\hbar(\mathbf{R}^2)$ for which $(s,t) \mapsto \theta_{s,t}(A)$ is norm continuous. It is straightforward to check that this is a C*-algebra, and in some respects it is nicer than $C_0^\hbar(\mathbf{R}^2)$. Its principal drawback is that it is not separable.

Finally, we examine the structure of the algebras $L^\infty_\hbar(\mathbf{R}^2)$ and $L^\infty_\hbar(\mathbf{T}^2)$ when $\hbar \neq 0$. The former is quite simple:

PROPOSITION 6.6.6
If $\hbar \neq 0$ *then* $L^\infty_\hbar(\mathbf{R}^2) \cong B(L^2(\mathbf{R}))$.

PROOF We showed in Theorem 5.4.13 that $\tilde{C}_0^\hbar(\mathbf{R}^2)$, realized as a subalgebra of $B(L^2(\mathbf{R}))$, is precisely the ideal $K(L^2(\mathbf{R}))$ of compact operators. Thus its weak* closure equals $B(L^2(\mathbf{R}))$. Conjugation by the unitary W in Proposition 4.2.2 yields $L^\infty_\hbar(\mathbf{R}^2) \cong B(L^2(\mathbf{R})) \otimes I \cong B(L^2(\mathbf{R}))$. ∎

The algebras $L^\infty_\hbar(\mathbf{T}^2)$ are mutually $*$-isomorphic for all positive values of $\hbar$, despite the fact that the corresponding algebras $C^\hbar(\mathbf{T}^2)$ are not. We will not prove this result, but we will show that $L^\infty_\hbar(\mathbf{T}^2)$ is not isomorphic to any $B(\mathcal{H})$. Interestingly, however, like $B(\mathcal{H})$ it is a factor: any operator $A \in B(l^2(\mathbf{Z}^2))$ that commutes with $\hat{U}$ and $\hat{V}$ must satisfy

$$\langle Ae_{m,n}, e_{m+k,n+l}\rangle = e^{-i\hbar(ml-nk)/2}\langle Ae_{0,0}, e_{k,l}\rangle$$

for all $k, l, m, n \in \mathbf{Z}$, and it follows from Proposition 6.6.4 that the center of $\hat{L}^\infty_\hbar(\mathbf{T}^2)$ is trivial. So the same must be true of $L^\infty_\hbar(\mathbf{T}^2)$.

PROPOSITION 6.6.7
$L^\infty_\hbar(\mathbf{T}^2)$ is not $*$-isomorphic to any $B(\mathcal{H})$ for any value of $\hbar$.

PROOF Since $L^\infty_\hbar(\mathbf{T}^2) \cong \hat{L}^\infty_\hbar(\mathbf{T}^2)$ we can work with the latter. Consider the weak* continuous linear functional $\tau : \hat{L}^\infty_\hbar(\mathbf{T}^2) \to \mathbf{C}$ defined by

$$\tau(A) = \langle Ae_{0,0}, e_{0,0}\rangle.$$

If A and B are polynomials in $\hat{U}$ and $\hat{V}$ then $\tau(AB) = \tau(BA)$, so by continuity this equality holds for all $A, B \in \hat{L}^\infty_\hbar(\mathbf{T}^2)$. Also $\tau(I) = 1$.

If $\mathcal{H}$ is infinite dimensional there is no weak* continuous linear functional on $B(\mathcal{H})$ with the above properties. To see this let P and Q be rank one projections in $B(\mathcal{H})$. Then there is a unitary W such that $W^*PW = Q$. So if $\tau \in B(\mathcal{H})^*$ satisfies $\tau(AB) = \tau(BA)$ for all $A, B \in B(\mathcal{H})$ then $\tau(P) = \tau(Q)$ for all rank one projections P and Q.

Let Λ be an orthonormal basis of $\mathcal{H}$ and for any finite subset $S \subset \Lambda$ let P_S be the projection onto its span. Then (P_S) is a net of finite rank projections which converges weak* to the identity operator. But $\tau(P_S) = an$ where n is the cardinality of S and a is the value of τ on any rank one projection. So weak* continuity is inconsistent with the condition $\tau(I) = 1$. ∎

6.7 Notes

[19] and [66] are good references on von Neumann algebras. Most of the material in Section 6.6 was taken from [71].

Chapter 7

Quantum Field Theory

7.1 Fock space

The subject matter of this chapter lies outside the main line of the book. Elsewhere we use physics only as motivation for our treatment of mathematical topics, but here we take the reverse attitude. In this chapter we will apply the ideas of previous chapters to the problem of modelling free quantum fields. We choose this topic because it is here that the best argument can be made for the relevance of C*-algebras to physics. As we will explain in Section 7.5, C*-algebras are needed to construct relativistically invariant models of free quantum fields in curved spacetime.

The construction of successful models of quantum fields is predicated on a fundamental understanding of field dynamics. Unfortunately, relativistic dynamical problems are extremely difficult even in the simplest field systems, and despite half a century of heroic mathematical efforts, this area still seems not to be fully understood at a basic level.[1] For this reason we will discuss only systems with trivial dynamics, i.e., the case of free (noninteracting) fields.

We begin with nonrelativistic fields. Consider a classical real scalar field on $\mathbf{R}^3$: such a field is described by a real value at each point of space. By contrast, for example, the electromagnetic field is a vector field and is described at each point by a vector with six real dimensions, three electric and three magnetic. Fields with more than one component can be built up from the scalar case without any fundamental obstruction, so we will consider only this simplest case.

It is easier to conceptualize the transition to quantum mechanics if we temporarily replace $\mathbf{R}^3$ with a discrete set X. We can model a classical field on X by a collection of one-dimensional particles indexed by the points of X, where the "position" of the nth particle is the field

[1] In the words of [48] it is a "deep gloomy mess."

strength at the nth point. Thus, the quantum mechanical version of the field is modelled at each point of X by the Hilbert space $L^2(\mathbf{R})$ of a one-dimensional quantum mechanical particle (see Section 4.1), and described as a whole by the tensor product of a set of copies of $L^2(\mathbf{R})$ indexed by the set X.

If X is infinite, taking this tensor product requires the selection of a distinguished unit vector in each factor (Definition 2.5.1). This suggests that at each point we use the $l^2(\mathbf{N})$ model of a one-dimensional particle discussed in Section 4.3. We can then take the tensor power of $l^2(\mathbf{N})$ as in Example 2.5.7, and the result is the symmetric Fock space over $l^2(X)$. Thus, the Hilbert space of a quantum real scalar field over the discrete set X is $\mathcal{F}_s l^2(X)$.

When the underlying space is not discrete (e.g., $X = \mathbf{R}^3$), we can, as discussed in Section 2.5, retain the intuition of a "measurable tensor product over X" by taking the Hilbert space to be $\mathcal{F}_s L^2(X)$. This is the Hilbert space of a free scalar Boson field over X. Similarly, the antisymmetric Fock space $\mathcal{F}_a L^2(X)$ is the Hilbert space of a free scalar Fermion field; here the classical version involves a two-state system at each point in space, so that the Hilbert space of the quantum system is intuitively a measurable tensor product of two-dimensional Hilbert spaces. For the sake of simplicity we will discuss only Boson fields in this chapter; much of what we do can be transferred to the Fermionic case by a simple substitution of antisymmetric Fock spaces for symmetric Fock spaces.

Fock spaces are constructed as direct sums of symmetric and antisymmetric tensor powers of a Hilbert space, in our case $L^2(\mathbf{R}^3)$ (Definition 2.5.8). This leads to a particle interpretation of quantum fields which has no classical analog. Namely, we interpret an element of $L^2(\mathbf{R}^3)^{\otimes n}_s \subset \mathcal{F}_s L^2(\mathbf{R}^3)$ as describing a system of n identical particles. $L^2(\mathbf{R}^3)^{\otimes n}_s$ is called the n-particle space.

To complete the basic model of a free nonrelativistic field, we must describe its dynamics. This can be done by going back to the discrete model and letting X be a lattice in $\mathbf{R}^3$. Then we can treat the classical field as a discrete family of one-dimensional particles, write an expression for their total energy, follow the standard prescription from physics for obtaining the quantum Hamiltonian (i.e., $i\hbar$ times the generator of time evolution) from the classical energy, and finally pass to a continuous limit. Of course some care must be taken to ensure that the limit exists; we will not go into the details. The result can be expressed in terms of the unitary operator $U_t \in B(L^2(\mathbf{R}^3))$ defined by

$$(U_t f)\hat{}\,(p) = e^{-i\hbar\nu^2|p|^2 t}\hat{f}(p),$$

where $\hat{f}$ is the three-dimensional Fourier transform; namely, the n-

particle subspace $L^2(\mathbf{R}^3)\otimes_s\cdots\otimes_s L^2(\mathbf{R}^3)\subset \mathcal{F}_s l^2(\mathbf{R}^3)$ evolves according to $U_t\otimes\cdots\otimes U_t$. Here ν is a constant which arises from the strength of the coupling between adjacent particles in the discrete approximation and is interpreted as wave velocity in the continuous limit.

By comparison, a single free nonrelativistic particle in $\mathbf{R}^3$ evolves according to

$$(U_t f)\hat{}\,(p) = e^{-i\hbar|p|^2 t/2m}\hat{f}(p),$$

where m is its mass. Thus the free nonrelativistic field behaves dynamically like an ensemble of free particles, with the wave velocity ν playing the role of $(2m)^{-1/2}$.

This completes our initial discussion of free nonrelativistic quantum fields. Now we want to introduce a different model of symmetric Fock space which will be useful in the sequel. First we record some basic facts about $\mathcal{F}_s\mathcal{H}$ (similar statements hold for $\mathcal{F}_a\mathcal{H}$).

PROPOSITION 7.1.1

Let $\mathcal{H}$ be a Hilbert space.

(a) If E is a closed subspace of $\mathcal{H}$ then there is a natural isometric isomorphism between $\mathcal{F}_s\mathcal{H}$ and $\mathcal{F}_s E\otimes\mathcal{F}_s E^\perp$, and $\mathcal{F}_s E$ naturally embeds in $\mathcal{F}_s\mathcal{H}$ by the map $v\mapsto v\otimes 1$. More generally, if $\mathcal{H}=\bigoplus E_\kappa$ then $\mathcal{F}_s\mathcal{H}\cong\bigotimes\mathcal{F}_s E_\kappa$.

(b) We can identify $\mathcal{F}_s\mathcal{H}$ with the completion of the directed union $\bigcup\mathcal{F}_s E$ of the symmetric Fock spaces over all finite dimensional subspaces E of $\mathcal{H}$.

The infinite tensor product in part (a) is taken with respect to the unit vectors $1\in\mathbf{C}\subset\mathcal{F}_s E_\kappa$ (see Definition 2.5.1). Proposition 7.1.1 is a routine consequence of Proposition 2.5.3 and Definition 2.5.8. Notice that part (a) implies $\mathcal{F}_s L^2(X)\cong\mathcal{F}_s L^2(S)\otimes\mathcal{F}_s L^2(X-S)$ for any $S\subset X$, in keeping with the idea that $\mathcal{F}_s L^2(X)$ is a measurable tensor product of Hilbert spaces indexed by X.

Now $\mathcal{F}_s\mathbf{C}$ can be identified with $l^2(\mathbf{N})$ and subsequently with $L^2(\mathbf{R})$ (Proposition 4.3.1). By Proposition 7.1.1 (a), taking finite tensor powers then yields a natural isometric isomorphism of $\mathcal{F}_s\mathbf{C}^n$ with $L^2(\mathbf{R}^n)$. Our next goal is generalize this to infinite dimensions and model any symmetric Fock space as a kind of space of L^2 functions. In general, let $\mathcal{H}_\mathbf{R}$ be a real Hilbert space and let $\mathcal{H}=\mathcal{H}_\mathbf{R}\oplus i\mathcal{H}_\mathbf{R}$ be its complexification. We want to say $\mathcal{F}_s\mathcal{H}\cong L^2(\mathcal{H}_\mathbf{R})$. Of course $L^2(\mathcal{H}_\mathbf{R})$ does not immediately make sense since $\mathcal{H}_\mathbf{R}$ does not carry a natural measure, but it can be defined as a limit over finite dimensional subspaces of $\mathcal{H}_\mathbf{R}$. This requires us to coordinate the L^2 functions on different finite dimensional subspaces.

Let $\hbar$ be a fixed positive real number.

DEFINITION 7.1.2 *Let $\mathcal{H}_{\mathbf{R}}$ be a real Hilbert space. On each finite dimensional subspace $E_{\mathbf{R}}$ of $\mathcal{H}_{\mathbf{R}}$ define the function $h_{E_{\mathbf{R}}}(v) = e^{-\|v\|^2/2\hbar}$ and let $\mu = \mu_{E_{\mathbf{R}}}$ be $(\hbar\pi)^{-n/2}$ times Lebesgue measure, where n is the dimension of $E_{\mathbf{R}}$. Then $\|h_{E_{\mathbf{R}}}\| = 1$ in $L^2(E_{\mathbf{R}}, \mu)$, and if $E_{\mathbf{R}} = E^1_{\mathbf{R}} \oplus E^2_{\mathbf{R}}$ then $h_{E_{\mathbf{R}}} = h_{E^1_{\mathbf{R}}} \otimes h_{E^2_{\mathbf{R}}}$. Thus the map $f \mapsto f \otimes h_{E^2_{\mathbf{R}}}$ isometrically embeds $L^2(E^1_{\mathbf{R}}, \mu)$ in $L^2(E_{\mathbf{R}}, \mu)$ and these embedding maps are consistent. We therefore define $L^2(\mathcal{H}_{\mathbf{R}})$ to be the completion of the directed union $\bigcup L^2(E_{\mathbf{R}})$ over all finite dimensional subspaces $E_{\mathbf{R}}$ of $\mathcal{H}_{\mathbf{R}}$.*

Observe that $h_{E_{\mathbf{R}}}$ can actually be defined on any subspace of $\mathcal{H}_{\mathbf{R}}$. Thus, for any $f \in L^2(E_{\mathbf{R}}, \mu)$ ($E_{\mathbf{R}}$ finite dimensional) let $\tilde{f} : \mathcal{H}_{\mathbf{R}} \to \mathbf{C}$ be the function $\tilde{f} = f \otimes h_{E^{\perp}_{\mathbf{R}}}$. The map $f \mapsto \tilde{f}$ is consistent with the embeddings in Definition 7.1.2, so it is possible to regard $L^2(\mathcal{H}_{\mathbf{R}})$ as the completion of the set of functions of the form $\tilde{f}$ for $f \in L^2(E_{\mathbf{R}}, \mu)$ with $E_{\mathbf{R}}$ a finite dimensional subspace of $\mathcal{H}_{\mathbf{R}}$. Thus, although $L^2(\mathcal{H}_{\mathbf{R}})$ is not actually a space of functions of $\mathcal{H}_{\mathbf{R}}$, it has a dense subspace which is.

Also note that each $L^2(E_{\mathbf{R}}, \mu)$ is naturally isometric to $L^2(E_{\mathbf{R}}, m)$, where m is Lebesgue measure, by the map $f \mapsto (\hbar\pi)^{-n/4} f$. By the comment made after Proposition 7.1.1, we see that there is a natural isomorphism $\mathcal{F}_s E \cong L^2(E_{\mathbf{R}}, \mu)$ for any finite dimensional real Hilbert space $E_{\mathbf{R}}$, where $E = E_{\mathbf{R}} \oplus iE_{\mathbf{R}}$. Taking direct limits and applying Proposition 7.1.1 (b) yields the following.

PROPOSITION 7.1.3
Let $\mathcal{H}_{\mathbf{R}}$ be a real Hilbert space and let $\mathcal{H} = \mathcal{H}_{\mathbf{R}} \oplus i\mathcal{H}_{\mathbf{R}}$. Then there is a natural isomorphism between $\mathcal{F}_s\mathcal{H}$ and $L^2(\mathcal{H}_{\mathbf{R}})$.

7.2 CCR algebras

The $L^2(\mathcal{H}_{\mathbf{R}})$ model of Fock space allows us to define field observables which are analogous to the position and momentum operators of the one-dimensional particle. Namely, for each $v \in \mathcal{H}_{\mathbf{R}}$ define operators $\mathcal{Q}_v$ and $\mathcal{P}_v$ on $L^2(\mathcal{H}_{\mathbf{R}})$ by

$$\mathcal{Q}f(w) = \langle w, v\rangle f(w) \qquad \mathcal{P}f(w) = -i\hbar\frac{\partial f}{\partial v}(w),$$

where $f \in L^2(\mathcal{H}_{\mathbf{R}})$ is a function on $\mathcal{H}_{\mathbf{R}}$ of the type described in the remark following Definition 7.1.2. Equivalently, given a complex Hilbert space $\mathcal{H}$ and $v \in \mathcal{H}$, let E be the complex span of v and write $\mathcal{F}_s(\mathcal{H}) \cong \mathcal{F}_s(E) \otimes \mathcal{F}_s(E^{\perp})$; then $\mathcal{Q}_v = \mathcal{Q} \otimes I$ and $\mathcal{P}_v = \mathcal{P} \otimes I$, after the usual identification of $\mathcal{F}_s(E) \cong \mathcal{F}_s(\mathbf{C})$ with $L^2(\mathbf{R})$. This second definition makes sense for any v in $\mathcal{H}$, not just $\mathcal{H}_{\mathbf{R}}$, and is clearly independent of the decomposition $\mathcal{H} = \mathcal{H}_{\mathbf{R}} \oplus i\mathcal{H}_{\mathbf{R}}$. However, the first definition is probably easier to visualize. We think of $\mathcal{Q}_v$ as measuring the strength of the v component of the field and we think of $\mathcal{P}_v$ as measuring its rate of change.

For the one-dimensional particle we went beyond $\mathcal{Q}$ and $\mathcal{P}$ and constructed C*- and von Neumann algebras of observables (Sections 5.4 and 6.6). (This is a slight abuse of language. Only the self-adjoint elements of the algebras, at most, should be regarded as genuine observables.) Now we would like to do the same thing for a scalar field, but we encounter the following difficulty. In the case of a one-dimensional particle the classical phase space is $\mathbf{R}^2$, and classical observables are functions on the plane. For the corresponding quantum mechanical system we have a C*-algebra $C_0^\hbar(\mathbf{R}^2)$ of observables which reduces, in the case $\hbar = 0$, to the continuous functions on $\mathbf{R}^2$ which vanish at infinity. But for fields, the classical phase space is infinite dimensional and hence not locally compact. So there is no sensible notion of "continuous functions vanishing at infinity."

Thus, we need to find a substitute for the classical algebra $C_0(X)$. In order to do this we must take into account the structure of the phase space of a free classical field. Namely, it is always a <u>symplectic space</u>. That is, it is a real vector space $\mathcal{V}$ equipped with a <u>symplectic form</u>, an antisymmetric bilinear map $\{\cdot,\cdot\} : \mathcal{V} \times \mathcal{V} \to \mathbf{R}$. We topologize $\mathcal{V}$ with the weakest topology that makes the map $w \mapsto \{v, w\}$ continuous for all $v \in \mathcal{V}$.

Generally, in the nonrelativistic case we can take $\mathcal{V}$ to be a complex Hilbert space, with symplectic form $\{v, w\} = \mathrm{Im}\langle v, w\rangle$. For the free scalar field $\mathcal{V}$ would just be $L^2(\mathbf{R}^3)$, where we identify the real functions in $L^2(\mathbf{R}^3)$ with the possible configurations of the classical field and the purely imaginary functions with their time derivatives. This is analogous to identifying the phase space of a one-dimensional particle with the complex plane, taking position to be the real axis and momentum to be the imaginary axis.

However, this construction is not canonical, and as we will see in Section 7.5, in curved spacetime it breaks down completely. At that point $\mathcal{V}$ will really have only a symplectic structure, but even then we will be able to embed it as a real linear subspace of a complex Hilbert space, in such a way that the symplectic form agrees with the imaginary part of the inner product. So this is the most general class of symplectic spaces we will need to consider. In this situation we always take $\{v, w\} = \mathrm{Im}\langle v, w\rangle$.

Now on any symplectic space there is a rich supply of well-behaved exponential functions $\varepsilon_v : w \mapsto e^{i\{v,w\}}$. We can use these functions to build a nice algebra of continuous functions which can then be deformed to yield a C*-algebra of observables for a quantum field.

The functions ε_v do not vanish at infinity, but they are periodic. Moreover, linear combinations of the ε_v are still "almost" periodic in the following sense.

DEFINITION 7.2.1 *Let $\mathcal{V}$ be a symplectic space. For $v \in \mathcal{V}$ define $T_v : l^\infty(\mathcal{V}) \to l^\infty(\mathcal{V})$ by $(T_v f)(w) = f(w - v)$. A function $f \in l^\infty(\mathcal{V})$ is almost periodic if the set $\{T_v f : v \in \mathcal{V}\}$ is precompact in sup norm. The set of continuous almost periodic functions is denoted $AP(\mathcal{V})$.*

The point is that when $\mathcal{V}$ is infinite dimensional we can use $AP(\mathcal{V})$ in place of (the nonexistant) $C_0(\mathcal{V})$.

Observe that $T_u\varepsilon_v = e^{-i\{v,u\}}\varepsilon_v$, so the set $\{T_u\varepsilon_v : u \in \mathcal{V}\}$ is homeomorphic to a circle, and hence is compact. Thus ε_v is indeed almost periodic. The fact that linear combinations of the ε_v are also almost periodic follows from the next result.

PROPOSITION 7.2.2
Let $\mathcal{V}$ be a symplectic space. Then $AP(\mathcal{V})$ is a unital C-subalgebra of $l^\infty(\mathcal{V})$.*

PROOF First we show that $AP(\mathcal{V})$ is a $*$-algebra. It is clear that any scalar multiple of an almost periodic function is almost periodic, as is its complex conjugate. If f and g are almost periodic then the set

$$\{(T_v f, T_w g) : v, w \in \mathcal{V}\} \subset l^\infty(\mathcal{V}) \oplus l^\infty(\mathcal{V})$$

is precompact, so its images under the sum and product maps, which contain $\{T_v(f+g) : v \in \mathcal{V}\}$ and $\{T_v(fg) : v \in \mathcal{V}\}$, are also precompact. We conclude that $f + g$ and fg are also almost periodic, and this shows that $AP(\mathcal{V})$ is a $*$-algebra.

Next we prove norm closure. Let (f_n) be a sequence of almost periodic functions which converges in norm to $f \in l^\infty(\mathcal{V})$. Given $\epsilon > 0$, choose n such that $\|f - f_n\| \leq \epsilon/3$ and find $v_1, \ldots, v_k \in \mathcal{V}$ such that every $T_v f_n$ $(v \in \mathcal{V})$ is within $\epsilon/3$ of $T_{v_i} f_n$ for some i. Then we have

$$\|T_v f - T_{v_i} f\| \leq \|T_v(f - f_n)\| + \|T_v f_n - T_{v_i} f_n\| + \|T_{v_i}(f_n - f)\| \leq \epsilon.$$

This shows that $\{T_v f : v \in \mathcal{V}\}$ is totally bounded and hence precompact. Thus $AP(\mathcal{V})$ is closed in norm, so it is a C*-subalgebra of $l^\infty(\mathcal{V})$. It is clear that $AP(\mathcal{V})$ is unital. ∎

In fact, if $\mathcal{H}$ is a complex Hilbert space with the standard symplectic form $\mathrm{Im}\langle\cdot,\cdot\rangle$, then $AP(\mathcal{H})$ is precisely the closed linear span of the functions ε_v. One direction of this assertion is clear, as every ε_v belongs to $AP(\mathcal{H})$. For the converse direction one needs to develop a theory of Fourier expansion of almost periodic functions. It turns out that any almost periodic function has a countably supported Fourier transform defined on $\mathcal{H}$, and can be approximated in norm by functions with finitely supported Fourier transforms, i.e., linear combinations of the ε_v.

Now suppose $\mathcal{V}$ is only a real linear subspace of a complex Hilbert space $\mathcal{H}$ with the inherited symplectic form. We next define an algebra of quantum observables, analogous to $AP(\mathcal{V})$, which acts on $\mathcal{F}_s\mathcal{H}$, and then we show that these algebras depend only on $\mathcal{V}$, not on $\mathcal{H}$.

DEFINITION 7.2.3 *Let $\mathcal{H}$ be a complex Hilbert space and let $\mathcal{V}$ be a real linear subspace of $\mathcal{H}$. For each $v \in \mathcal{V}$ let $W_v = e^{i\mathcal{Q}_v}$ where $\mathcal{Q}_v$ is the field operator introduced at the beginning of this section. Equivalently, let E be the complex span of v, let $E_{\mathbf{R}}$ be the real span of v, and identify $\mathcal{F}_sE$ with $L^2(E_{\mathbf{R}})$; then W_v is defined on $\mathcal{F}_s\mathcal{H} \cong \mathcal{F}_sE \otimes \mathcal{F}_sE^\perp$ by $W_v = M_f \otimes I$ where $f(tv) = e^{it\|v\|}$ and M_f is the multiplication operator acting on $L^2(E_{\mathbf{R}})$. These are called <u>Weyl operators</u>. The <u>CCR algebra</u> is the C*-algebra $CCR(\mathcal{V})$ generated by the operators W_v for $v \in \mathcal{V}$.*

The preceding construction is more subtle than it first appears. Suppose $\mathcal{H} = \mathbf{C}$ is one-dimensional and let $\{e_1, e_2\}$ be its canonical real basis, so that $e_2 = ie_1$. Then W_{e_1} and W_{e_2} can both be represented as the multiplication operator $M_{e^{ix}}$, but this is by way of different identifications of $\mathcal{F}_s\mathbf{C}$ with $L^2(\mathbf{R})$. The Hermite function h_n in the e_1 picture corresponds to the vector $e_1^{\otimes n} \in \mathcal{F}_s\mathbf{C}$, whereas h_n in the e_2 picture corresponds to $e_2^{\otimes n} = i^n e_1^{\otimes n}$. So the two identifications are related by the unitary map $U : L^2(\mathbf{R}) \to L^2(\mathbf{R})$ which takes the nth Hermite function h_n to $i^n h_n$.

By considering the actions of $\mathcal{Q}$ and $\mathcal{P}$ on the Hermite functions (which can be deduced from the expressions for $(\mathcal{Q} + i\mathcal{P})h_n$ and $(\mathcal{Q} - i\mathcal{P})h_n$ given after Proposition 4.3.1), it is easily seen that $U^{-1}\mathcal{Q}U = \mathcal{P}$. Thus in the e_1 picture we have $W_{e_1} = e^{i\mathcal{Q}}$ and $W_{e_2} = U^{-1}e^{i\mathcal{Q}}U = e^{i\mathcal{P}}$.

This analysis allows us to deduce commutation relations for the Weyl operators. If v and w are (complex) orthogonal, or if w is a real multiple of v, it is easy to see that W_v and W_w commute. On the other hand, if $\|v\| = 1$ then the preceding comment (together with Theorem 4.1.3 (c)) shows that $W_{sv}W_{itv} = e^{-i\hbar st}W_{itv}W_{sv}$. By linearity this generalizes to the following result:

PROPOSITION 7.2.4
Let $\mathcal{V}$ be as in Definition 7.2.3. For any $v, w \in \mathcal{V}$ we have $W_v W_w = e^{i\hbar\{v,w\}} W_w W_v$.

This is a generalization of the Weyl form of the canonical commutation relations presented in Theorem 4.1.3 (c).

Next we show that the algebra $CCR(\mathcal{V})$ is uniquely determined by the Weyl relations. This fact requires that $\mathcal{V}$ be a real linear subspace of a complex Hilbert space and that $\mathcal{V}$ be nondegenerate in the sense that for every nonzero $v \in \mathcal{V}$ there exists $w \in \mathcal{V}$ such that $\{v, w\} \neq 0$.

THEOREM 7.2.5
Let $\mathcal{V}$ be a nondegenerate real linear subspace of a complex Hilbert space and let $\tilde{W}_v$ ($v \in \mathcal{V}$) be unitary operators which act on some other Hilbert space and satisfy $\tilde{W}_v \tilde{W}_w = e^{i\hbar\{v,w\}} \tilde{W}_w \tilde{W}_v$. Then the C-algebra the $\tilde{W}_v$ generate is $*$-isomorphic to $CCR(\mathcal{V})$.*

PROOF Let $\mathcal{A}$ be the C*-algebra generated by the $\tilde{W}_v$. Our proof of uniqueness uses the notion of $\mathcal{A}$-valued almost periodic functions on $\mathcal{V}$. These can be defined as functions from $\mathcal{V}$ to $\mathcal{A}$ whose composition with any $\omega \in \mathcal{A}^*$ is almost periodic, or one can directly mimic Definition 7.2.1, replacing $l^\infty(\mathcal{V})$ with the set of bounded $\mathcal{A}$-valued functions on $\mathcal{V}$. The key fact that we need is that every continuous almost periodic function f possesses a mean value $\tau(f) \in \mathcal{A}$ with the following properties: (1) $\|\tau(f)\| \leq \|f\|$; (2) $f \geq 0$ implies $\tau(f) \geq 0$; (3) $\tau(af + bg) = a\tau(f) + b\tau(g)$ for any $a, b \in \mathbf{C}$; (4) $\tau(T_v f) = \tau(f)$ for all v; and (5) if $f(v) = A$ constantly then $\tau(f) = A$. This is a general fact which is true of almost periodic functions from any group into a Banach space.

Now define an action θ of $\mathcal{V}$ on $\mathcal{A}$ by letting θ_w be conjugation with $\tilde{W}_w$, i.e., $\theta_w(A) = \tilde{W}_w^{-1} A \tilde{W}_w$. For each $v \in \mathcal{V}$ the map $f_{\tilde{W}_v} : w \mapsto \theta_w(\tilde{W}_v)$ is periodic, and an argument like the one in Proposition 7.2.2 then shows that the map $f_A : w \mapsto \theta_w(A)$ is almost periodic for all $A \in \mathcal{A}$. So it has a mean value $\tau(f_A)$. We have $\|\tau(f_A)\| \leq \|A\|$, so the map $A \mapsto \tau(f_A)$ is continuous. Also, when $A = \tilde{W}_v$ we have $f_A = \varepsilon_{\hbar v} A$, and properties (3) and (4) of the mean value imply

$$\tau(f_A) = \tau(T_w f_A) = e^{i\hbar\{w,v\}} \tau(f_A),$$

so that nondegeneracy of $\mathcal{V}$ implies $\tau(f_A) = 0$ unless $v = 0$, when $\tau(f_A) = I$ by property (5). By continuity it follows that $\tau(f_A) \in \mathbf{C} \cdot I$ for any $A \in \mathcal{A}$. Also, $A \geq 0$ implies $\tau(f_A) \geq 0$. We can now prove that $\mathcal{A}$

has no proper C*-ideals by the reasoning used in the proof of Theorem 5.5.8, and conclude that $\mathcal{A} \cong CCR(\mathcal{V})$ just as in Corollary 5.5.9. ∎

This uniqueness result shows in particular that the construction of $CCR(\mathcal{V})$ in Definition 7.2.3 depends only on $\mathcal{V}$, and not on the ambient Hilbert space $\mathcal{H}$. However, it is important to note that embeddings of $\mathcal{V}$ in different Hilbert spaces will generally give rise, via Definition 7.2.3, to representations of $CCR(\mathcal{V})$ which are *not* unitarily equivalent. In particular, the von Neumann algebra generated by the Weyl operators does depend on $\mathcal{H}$.

We can now introduce a representation of $CCR(\mathcal{V})$ which is similar to the $L^2(\mathbf{R}^2)$ representation given in Section 4.2. For $v \in \mathcal{V} \subset \mathcal{H}$ define a unitary operator L_v on $L^2(\mathcal{H})$ (in the sense of Definition 7.1.2, treating $\mathcal{H}$ as a real Hilbert space) by

$$L_v f(w) = e^{i\mathrm{Im}\langle v,w\rangle/2} f(w - \hbar v).$$

These are analogous to the operators L_{t_1,t_2}, although our conventions here are slightly different. Since they obey the same commutation relations as the W_v, it follows from Theorem 7.2.5 that the C*-algebra they generate is $*$-isomorphic to $CCR(\mathcal{V})$. But, as in the case of the quantum plane, this representation allows a definition of $CCR(\mathcal{V})$ when $\hbar = 0$.

(The above formula also defines a representation of $CCR(\mathcal{V})$ on the non-separable Hilbert space $l^2(\mathcal{H})$; this is the GNS representation construction analogous to the one in Example 5.6.7 (a).)

If we set $\hbar = 0$ then L_v is just multiplication by $\varepsilon_{v/2}$. So the next result is a consequence of the comment following Proposition 7.2.2.

PROPOSITION 7.2.6
Let $\mathcal{H}$ be a complex Hilbert space. If $\hbar = 0$ then $CCR(\mathcal{H}) \cong AP(\mathcal{H})$.

Thus, for $\hbar > 0$ the C*-algebra $CCR(\mathcal{H})$ is a "deformation" of $AP(\mathcal{H})$ in the same way that $C_0^\hbar(\mathbf{R}^2)$ and $C^\hbar(\mathbf{T}^2)$ are deformations of $C_0(\mathbf{R}^2)$ and $C(\mathbf{T}^2)$.

7.3 Relativistic particles

In this section we address the question of how free particles behave in Minkowski spacetime. The corresponding field theory will be described in Section 7.4.

In special relativity, each inertial observer sees a three-dimensional slice of spacetime at each instant of time. Thus the state of a spinless particle at any moment is described by a normalized function in $L^2(\mathbf{R}^3)$. (Spin is accomodated by instead using a direct sum of some number of copies of $L^2(\mathbf{R}^3)$; for simplicity we will stick to the spinless case.) Furthermore, complete knowledge of the particle's state at any given moment determines its entire past and future. Thus $L^2(\mathbf{R}^3)$ models the phase space of the system, just as in nonrelativistic quantum mechanics. We must now describe how a state given at time $t = 0$ evolves as t changes, and also how it will appear to other inertial observers.

The time evolution of a free relativistic particle is similar to the nonrelativistic case (Section 7.1). As there, the evolution operators U_t are diagonal (i.e., appear as multiplication operators) in the Fourier transform picture; the appropriate relativistic formula is

$$(U_t f)\hat{}\,(p) = e^{-i(\hbar^2 c^2 |p|^2 + m^2 c^4)^{1/2} t/\hbar} \hat{f}(p) = e^{-ip_0 t} \hat{f}(p)$$

where c is the speed of light, m is the mass of the particle (assumed throughout to be strictly positive), and $p_0 = (c^2|p|^2 + m^2c^4\hbar^{-2})^{1/2}$ is $\hbar^{-1}$ times the relativistic energy of a quantum mechanical particle with mass m and momentum $\hbar p = \hbar(p_1, p_2, p_3)$.

Now we must say how states transform under Lorentz transformations. That is, if L is a 4×4 matrix which leaves the relativistic length $c^2x_0^2 - x_1^2 - x_2^2 - x_3^2$ invariant (here $x_0 = t$ is the time coordinate), then we must determine how the state of a particle given on the $t = 0$ slice in the old frame appears on the $t = 0$ slice in the new frame, i.e., the set of points $\{L^{-1}(0, x) : x = (x_1, x_2, x_3) \in \mathbf{R}^3\}$. We assume throughout that L preserves the direction of time and the orientation of space.

The way to proceed is partially clarified by considering "momentum eigenstates" of the form $e^{ip\cdot x}$ where $x = (x_1, x_2, x_3)$ is the space variable and $p = (p_1, p_2, p_3)$ is fixed. Of course, these functions do not belong to $L^2(\mathbf{R}^3)$, so they are not really states at all, but we can imagine them as being approximated by L^2 functions if we wish. In the Fourier transform picture these momentum eigenstates appear as delta functions concentrated at p. Now the point is that under time evolution they remain delta functions, and we expect that in *any* frame, at *any* time they will appear as momentum eigenstates. In particular, in a new frame related to the original one by a Lorentz transformation L, we expect an eigenstate of momentum $\hbar p$ to transform into one of momentum $\hbar p'$ such that $(p'_0, p') = \tilde{L}(p_0, p)$, where $\tilde{L} = (L^{-1})^T$ is the inverse transpose of L. (The transpose arises because we have taken the Fourier transform.)

It may seem that the preceding completely determines how states should transform: given a state f, one might think, we simply have to take the Fourier transform of f, transform each momentum value in the above manner, and finally apply the inverse Fourier transform. However, there is an ambiguity due to the fact that momentum eigenstates are not L^2 functions. If they were, we would have to normalize them before carrying out the above prescription. Thus, we actually expect only to transform $N(p)e^{ip\cdot x}$ to $N(p')e^{ip'\cdot x}$ where $N(p)$ is some "normalizing" factor.

The value of $N(p)$ is determined by the requirement that transformation of states must be unitary. It is helpful here to consider the mass shell

$$X_m = \{(p_0, p) : p_0 = (c^2|p|^2 + m^2c^4\hbar^{-2})^{1/2}\},$$

which is an orbit of the group of Lorentz transformations acting in momentum space in the inverse transpose manner described above. Projection onto the p coordinates defines a homeomorphism between X_m and $\mathbf{R}^3$, so Lebesgue measure dp on $\mathbf{R}^3$ can be transferred to X_m. However, this measure is not Lorentz invariant; the measure dp/p_0 is. Thus, composition with any $\tilde{L}$ defines a unitary map on $L^2(X_m, dp/p_0)$.

(The sense of the notation dp/p_0 is the following. If a function $f(p_0, p)$ is defined on X_m then its integral with respect to the measure dp/p_0 is $\int_{\mathbf{R}^3} f(p_0, p)\, dp/p_0$.)

Therefore, we define a unitary map $T : L^2(\mathbf{R}^3) \to L^2(X_m, dp/p_0)$ by

$$(Tf)(p_0, p) = \sqrt{p_0}\hat{f}(p),$$

we let $V_L : L^2(X_m) \to L^2(X_m)$ be the operator of composition with $\tilde{L}^{-1}$, and we define the action of L on the original $L^2(\mathbf{R}^3)$ by $U_L = T^{-1}V_LT$. That is, U_Lf is defined by passing to $L^2(X_m)$ via T, composing with $\tilde{L}^{-1}$, and then passing back to $L^2(\mathbf{R}^3)$. This is manifestly a unitary transformation, and it permutes momentum eigenstates in the desired manner. We see that the scalar $N(p)$ equals $p_0^{-1/2}$.

This completes our description of relativistic particles. But the time evolution operators U_t defined near the beginning of this section have a peculiar property which is worth discussing. Suppose $f \in L^2(\mathbf{R}^3)$ has unit norm and is supported on a compact set. Then $\hat{f} \in L^2(\mathbf{R}^3)$ is real analytic, meaning that it is the restriction to $\mathbf{R}^3$ of the function

$$\hat{f}(z) = \frac{1}{(2\pi)^{3/2}} \int f(x)e^{-ix\cdot z}\, dx$$

on $\mathbf{C}^3$, which is separately analytic in z_1, z_2, and z_3. (This follows from the fact that $f \in L^1(\mathbf{R}^3)$ and so we can differentiate under the integral sign.) Now if this is the case, then $(U_tf)\hat{}$ *cannot* be real analytic for any nonzero value of t, because of the square root in the exponent which defines U_t. Thus the state which equals f at time $t = 0$ will *not* be supported in any compact subset of position space for any $t \neq 0$.

This phenomenon has serious implications. Let K_1 and K_2 be two well-separated compact spatial regions and let P_1 and P_2 be the orthogonal projections of $L^2(\mathbf{R}^3)$ onto $L^2(K_1)$ and $L^2(K_2)$. Then fix t small enough that light cannot travel from K_1 to K_2 in time t, and define

$P_2' = U_t^{-1} P_2 U_t$. We claim that P_1 and P_2' do not commute in general. To see this, find a state f which is supported on K_1 at time 0 and has a nonzero component in K_2 at time t; if $P_1 P_2' f = P_2' P_1 f = P_2' f$ then $P_2' f$ would be supported on K_1 at time 0 and on K_2 at time t, violating the conclusion of the preceding paragraph.

One can show that an agent located in K_1 could use this effect to send a faster-than-light signal to an agent located in K_2, violating relativistic causality. This is because, given a particle whose prior state is known, merely measuring whether the particle is in K_1 at time 0 can alter the probability that it is in K_2 at time t in such a way that the second agent could infer that a measurement had been made, regardless of the result of the first measurement.

The moral is that one cannot make local measurements of individual particles in relativistic quantum mechanics! This point will become more clear in the next section when we discuss the quantum field observables which actually can be measured locally. It turns out that these local measurements, when applied to a quantum field in a single-particle state, always yield a result which has a nonzero many-particle component (though this component can be made arbitrarily small by making the measurement over a large enough region). The phenomenon is similar to what happens when one measures whether a diagonally polarized photon is horizontally polarized: the result of such a measurement is either a horizontally or vertically polarized photon, so the photon cannot remain in a diagonally polarized state. The difference is that in the photon example there is no locality requirement which prevents one from directly measuring diagonal polarization.

Thus, while we do have a single-particle special relativistic theory, it is incompatible with local observers. (*General* relativity is more severely inconsistent with single particles; see Section 7.5.)

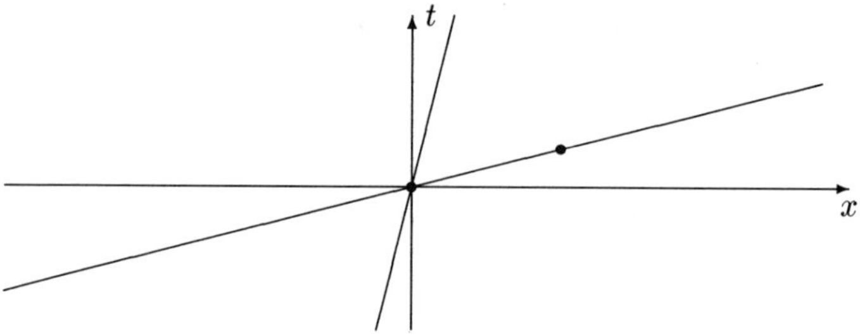

Figure 7.1 "A particle which travels faster than the speed of light appears twice in some frame"

Other, incorrect inferences have also been drawn. For instance, the argument is sometimes made that a particle that travels faster than light must appear twice in some reference frame, and hence one cannot have a single-particle relativistic model. A related argument claims to demonstrate that relativity implies the existence of antiparticles. The idea here is that if a particle localized in region K_1 travels faster than light to another region K_2, then in some other frame it would appear as an antiparticle travelling from K_2 to K_1. Although it undeniably stimulates the imagination, this argument too is wrong. The model presented in this section is an explicit counterexample to both of these suggestions: although it does exhibit the kind of superluminal travel at issue, it is manifestly a single-particle model, with no other particles or antiparticles. Thus, the most one can say is that relativity implies the existence of phenomena which are reminiscent of multiple particles or antiparticles. But even this seems misleading.

7.4 Flat spacetime

The free scalar field in flat (Minkowski) spacetime is easily described. As in the nonrelativistic case (Section 7.1), its Hilbert space is $\mathcal{F}_s L^2(\mathbf{R}^3)$. Its dynamics are based on the operators U_t and U_L given in Section 7.3. These give rise to unitary operators on $\mathcal{F}_s L^2(\mathbf{R}^3)$ in the natural way: time evolution and Lorentz transformations act on the n-particle space $L^2(\mathbf{R}^3)_s^{\otimes n}$ by $U_t^{\otimes n}$ and $U_L^{\otimes n}$.

Also as in the nonrelativistic case (Section 7.2), we have observables $\mathcal{Q}_v$ and $\mathcal{P}_v$ and Weyl operators W_v for $v \in L^2(\mathbf{R}^3)$. The intuition for $\mathcal{Q}_v$ and $\mathcal{P}_v$ mentioned in Section 7.2, that they respectively measure the strength and rate of change of the v component of the field, is still valid. However, this does not tell us which observables can be measured in which regions of space, an issue of some urgency given the paradox discussed at the end of Section 7.3. To answer this question we will need to correlate quantum observables with elements of classical phase space; then a quantum observable will be measurable within some region K if and only if the corresponding classical field state is supported in K.

Any state of the classical field can be described at a given time by the field strength f and its time derivative $f_\bullet$. For definiteness let us suppose f and $f_\bullet$ are real-valued functions in $C_c^\infty(\mathbf{R}^3)$. Thus the classical phase space $\mathcal{V}$ is the set of all such pairs $(f, f_\bullet)$.

Now given $(f, f_\bullet)$ we want to identify a vector $v = T^q(f, f_\bullet) \in L^2(\mathbf{R}^3)$ such that $\mathcal{Q}_v$ and $\mathcal{P}_v$ can be interpreted as the strength and rate of change of the classical mode $(f, f_\bullet)$ of the quantum field. In this way we will correlate quantum observables with elements of classical phase space much as the operators $\mathcal{Q}$ and $\mathcal{P}$ correspond to classical position and momentum.

There are three natural consistency conditions that can be placed on this classical labelling $T^q : \mathcal{V} \to L^2(\mathbf{R}^3)$ of quantum observables: it should respect time evolution, the action of the Lorentz group, and symplectic structure. It turns out that these requirements completely determine T^q.

The dynamics of the classical field are described by the Klein-Gordon equation

$$\frac{1}{c^2}\frac{\partial^2 f}{\partial x_0^2} - \frac{\partial^2 f}{\partial x_1^2} - \frac{\partial^2 f}{\partial x_2^2} - \frac{\partial^2 f}{\partial x_3^2} = -\left(\frac{mc}{\hbar}\right)^2 f.$$

It follows that in the Fourier transform picture the time derivative of $\hat{f}$ is $\hat{f}_\bullet$ and the time derivative of $\hat{f}_\bullet$ is $-p_0^2\hat{f}$. Lorentz transformations act in both the classical and quantum pictures by composition. Finally, the symplectic structure on $\mathcal{V}$ is given by

$$\{(f, f_\bullet), (g, g_\bullet)\} = \int_{\mathbf{R}^3} f_\bullet g - f g_\bullet$$

and on $L^2(\mathbf{R}^3)$ it is the imaginary part of the inner product.

The unique (up to multiplication by a scalar of modulus one) quantization map T^q which satisfies the three consistency conditions is

$$T^q(f, f_\bullet)\hat{} = p_0^{1/2}\hat{f} + i p_0^{-1/2}\hat{f}_\bullet.$$

Consistency with time evolution follows from the computations

$$T^q(f_t, (f_\bullet)_t)\hat{} = p_0^{1/2}\hat{f}_t + i p_0^{-1/2}(\hat{f}_\bullet)_t = -i p_0^{3/2}\hat{f} + p_0^{1/2}\hat{f}_\bullet$$

and

$$(T^q(f, f_\bullet)\hat{})_t = -i p_0 (p_0^{1/2}\hat{f} + i p_0^{-1/2}\hat{f}_\bullet).$$

Invariance under Lorentz transformations follows from the computation

$$T^q(f \circ L^{-1}, f_\bullet \circ L^{-1})\hat{} = T^q(f, f_\bullet)\hat{} \circ L^T = T^q(f, f_\bullet)\hat{} \circ \tilde{L}^{-1}.$$

Finally, preservation of symplectic structure follows from the computations

$$\{(f, f_\bullet), (g, g_\bullet)\} = \int f_\bullet g - f g_\bullet = \int \hat{f}_\bullet \bar{\hat{g}} - \hat{f} \bar{\hat{g}}_\bullet$$

and

$$\begin{aligned}\operatorname{Im}\langle T^q(f, f_\bullet), T^q(g, g_\bullet)\rangle &= \operatorname{Im}\int (p_0^{1/2}\hat{f} + i p_0^{-1/2}\hat{f}_\bullet)\overline{(p_0^{1/2}\hat{g} + i p_0^{-1/2}\hat{g}_\bullet)} \\ &= \operatorname{Im}\int (p_0 \hat{f}\bar{\hat{g}} + i\hat{f}_\bullet\bar{\hat{g}} - i\hat{f}\bar{\hat{g}}_\bullet + p_0^{-1}\hat{f}_\bullet\bar{\hat{g}}_\bullet).\end{aligned}$$

Thus T^q has the desired properties and this enables us to associate to any $(f, f_\bullet) \in \mathcal{V}$ an element of $L^2(\mathbf{R}^3)$ which then gives rise to a Weyl operator on $\mathcal{F}_s L^2(\mathbf{R}^3)$. The important point here is that if $(f, f_\bullet)$ and $(g, g_\bullet)$ are disjointly supported then $\{(f, f_\bullet), (g, g_\bullet)\} = 0$ and so the corresponding Weyl operators commute. Moreover, since classical solutions of the Klein-Gordon equation propagate at the speed of light, this remains true for Weyl operators associated to classical fields supported on spatial regions at different times that are sufficiently separated to prevent light travelling between them. The commutation of these "local" operators implies that one cannot use the measurement process to send a faster-than-light signal.

7.5 *Curved spacetime*

Free quantum field theory can also be formulated against a curved spacetime background. This means that we replace Minkowski spacetime with a four-manifold M satisfying the following conditions: (1) M is smooth; (2) M is Lorentzian, meaning that its tangent bundle is equipped with a smooth bilinear form $[\cdot,\cdot]$ which can be expressed in local coordinates as

$$[x, y] = c^2 x_0 y_0 - x_1 y_1 - x_2 y_2 - x_3 y_3;$$

and (3) the Lorentz form satisfies Einstein's field equations. We must also require that M be globally hyperbolic; this means that there exists a Cauchy surface, i.e., a smooth three-manifold $M_0 \subset M$ such that every point in M lies in the past or future of exactly one point on M_0. In other words, any point in M can be connected to a point in M_0 by a curve all of whose tangent vectors v satisfy $[v, v] > 0$. Global hyperbolicity is essential to our approach because Cauchy surfaces will play the role that constant t slices play in flat spacetime. However, it is a strong restriction: it implies that M is homeomorphic to $M_0 \times \mathbf{R}$ and is in fact foliated by Cauchy surfaces.

The restriction of the negation of the Lorentz form to any Cauchy surface M_0 makes M_0 a Riemannian manifold, so it has a unique volume measure and there is a canonical $L^2(M_0)$. Following the construction of free quantum fields in flat spacetime, it seems natural to take the Hilbert space of a free scalar field on M to be $\mathcal{F}_s L^2(M_0)$, with one-particle space $L^2(M_0)$. However, here we run into difficulties because this construction is not relativistically invariant. In particular, an element of the one-particle space on a Cauchy surface M_0 will in general *not* evolve into an element of the one-particle space on a different Cauchy surface M_0'. In other words, there simply is no relativistically invariant model of a single particle on general globally hyperbolic spacetimes (and it seems that this is the case even in Minkowski spacetime, if one allows noninertial observers). Even worse, the Fock space constructions on different Cauchy surfaces are in general not unitarily equivalent, in a sense that we will explain below. That is, general relativistic dynamics not only fail to take the one-particle space of a given Cauchy surface to the one-particle space of another Cauchy surface, they fail to take the Fock space of states on one surface to the Fock space of states on another!

For these reasons, it seems hopeless to try to find a canonical Hilbert space construction of free fields in curved spacetimes. However, it is still possible to formulate a general model of free fields if one adopts the point of view that the CCR algebras of observables are primary. We now describe this approach.

Let M be a globally hyperbolic spacetime and for any Cauchy surface M_0 let $\mathcal{V}_0$ be the set of pairs $(f, f_\bullet)$ with f and $f_\bullet$ real-valued functions in $C_c^\infty(M_0)$. The dynamics of classical fields are described by the Klein-Gordon equation

$$\nabla^2\phi = -\left(\frac{mc}{\hbar}\right)^2\phi,$$

where $\nabla^2 = \nabla\cdot\nabla$ is the four-dimensional divergence of the gradient, so that in local coordinates which diagonalize the Lorentz form at a point we have

$$\nabla^2 = \frac{1}{c^2}\frac{\partial^2}{\partial x_0^2} - \frac{\partial^2}{\partial x_1^2} - \frac{\partial^2}{\partial x_2^2} - \frac{\partial^2}{\partial x_3^2}.$$

The general theory of partial differential equations implies that for any Cauchy surface M_0, any pair $(f, f_\bullet) \in \mathcal{V}_0$ is the initial data of a unique solution ϕ of the Klein-Gordon equation such that $\phi|_{M_0} = f$ and the forward normal derivative of ϕ on M_0 is $f_\bullet$. Moreover, ϕ will be smooth and have <u>spatially compact support</u> in the sense that its restriction to any Cauchy surface has compact support. Thus, if we let the classical phase space $\mathcal{V}$ of a free scalar field be the set of all smooth solutions of the Klein-Gordon equation with spatially compact support, then for any M_0 there is a natural bijection between $\mathcal{V}$ and $\mathcal{V}_0$.

There is a natural symplectic form on $\mathcal{V}$ defined by

$$\{(f, f_\bullet), (g, g_\bullet)\} = \int_{M_0} f_\bullet g - f g_\bullet.$$

It is independent of M_0. To see this let M_0' be another Cauchy surface that lies in the future of M_0 and let M_1 be the four-dimensional region bounded by M_0 and M_0'. Let ϕ and ψ be solutions of the Klein-Gordon equation with respective data $(f, f_\bullet)$ and $(g, g_\bullet)$ on M_0 and $(f', f_\bullet')$ and $(g', g_\bullet')$ on M_0'. Then

$$\int_{M_0'} (f_\bullet' g' - f' g_\bullet') - \int_{M_0} (f_\bullet g - f g_\bullet) = \int_{\partial M_1} (\psi\nabla\phi - \phi\nabla\psi)\cdot\mathbf{n}$$

where $\mathbf{n}$ is the outward normal vector. By the divergence theorem this equals

$$\int_{M_1} \nabla \cdot (\psi\nabla\phi - \phi\nabla\psi) = \int_{M_1} (\psi\nabla^2\phi - \phi\nabla^2\psi) = 0$$

since $\nabla^2\psi = -m^2c^2\hbar^{-2}\psi$ and $\nabla^2\phi = -m^2c^2\hbar^{-2}\phi$. Thus $\int_{M_0}(f_\bullet g - fg_\bullet) = \int_{M_0'}(f_\bullet' g' - f' g_\bullet')$. This was shown for a surface M_0' lying in the future of M_0, but given any two Cauchy surfaces we can find a third lying in both of their futures, so we conclude that the symplectic form is the same on every Cauchy surface. Thus $\mathcal{V}$ is equipped with a well-defined symplectic form.

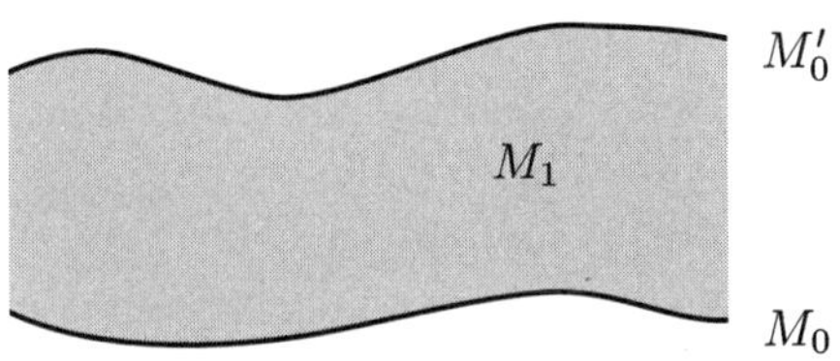

Figure 7.2 M_0, M_0', and M_1

Now it is possible to embed $\mathcal{V}$ in a complex Hilbert space in such a way that its symplectic form agrees with the imaginary part of the inner product. For example, mimicking the corresponding construction in flat spacetime, we can embed $\mathcal{V} \cong \mathcal{V}_0$ in $L^2(M_0)$ by

$$T^q(f, f_\bullet) = A^{1/4}f + iA^{-1/4}f_\bullet$$

where $A = -c^2\nabla^2_{M_0} + m^2c^4\hbar^{-2}$, $\nabla^2_{M_0}$ being the three-dimensional Riemannian Laplacian on M_0. However, the consistency conditions which provided motivation for this definition in flat spacetime are no longer meaningful, and the resulting $CCR(\mathcal{V})$ representations (Definition 7.2.3) for different Cauchy surfaces are not unitarily equivalent. Nonetheless, this does allow us to define $CCR(\mathcal{V})$ and to know that it has at least one Hilbert space representation. Recall that by Theorem 7.2.5 the C*-algebra $CCR(\mathcal{V})$ is independent of M_0.

Let us take stock of the situation. We now have a well-defined algebra of observables $CCR(\mathcal{V})$, but we do not have a canonical embedding of $CCR(\mathcal{V})$ in a "one-particle" Hilbert space because the latter concept does not have a relativistically invariant meaning. But suppose we are given *any* representation of $CCR(\mathcal{V})$ on a Hilbert space $\mathcal{H}$. Then for any

Cauchy surface M_0 and any $(f, f_\bullet) \in \mathcal{V}_0$, there is a unique $\phi \in \mathcal{V}$ with initial data $(f, f_\bullet)$ and a corresponding Weyl operator $W_\phi \in CCR(\mathcal{V})$ which acts on $\mathcal{H}$. We can therefore do quantum mechanics in the usual way, and by correlating the Weyl operators with various Cauchy surfaces we are able to interpret the results of observations made by different observers. In particular, for any region $S \subset M_0$ the observables that can be measured in S are precisely those in the sub-C*-algebra of $CCR(\mathcal{V})$ generated by the operators W_ϕ such that the corresponding f and $f_\bullet$ on M_0 are supported in S.

Thus, if we accept that there simply is no canonical Hilbert space construction, we can still be satisfied with interpreting different representations of $CCR(\mathcal{V})$ as different realizations of a free field on M. There is not even any good reason to restrict ourselves to Fock space constructions, although there are various other physically and mathematically motivated restrictions on which representations ought to be allowed. On this view, among the allowed representations of $CCR(\mathcal{V})$ none is accorded a fundamental status. The cleanest mathematical expression of this approach is to take C*-algebraic states in the sense of Definition 5.6.3 as primary, giving rise to Hilbert space representations via the GNS construction. The drawback is that a particle interpretation becomes difficult in general. But this seems to be a feature of general relativity that one just has to accept. Presumably one should still be able to define particles locally via a flat spacetime approximation.

The fact that representations of infinite dimensional CCR algebras are not unique up to unitary equivalence has been put forward as an argument for the value of C*-algebraic methods in quantum field theory. Using the CCR algebra point of view, one is able to recognize different representations of the same algebra as being realizations of the "same" quantum field; moreover, the language of C*-algebraic states is convenient for constructing representations. Quantum field theory in curved spacetime provides an even stronger version of this argument because here the Hilbert space constructions of different observers are generally inequivalent, and it is only through the C*-algebra $CCR(\mathcal{V})$ that one is able to relate the experience of states on different Cauchy surfaces. That is, the theory cannot be formulated in a relativistically invariant manner at the level of Hilbert spaces, only at the level of C*-algebras.

7.6 Notes

A thorough treatment of nonrelativistic fields is given in [36]. For a Fermionic version of the $L^2(\mathcal{H})$ construction see [6]. [2] is another good mathematical reference on quantum field theory.

Standard references on CCR and CAR algebras (the Fermionic version of CCR algebras) are [10] and [32]. Almost periodic functions are treated in [14], and their relation to CCR algebras is discussed in [72].

Relativistic single-particle systems, including spin, are treated in [68]. For more on localizability and superluminal signalling see [76] and [35].

Special relativistic quantum field theory is treated in [33], with emphasis on the use of C*- and von Neumann algebras to describe the observables which can be measured locally.

Free fields in curved spacetime are extensively discussed in [30] and [69]. The embedding of $T^q : \mathcal{V} \to L^2(M_0)$ described in Section 7.5 was given in [4]; the fact that there exist spacetimes in which it has undesirable properties follows from [42]. Also see [4] for further discussion of the conceptual basis of the Cauchy surface approach to quantum mechanics in curved spacetime. The argument that shows the symplectic form is independent of the Cauchy surface was shown to me by Renato Feres.

Chapter 8

Operator Spaces

8.1 The spaces $V(K)$

The topics in Chapters 2, 3, 5, and 6 constitute the "elementary" theory of mathematical quantization. In the remaining chapters we will introduce some more advanced topics. This chapter is about operator spaces, which are norm closed linear subspaces of $B(\mathcal{H})$ and are thought of as a quantum version of Banach spaces.

In order to make sense of this interpretation, we need to take a concrete view of Banach spaces. We will accomplish this by introducing a class of objects called "dual unit balls" and defining, for each dual unit ball K, a space of functions $V(K) \subset l^\infty(K)$, such that the spaces $V(K)$ are concrete realizations of Banach spaces. Then operator spaces will be the corresponding subspaces of $B(\mathcal{H})$.

DEFINITION 8.1.1 *A subset K of a topological vector space (TVS) is balanced if $x \in K$ and $a \in \mathbf{C}$, $|a| = 1$, implies $ax \in K$. A dual unit ball is then a compact, convex, balanced subset of a locally convex TVS.*

For any dual unit ball K, let $V(K)$ be the space of continuous functions f from K into $\mathbf{C}$ which are linear in the sense that $f(ax + by) = af(x) + bf(y)$ whenever $x, y, ax + by \in K$. Give $V(K)$ the supremum norm it inherits from $l^\infty(K)$.

Note that this definition of linearity is equivalent to asserting that f extends to a linear function on the span of K.

It is easy to see that $V(K)$ is a norm closed subspace of $l^\infty(K)$, so it is a Banach space.

We write $(\mathcal{V})_1$ for the closed unit ball of a Banach space $\mathcal{V}$.

Example 8.1.2

Let $\mathcal{V}$ be a Banach space and let $K = (\mathcal{V}^*)_1$ be the unit ball of the dual space $\mathcal{V}^*$. Then K is a dual unit ball.

PROPOSITION 8.1.3
Let K be a dual unit ball. Then K is linearly homeomorphic to the unit ball of $V(K)^*$.

PROOF Let $\mathcal{V} = V(K)$ and for $x \in K$ define $\hat{x} \in (\mathcal{V}^*)_1$ by $\hat{x}(f) = f(x)$. It is easy to check that the map $x \mapsto \hat{x}$ is linear and that it is continuous going into the weak* topology on $\mathcal{V}^*$. It is also one-to-one since local convexity implies that the continuous linear functionals separate points of K. Thus K is linearly homeomorphic to a weak* compact, convex subset of $(\mathcal{V}^*)_1$.

To show that the map $x \mapsto \hat{x}$ is onto, suppose $\omega \in \mathcal{V}^*$ is not in its image. Then by a standard separation theorem there exists a weak* continuous linear functional T on $\mathcal{V}^*$ such that

$$\operatorname{Re} T(\hat{x}) \leq 1 < \operatorname{Re} T(\omega)$$

for all $x \in K$. It is also standard that there exists $f \in \mathcal{V}$ such that $T(\rho) = \rho(f)$ for all $\rho \in \mathcal{V}^*$, so we have

$$\operatorname{Re} f(x) \leq 1 < \operatorname{Re} \omega(f)$$

for all $x \in K$. Since K is balanced and f is linear, this implies that $\|f\| \leq 1$, and hence $\|\omega\| > 1$. We conclude that $K \cong (\mathcal{V}^*)_1$. ∎

The preceding result shows that every abstract dual unit ball gives rise to a Banach space of which it is the dual unit ball. Conversely, we have the following result.

PROPOSITION 8.1.4
Let $\mathcal{V}$ be a Banach space and let $K = (\mathcal{V}^*)_1$. Then $\mathcal{V}$ is isometrically isomorphic to $V(K)$.

PROOF The weak* continuous linear functionals on $\mathcal{V}^*$ are precisely the maps $\hat{\xi} : \omega \mapsto \omega(\xi)$ for $\xi \in \mathcal{V}$. Each $\hat{\xi}$ restricts to a continuous linear function on K, and conversely, by the Krein-Smulian theorem every continuous linear function on K extends to a weak* continuous linear functional on $\mathcal{V}^*$. Thus the map $\xi \mapsto \hat{\xi}|_K$ is a linear isomorphism of $\mathcal{V}$ onto $V(K)$. It is isometric by the Hahn-Banach theorem. ∎

Next we observe that maps between Banach spaces correspond to maps between their dual unit balls.

Example 8.1.5

Let K and L be dual unit balls and let $\phi : K \to L$ be a continuous linear map (in the same sense as in Definition 8.1.1). Then composition with ϕ defines a linear contraction from $V(L)$ into $V(K)$.

Example 8.1.6

Let $\mathcal{V}$ and $\mathcal{W}$ be Banach spaces and let $T : \mathcal{V} \to \mathcal{W}$ be a linear contraction. Then composition with T defines a weak* continuous linear map from $(\mathcal{W}^*)_1$ into $(\mathcal{V}^*)_1$.

The verification of both examples is routine.

8.2 Matrix norms and convexity

We found in Section 8.1 that every Banach space can be realized concretely as a space of continuous linear functions on a dual unit ball, and thus Banach spaces are the abstract version of the spaces $V(K)$.

Each $V(K)$ is a norm closed linear subspace of $l^\infty(K)$, and conversely, every norm closed linear subspace of any $l^\infty(X)$ is a Banach space. This motivates the following definition.

DEFINITION 8.2.1 *An* operator space *is a norm closed linear subspace of some* $B(\mathcal{H})$.

Operator spaces are the quantum version of dual unit balls in the same way that C*-algebras are the quantum version of topological spaces and von Neumann algebras are the quantum version of measure spaces. In each case we have a classical object (topological space, measure space, dual unit ball), a function theoretic parallel ($C(X)$, $L^\infty(X)$, $V(K)$), and an operator analog (C*-algebra, von Neumann algebra, operator space).

Now every operator space is a Banach space, so at first it appears that there is no difference between operator spaces and Banach spaces. However, $B(\mathcal{H})$ has an extra level of structure which we have not discussed yet, and this extra structure is inherited by operator spaces. Namely, any $n \times n$ matrix of bounded operators $[A_{ij}]$, with each A_{ij} in $B(\mathcal{H})$, can be viewed as operating on the n-fold direct sum Hilbert space $\mathcal{H}^n$. Thus the matrix $[A_{ij}]$ has a natural norm, and this norm is compatible with matrices of different sizes in a certain way. This suggests the following definition.

To simplify notation, we will write M_n for $M_n(\mathbf{C})$. We will also write $M_{m,n}$ for the space of all $m \times n$ complex matrices. Note that $M_{m,n}$ can be identified with the set of linear maps from $\mathbf{C}^n$ to $\mathbf{C}^m$; we therefore take the norm of an element in $M_{m,n}$ to be its operator norm.

DEFINITION 8.2.2 *For any vector space $\mathcal{V}$ and any $n \in \mathbf{N}$, let $M_n(\mathcal{V})$ be the vector space of all $n \times n$ matrices with entries in $\mathcal{V}$. We can multiply matrices over $\mathcal{V}$ with scalar matrices using the usual formula for matrix products.*

A matrix norm on $\mathcal{V}$ is a sequence of norms defined on each $M_n(\mathcal{V})$ with the property that

$$||A\Xi B|| \leq ||A||||\Xi||||B||$$

for all $A \in M_{m,n}$, $B \in M_{n,m}$, and $\Xi \in M_n(\mathcal{V})$.

Any linear map $T : \mathcal{V} \to \mathcal{W}$ induces a linear map $T^{(n)} : M_n(\mathcal{V}) \to M_n(\mathcal{W})$ defined entrywise. If $\mathcal{V}$ and $\mathcal{W}$ are matrix normed spaces, then a linear map $T : \mathcal{V} \to \mathcal{W}$ is completely bounded if its completely bounded (CB) norm $||T||_{cb} = \sup_n ||T^{(n)}||$ is finite. T is completely contractive if $||T||_{cb} \leq 1$ and completely isometric if each $T^{(n)}$ is isometric.

We present two examples of matrix norms.

Example 8.2.3

Let $\mathcal{V} \subset B(\mathcal{H})$ be an operator space. Give $M_n(\mathcal{V})$ the norm it inherits from $M_n(B(\mathcal{H})) \cong B(\mathcal{H}^n)$. This is a matrix norm on $\mathcal{V}$.

Example 8.2.4

Let $\mathcal{V}$ be a Banach space. Then $\mathcal{V} \cong V(K)$ where $K = (\mathcal{V}^*)_1$, and $V(K) \subset l^\infty(K)$. Define the norm of any matrix $F = [f_{ij}] \in M_n(V(K))$ by

$$||F|| = \sup_{x \in K} ||F(x)|| = \sup_{x \in K} ||[f_{ij}(x)]||,$$

where $||[f_{ij}(x)]||$ is the usual operator norm of a scalar matrix. This is a matrix norm on $V(K) \cong \mathcal{V}$.

The matrix norm in Example 8.2.4 agrees with the matrix norm in Example 8.2.3 when we embed $l^\infty(K)$ in $B(l^2(K))$ in the usual way as multiplication operators.

Next we give examples of complete contractions.

Example 8.2.5

Let $\mathcal{A}$ and $\mathcal{B}$ be C*-algebras and let $\pi : \mathcal{A} \to \mathcal{B}$ be a $*$-homomorphism. Then π is a contraction by the comment following Definition 5.3.1. But $M_n(\mathcal{A})$ and $M_n(\mathcal{B})$ are also C*-algebras and $\pi^{(n)} : M_n(\mathcal{A}) \to M_n(\mathcal{B})$ is a $*$-homomorphism for all n, so π is in fact completely contractive.

This example shows why matrix norms have not played a role in earlier chapters. When one is dealing with $*$-homomorphisms, the higher matrix levels are typically irrelevant. However, in the setting of general linear maps between operator spaces the CB norm is usually more important than the first level (Banach space) norm.

The simplest example where the Banach space and CB norms differ is the transpose map on M_2. The map

$$\begin{bmatrix} a & b \\ c & d \end{bmatrix} \mapsto \begin{bmatrix} a & c \\ b & d \end{bmatrix}$$

is isometric, but its completely bounded norm is 2.

We have now defined the basic concepts surrounding operator spaces. Next we want to characterize these spaces abstractly. This requires a separation theorem whose proof is a bit technical but which is quite useful. It involves separating points from compact convex sets. However, the desired result does not just accomplish this on a single level (that could be done using classical separation theorems); rather, we consider convex sets on all levels. We need the following definition.

DEFINITION 8.2.6 *Let $\mathcal{V}$ be a locally convex TVS. A* <u>balanced matrix convex set</u> *over $\mathcal{V}$ is a sequence $K = (K_n)$ of subsets of $M_n(\mathcal{V})$ such that*

(a) $X \in K_m$ and $Y \in K_n$ implies $\begin{bmatrix} X & 0 \\ 0 & Y \end{bmatrix} \in K_{m+n}$, and

(b) if $A \in (M_{m,n})_1$, $B \in (M_{n,m})_1$, and $X \in K_n$, then $AXB \in K_m$.

K is <u>closed</u> *if each K_n is closed in $M_n(\mathcal{V})$, giving $M_n(\mathcal{V})$ the natural topology arising from the topology on $\mathcal{V}$ (nets in $M_n(\mathcal{V})$ converge if and only if they converge entrywise).*

Observe that each K_n must be convex, for if $X, Y \in K_n$ and $s + t = 1$ $(s, t \geq 0)$ then

$$sX + bY = [\sqrt{s}I_n \quad \sqrt{t}I_n] \begin{bmatrix} X & 0 \\ 0 & Y \end{bmatrix} \begin{bmatrix} \sqrt{s}I_n \\ \sqrt{t}I_n \end{bmatrix},$$

where I_n is the $n \times n$ identity matrix, belongs to K_n as well. Property (b) also implies that each K_n is balanced.

We now proceed to prove the desired separation theorem. A function f on a convex set K is <u>affine</u> if

$$f(tx + (1-t)y) = tf(x) + (1-t)f(y)$$

for all $x, y \in K$ and all $t \in [0, 1]$. A <u>cone</u> is a subset of a vector space that is closed under addition of vectors and multiplication by nonnegative real numbers.

LEMMA 8.2.7
Let K be a compact convex subset of a TVS and let E be a cone of real continuous affine functions on K. Suppose that for each $\phi \in E$ there exists $x \in K$ such that $\phi(x) \geq 0$. Then there exists $x_0 \in K$ such that $\phi(x_0) \geq 0$ for all $\phi \in E$.

PROOF By compactness it will suffice to show that for any finite set of functions $\phi_1, \ldots, \phi_n \in E$ there exists $x \in K$ such that $\phi_i(x) \geq 0$ for $1 \leq i \leq n$. Suppose this fails for some $\phi_1, \ldots, \phi_n$. Let $\phi : K \to \mathbf{R}^n$ be the direct sum map $\phi = \bigoplus \phi_i$. Then $\phi(K)$ is a compact convex subset of $\mathbf{R}^n$, and by assumption it does not intersect the closed convex set $[0, \infty)^n$. So by a standard separation theorem there is a linear function $\omega : \mathbf{R}^n \to \mathbf{R}$ such that $\omega(\phi(K)) < 0$ and $\omega([0, \infty)^n) \geq 0$. We have $\omega \circ \phi = \sum a_i \phi_i$ where $a_i = \omega(e_i) \geq 0$ (e_i being the ith basis vector in $\mathbf{R}^n$), so that $\omega \circ \phi \in E$. But then $\omega(\phi(K)) < 0$ contradicts the hypothesis of the lemma, so we have reached a contradiction. ∎

LEMMA 8.2.8
Let $K = (K_n)$ be a closed balanced matrix convex set over a locally convex TVS $\mathcal{V}$. Let $X_0 \in M_n(\mathcal{V})$ and suppose $X_0 \notin K_n$. Then there is a continuous linear map $F : M_n(\mathcal{V}) \to \mathbf{C}$ and a pair of states ω and ρ on M_n such that

$$|F(X)| \leq 1 < |F(X_0)|$$

for all $X \in K_n$ and

$$|F(AXB)| \leq (\omega(AA^*)\rho(B^*B))^{1/2}$$

for all $A \in M_{n,m}$, $B \in M_{m,n}$, and $X \in K_m$ ($m \in \mathbf{N}$).

PROOF The existence of F such that $|F(X)| \leq 1 < |F(X_0)|$ for all $X \in K_n$ follows from a standard separation theorem and the fact that K_n is a closed balanced convex set in $M_n(\mathcal{V})$.

Let S_n be the set of states on M_n. Then $S_n^2 = S_n \times S_n$ is a compact convex set. Given any $X \in K_m$ and any $A \in M_{n,m}$ and $B \in M_{m,n}$, define a real continuous affine function $\phi = \phi_{A,X,B}$ on S_n^2 by

$$\phi(\omega, \rho) = \omega(AA^*) + \rho(B^*B) - 2\mathrm{Re}\, F(AXB).$$

Let E be the set of all such functions $\phi_{A,X,B}$. We have $aE \subset E$ for all $a \geq 0$ since $a\phi_{A,X,B} = \phi_{A',X,B'}$ where $A' = a^{1/2}A$ and $B' = a^{1/2}B$. We also have $E + E \subset E$ because $\phi_{A_1,X_1,B_1} + \phi_{A_2,X_2,B_2} = \phi_{A,X,B}$ where $A = [A_1 \quad A_2]$, $B = \begin{bmatrix} B_1 \\ B_2 \end{bmatrix}$, and $X = X_1 \oplus X_2 \in K_{m_1+m_2}$. Thus E is a cone.

Moreover, for each $\phi \in E$ there exists a point $(\omega, \rho) \in S_n^2$ such that $\phi(\omega, \rho) \geq 0$. To see this fix A, X, B and find states ω and ρ on M_n such that

$$\omega(AA^*) = ||AA^*|| = ||A||^2 \qquad \text{and} \qquad \rho(B^*B) = ||B^*B|| = ||B||^2.$$

Then

$$\phi_{A,X,B}(\omega, \rho) = ||A||^2 + ||B||^2 - 2\mathrm{Re}\, F(AXB) \geq 0$$

since

$$\mathrm{Re}\, F(AXB) \leq |F(AXB)| \leq ||A|| ||B||.$$

Thus Lemma 8.2.7 applies, and we conclude that there exist states ω and ρ on M_n such that $\phi_{A,X,B}(\omega, \rho) \geq 0$ for any $A \in M_{n,m}$, $B \in M_{m,n}$, and $X \in K_m$.

For this choice of ω and ρ we have

$$2\mathrm{Re}\, F(AXB) \leq \omega(AA^*) + \rho(B^*B)$$

for all $A \in M_{n,m}$, $B \in M_{m,n}$, and $X \in K_m$. Replacing A by $b^{1/2}A$ and B by $b^{-1/2}B$ where $b = \omega(AA^*)^{-1/2}\rho(B^*B)^{1/2}$ then yields

$$\mathrm{Re}\, F(AXB) \leq \omega(AA^*)^{1/2}\rho(B^*B)^{1/2}.$$

Finally, replacing A with $(|F(AXB)|/F(AXB))A$ yields the desired inequality. ∎

A state ω on a C*-algebra is faithful if $\omega(A^*A) = 0$ implies $A = 0$.

LEMMA 8.2.9

Under the hypotheses of Lemma 8.2.8, we can obtain the same conclusion with ω and ρ faithful.

PROOF Let F, ω, and ρ be as in Lemma 8.2.8. The normalized trace $\tau = \frac{1}{n}\mathrm{tr}$ on M_n is faithful, so for any $\epsilon \in (0,1)$ the states $\omega' = (1-\epsilon)\omega + \epsilon\tau$ and $\rho' = (1-\epsilon)\rho + \epsilon\tau$ are faithful. Let $F' = (1-\epsilon)F$; for sufficiently small ϵ we still have

$$|F'(X)| \leq 1 < |F'(X_0)|$$

for all $X \in K_n$, and for $A \in M_{n,m}$, $B \in M_{m,n}$, and $X \in K_m$ we also have

$$\begin{aligned} |F'(AXB)| &\leq (1-\epsilon)\omega(AA^*)^{1/2}\rho(B^*B)^{1/2} \\ &\leq \frac{1}{2}(1-\epsilon)(\omega(AA^*) + \rho(B^*B)) \end{aligned}$$

$$\leq \frac{1}{2}(\omega'(AA^*) + \rho'(B^*B)).$$

Then replacing A and B by $b^{1/2}A$ and $b^{-1/2}B$ as in the proof of Lemma 8.2.8 yields

$$|F'(AXB)| \leq (\omega'(AA^*)\rho'(B^*B))^{1/2},$$

as desired. ∎

THEOREM 8.2.10

Let $K = (K_n)$ be a closed balanced matrix convex set over a locally convex TVS $\mathcal{V}$. Let $X_0 \in M_n(\mathcal{V})$ and suppose $X_0 \notin K_n$. Then there is a continuous linear map $\Omega : \mathcal{V} \to M_n$ such that $\|\Omega^{(m)}(X)\| \leq 1$ for all $m \in \mathbf{N}$ and $X \in K_m$ but $\|\Omega^{(n)}(X_0)\| > 1$.

PROOF Let F, ω, and ρ be as in Lemma 8.2.8 with ω and ρ faithful. Apply the GNS construction (Theorem 5.6.6) to ω and ρ to get representations $\pi : M_n \to B(\mathcal{H}_\omega)$ and $\theta : M_n \to B(\mathcal{H}_\rho)$. Faithfulness of ω and ρ implies that π and θ are $*$-isomorphisms. Let $v_0 = \bar{I}_n \in \mathcal{H}_\omega$ and $w_0 = \bar{I}_n \in \mathcal{H}_\rho$, so that $\omega(A) = \langle \pi(A)v_0, v_0 \rangle$ and $\rho(A) = \langle \theta(A)w_0, w_0 \rangle$ for all $A \in M_n$.

For any $1 \times n$ scalar matrix $A = [a_1 \quad \ldots \quad a_n]$ let $\tilde{A}$ be the $n \times n$ scalar matrix

$$\tilde{A} = \begin{bmatrix} a_1 & a_2 & \cdots & a_n \\ 0 & 0 & \cdots & 0 \\ \vdots & \vdots & \ddots & \vdots \\ 0 & 0 & \cdots & 0 \end{bmatrix}.$$

Let E be the set of all $n \times n$ matrices of this form, and define $\mathcal{H} = \pi(E)v_0 \subset \mathcal{H}_\omega$ and $\mathcal{K} = \theta(E)w_0 \subset \mathcal{H}_\rho$. We have $\dim(\mathcal{H}) = \dim(\mathcal{K}) = n$.

For $x \in \mathcal{V}$ define a sesquilinear form on $\mathcal{K} \times \mathcal{H}$ by

$$\{\theta(\tilde{B})w_0, \pi(\tilde{A})v_0\} = F(A^*xB).$$

Then let $\Omega(x) : \mathcal{K} \to \mathcal{H}$ be the linear map for which $\{w, v\} = \langle \Omega(x)w, v \rangle$. The set of linear maps from $\mathcal{K}$ into $\mathcal{H}$ can be completely isometrically identified with M_n, so we may regard Ω as a map from $\mathcal{V}$ into M_n.

Let $\{e_i\}$ be the standard basis of $\mathbf{C}^n \cong M_{1,n}$. For $X = [x_{ij}] \in M_n(\mathcal{V})$ we have $X = \sum_{ij} e_i^* x_{ij} e_j$, so

$$F(X) = \sum \langle \Omega(x_{ij})\theta(\tilde{e}_j)w_0, \pi(\tilde{e}_i)v_0 \rangle = \langle \Omega^{(n)}(X)w_1, v_1 \rangle$$

where

$$v_1 = \begin{bmatrix} \pi(\tilde{e}_1)v_0 \\ \vdots \\ \pi(\tilde{e}_n)v_0 \end{bmatrix} \quad \text{and} \quad w_1 = \begin{bmatrix} \theta(\tilde{e}_1)w_0 \\ \vdots \\ \theta(\tilde{e}_n)w_0 \end{bmatrix}$$

(both in $\mathbf{C}^{n^2}$). A short computation shows that $\|v_1\| = \|w_1\| = 1$. Since $|F(X_0)| > 1$, this shows that $\|\Omega^{(n)}(X_0)\| > 1$ also.

Finally, let $X \in K_m$ and fix v and w in $\mathbf{C}^{mn}$. Write

$$v = \begin{bmatrix} \pi(\tilde{A}_1)v_0 \\ \vdots \\ \pi(\tilde{A}_m)v_0 \end{bmatrix} \qquad \text{and} \qquad w = \begin{bmatrix} \theta(\tilde{B}_1)w_0 \\ \vdots \\ \theta(\tilde{B}_m)w_0 \end{bmatrix}$$

for some $1 \times n$ matrices A_i and B_i. Then $\|v\|^2 = \omega(A^*A)$ and $\|w\|^2 = \rho(B^*B)$ where

$$A = \begin{bmatrix} A_1 \\ \vdots \\ A_m \end{bmatrix} \qquad \text{and} \qquad B = \begin{bmatrix} B_1 \\ \vdots \\ B_m \end{bmatrix}$$

are $m \times n$ matrices. Thus

$$\begin{aligned} |\langle \Omega^{(m)}(X)w, v\rangle| &= |F(A^*XB)| \\ &\leq (\omega(A^*A)\rho(B^*B))^{1/2} \\ &= \|v\|\|w\|. \end{aligned}$$

As v and w were arbitrary, this shows that $\|\Omega^{(m)}(X)\| \leq 1$. ∎

The abstract characterization of operator spaces is an easy consequence of the preceding result. An <u>L^∞-matrix norm</u> on a vector space $\mathcal{V}$ is a matrix norm which satisfies

$$\left\| \begin{bmatrix} \Xi & 0 \\ 0 & H \end{bmatrix} \right\| = \max(\|\Xi\|, \|H\|)$$

for $\Xi \in M_m(\mathcal{V})$ and $H \in M_n(\mathcal{V})$. It is easy to see that the matrix norm on $B(\mathcal{H})$ is an L^∞-matrix norm, and hence the same is true of any operator space. The converse is also true; this is known as Ruan's theorem.

THEOREM 8.2.11

Let $\mathcal{V}$ be a complete L^∞-matrix normed space. Then $\mathcal{V}$ is completely isometric to an operator space.

PROOF Let T be the direct sum of all complete contractions from $\mathcal{V}$ into M_n ($n \in \mathbf{N}$). It is clear that T is a complete contraction, and to show that it is a complete isometry we must find, for each $\Xi_0 \in M_n(\mathcal{V})$ with $\|\Xi_0\| > 1$, a complete contraction $\Omega : \mathcal{V} \to M_m$ for some $m \in \mathbf{N}$ such that $\|\Omega^{(n)}(\Xi_0)\| > 1$. But the sequence of unit balls $K_m =$

$(M_m(\mathcal{V}))_1$ is a closed balanced matrix convex set over $\mathcal{V}$, so Theorem 8.2.10 implies the existence of a continuous linear map $\Omega : \mathcal{V} \to M_n$ such that $\|\Omega^{(m)}(\Xi)\| \leq 1$ for all $\Xi \in K_m$ ($m \in \mathbf{N}$) but $\|\Omega^{(n)}(\Xi_0)\| > 1$. That is, Ω is a complete contraction, and we conclude that T is a complete isometry. Finally, we have assumed that $\mathcal{V}$ is a Banach space, so $T(\mathcal{V})$ is closed and hence is an operator space. ∎

8.3 Duality

Every operator space $\mathcal{V}$ has a dual $\mathcal{V}^*$ which is also an operator space. These behave much like Banach space duals; for example, $\mathcal{V}$ completely isometrically embeds in $\mathcal{V}^{**}$ (Proposition 8.3.3). We will use operator space duals to formulate a definition of "dual matrix unit balls" which are the operator space analog of the dual unit balls discussed in Section 8.1.

DEFINITION 8.3.1 *Let $\mathcal{V}$ be an operator space. Its* dual *is the Banach space dual $\mathcal{V}^*$, with matrix norms defined by taking the norm of an element of $M_n(\mathcal{V}^*)$ to be the completely bounded norm of the map it induces from $\mathcal{V}$ into M_n. Thus, for $X = [x_{ij}] \in M_n(\mathcal{V}^*)$ we have*

$$\begin{aligned}\|X\| &= \sup\{\|X^{(m)}(\Xi)\| : m \in \mathbf{N} \text{ and } \Xi \in (M_m(\mathcal{V}))_1\} \\ &= \sup\{\|[x_{ij}(\xi_{kl})]\| : m \in \mathbf{N} \text{ and } \Xi = [\xi_{kl}] \in (M_m(\mathcal{V}))_1\}.\end{aligned}$$

When $n = 1$ this norm agrees with the Banach space norm on $\mathcal{V}^*$. To see this let $\Xi \in (M_n(\mathcal{V}))_1$, let $x \in \mathcal{V}^*$, and let $v, w \in \mathbf{C}^n \cong M_{n,1}$. Then

$$|\langle x^{(n)}(\Xi)v, w\rangle| = |x(w^*\Xi v)| \leq \|x\|\|v\|\|w\|,$$

which shows that $\|x^{(n)}(\Xi)\| \leq \|x\|$. Thus the operator space norm of x is at most its Banach space norm, and the reverse inequality is clear.

The dual of an operator space is itself an operator space. One can show this by means of Ruan's theorem (Theorem 8.2.11), but it is easier to give a direct proof.

PROPOSITION 8.3.2
Let $\mathcal{V}$ be an operator space. Then there is a completely isometric and weak homeomorphic isomporphism of its dual space $\mathcal{V}^*$ with a weak* closed subspace of some $B(\mathcal{H})$.*

PROOF Let Λ be the set of all pairs (m, Ξ) such that $m \in \mathbf{N}$ and

$\Xi \in (M_m(\mathcal{V}))_1$, and let

$$\mathcal{H} = \bigoplus_{(m,\Xi)\in\Lambda} \mathbf{C}^m.$$

Then define $T : \mathcal{V}^* \to B(\mathcal{H})$ by $T(x) = \bigoplus x^{(m)}(\Xi)$. For $X \in M_n(\mathcal{V}^*)$ we have

$$\|T^{(n)}(X)\| = \left\| \bigoplus_{(m,\Xi)\in\Lambda} X^{(m)}(\Xi) \right\| = \sup_{(m,\Xi)\in\Lambda} \|X^{(m)}(\Xi)\| = \|X\|_{cb},$$

so the matrix norms on $\mathcal{V}^*$ and $T(\mathcal{V}^*)$ agree, and hence T is a complete isometry.

If (x_κ) is a bounded net in $\mathcal{V}^*$ and $x_\kappa \to x$ weak*, then $x_\kappa^{(m)}(\Xi) \to x^{(m)}(\Xi)$ in each entry, for each $\Xi \in (M_m(\mathcal{V}))_1$. It follows that $Tx_\kappa \to Tx$ weak* in $B(\mathcal{H})$. Therefore, since T is an isometry, the unit ball of $T(\mathcal{V}^*)$ is weak* compact, and hence by the Krein-Smulian theorem $T(\mathcal{V}^*)$ is a weak* closed subspace of $B(\mathcal{H})$. Moreover, this shows that for bounded nets the weak* topology on $\mathcal{V}^*$ agrees with the induced weak* topology on $T(\mathcal{V}^*)$, which implies that the two weak* topologies are equal. ∎

Next we prove that any operator space completely isometrically embeds in its double dual. This result is an easy consequence of the separation theorem proven in Section 8.2.

PROPOSITION 8.3.3
Let $\mathcal{V}$ be an operator space. Then the natural map $\xi \mapsto \hat{\xi}$ completely isometrically embeds $\mathcal{V}$ in $\mathcal{V}^{**}$.

PROOF A straightforward computation shows that the map $\xi \mapsto \hat{\xi}$ is a complete contraction. To prove that it is a complete isometry, let $\Xi = [\xi_{ij}] \in M_m(\mathcal{V})$ and suppose $\|\Xi\| > 1$; we must show that $\|\hat{\Xi}\| = \|[\hat{\xi}_{ij}]\| > 1$. To do this it will suffice to find $X \in M_m(\mathcal{V}^*)$ such that $\|\hat{\Xi}^{(m)}(X)\| = \|X^{(m)}(\Xi)\| > 1$. But taking $K_n = (M_n(\mathcal{V}))_1$ in Theorem 8.2.10 yields precisely this. ∎

Next we consider the abstract characterization of dual operator spaces. Concretely, Proposition 8.3.2 shows that these are precisely the weak* closed subspaces of $B(\mathcal{H})$. Abstractly, one might expect that any operator space which is a dual Banach space should be a dual operator space; however, a slightly stronger assumption is needed.

PROPOSITION 8.3.4

Let $\mathcal{V}$ be an operator space which is a dual Banach space and suppose that the unit ball of $M_n(\mathcal{V})$ is weak closed for all n. Then $\mathcal{V}$ is a dual operator space.*

PROOF We may assume that $\mathcal{V} = \mathcal{W}^*$ for some Banach space $\mathcal{W}$. Then $\mathcal{W}$ isometrically embeds in $\mathcal{V}^*$, and inherits an operator space structure from this space. We must show that $\mathcal{V} = \mathcal{W}^*$ as operator spaces.

Let Λ be the set of all pairs (m, Ξ) such that $m \in \mathbf{N}$ and $\Xi \in (M_m(\mathcal{W}))_1$, and apply the construction of Proposition 8.3.2 to get a weak* continuous complete contraction $T : \mathcal{V} \to B(\mathcal{H})$. It will suffice to show that T is a complete isometry. To do this, let $X \in M_n(\mathcal{V})$ and suppose $\|X\| > 1$; we must find $\Xi \in M_m(\mathcal{W})$ such that $\|\Xi\| \leq 1$ but $\|X^{(m)}(\Xi)\| > 1$. Define $K_n = (M_n(\mathcal{V}))_1$ and apply Theorem 8.2.10 to $\mathcal{V}$ equipped with the weak* topology. Then we get a matrix Ω of weak* continuous maps from $\mathcal{V}$ into $\mathbf{C}$ such that $\|\Omega\| \leq 1$ but $\|\Omega^{(n)}(X)\| > 1$. However, every weak* continuous map from $\mathcal{V}$ into $\mathbf{C}$ is obtained by pairing with an element of $\mathcal{W}$, so we have $\Omega = \hat{\Xi}$ for some $\Xi \in (M_m(\mathcal{W}))_1$, as desired. ∎

Now we are ready to present the matrix version of the dual unit balls discussed in Section 8.1.

DEFINITION 8.3.5 *A <u>dual matrix unit ball</u> is a balanced matrix convex set $K = (K_n)$ such that each K_n is compact.*

Since K is a balanced matrix convex set, each K_n must be a balanced convex set (see the comment following Definition 8.2); as we are now also assuming that the K_n are compact, it follows that each K_n is a dual unit ball in the sense of Definition 8.1.1. The prototypical example of a dual matrix unit ball is the sequence of matrix unit balls over a dual operator space.

Example 8.3.6

Let $\mathcal{V}$ be an operator space and let $K_n = (M_n(\mathcal{V}^*))_1$ be the sequence of matrix unit balls of the dual operator space. Giving $\mathcal{V}^*$ the weak* topology, each K_n is compact, and $K = (K_n)$ is a dual matrix unit ball.

If K and L are dual matrix unit balls, we will use the term <u>completely linear map</u> from K to L to mean a linear map $\phi : K_1 \to L_1$

such that $\phi^{(n)}(K_n) \subset L_n$ for all n. We will say that ϕ is a complete homeomorphism if each $\phi^{(n)}$ is a homeomorphism.

PROPOSITION 8.3.7

Let $K = (K_n)$ be a dual matrix unit ball. Then K is completely linearly homeomorphic to the sequence of matrix unit balls of some dual operator space.

PROOF As we noted above, each K_n is a dual unit ball. Let $\mathcal{V} = \text{span}(K)$. Condition (b) of balanced matrix convexity implies that each entry of any element in K_n belongs to K_1, so $M_n(\mathcal{V}) = \text{span}(K_n)$ for all n. Define a matrix norm on $\mathcal{V}$ by letting the unit ball of $M_n(\mathcal{V})$ be K_n.

This does define a norm on each $M_n(\mathcal{V})$ by Proposition 8.1.3, and it makes $\mathcal{V}$ into an operator space by Theorem 8.2.11. As the unit ball of each $M_n(\mathcal{V})$ is K_n, which is weak* compact, it follows from Proposition 8.3.4 that $\mathcal{V}$ is a dual operator space. So K is identified with the sequence of matrix unit balls over a dual operator space. ∎

We now have a way of passing from operator spaces to dual matrix unit balls, and vice versa, just as we can pass between Banach spaces and dual unit balls. As in the Banach space case, morphisms between the two types of objects also correspond.

Example 8.3.8

Let $\mathcal{V}$ and $\mathcal{W}$ be operator spaces and let $K_n = (M_n(\mathcal{V}^*))_1$ and $L_n = (M_n(\mathcal{W}^*))_1$ be the corresponding dual matrix unit balls. If $T : \mathcal{V} \to \mathcal{W}$ is a complete contraction then composition with T defines a continuous completely linear map from L to K.

PROPOSITION 8.3.9

Let $\mathcal{V}$ and $\mathcal{W}$ be operator spaces and let K and L be the corresponding dual matrix unit balls. Then any continuous completely linear map $\phi : L \to K$ is given by composition with a complete contraction from $\mathcal{V}$ to $\mathcal{W}$.

PROOF By Proposition 8.1.4 and Example 8.1.5 there is a contraction $T : \mathcal{V} \to \mathcal{W}$ such that ϕ is composition with T. If T fails to be a complete contraction then for some n it does not take the unit ball of $M_n(\mathcal{V})$ into the unit ball of $M_n(\mathcal{W})$ for some n, and then composition with T cannot take L_n into K_n, contradicting complete linearity of ϕ. So T is a complete contraction. ∎

8.4 Matrix-valued functions

The notion of a dual matrix unit ball, introduced in the previous section, provides the basis for a different approach to operator spaces which relates them to C*- and von Neumann algebras. The basic idea is this. If $\mathcal{V}$ is a Banach space then we can identify $\mathcal{V}$ with $V(K)$, and as the dual unit ball K is a compact Hausdorff space we can write $V(K) \subset C(K) \subset l^\infty(K)$. For operator spaces there will be an analogous construction in which $C(K)$ and $l^\infty(K)$ are replaced by noncommutative C*- and von Neumann algebras. We give the von Neumann algebra construction first.

DEFINITION 8.4.1 *Let K be a dual matrix unit ball. We define the* <u>matrix l^∞-space</u> *$l^\infty_{mat}(K)$ to be the set of sequences $f = (f_n)$ of bounded functions from K_n into M_n such that*

(a) $\sup_n \|f_n\|_\infty \leq \infty$;
(b) $X \in K_n$ and $Y \in K_m$ implies

$$f_{n+m}\left(\begin{bmatrix} X & 0 \\ 0 & Y \end{bmatrix}\right) = \begin{bmatrix} f_n(X) & 0 \\ 0 & f_m(Y) \end{bmatrix};$$

and
(c) if $X \in K_n$ and $A \in M_n$ is unitary then $f_n(A^*XA) = A^*f_n(X)A$.

This is supposed to be a matrix version of l^∞ appropriate to the context of dual matrix unit balls. In fact, the functions at the first level do constitute precisely $l^\infty(K_1)$. The higher level functions take values in matrix algebras rather than $\mathbf{C}$ and are also required to satisfy the compatibility conditions (b) and (c) which relate to the structure of K.

Observe that $l^\infty_{mat}(K)$ has a natural algebraic structure with operations defined pointwise. It is not commutative, of course, because the functions f_n take values in matrix algebras.

DEFINITION 8.4.2 *For any dual matrix unit ball K let $V_{mat}(K)$ be the subspace of $l^\infty_{mat}(K)$ consisting of those sequences $f = (f_n)$ such that*

(a) $f_1 : K_1 \to \mathbf{C}$ is linear and continuous and
(b) $f_n[x_{ij}] = [f_1(x_{ij})]$ for all $n \in \mathbf{N}$ and $[x_{ij}] \in K_n$, i.e., $f_n = f_1^{(n)}$.

We think of $V_{mat}(K)$ as the space of "continuous linear functions" in $l^\infty_{mat}(K)$.

Now we present the basic facts about $V_{mat}(K)$ and $l^\infty_{mat}(K)$.

LEMMA 8.4.3

Let K be a dual matrix unit ball and suppose $X \in K_n$ is nonzero. Then there exists $f \in V_{mat}(K)$ such that $f_n(X) \neq 0$.

PROOF Since X is nonzero it must have a nonzero entry x_{ij}. Let $f_1 : K_1 \to \mathbf{C}$ be a continuous linear map such that $f_1(x_{ij}) \neq 0$. Then defining $f = (f_n)$ by $f_n = f_1^{(n)}$ immediately yields $f_n(X) \neq 0$, and it is straightforward to verify that $f \in l^\infty_{mat}(K)$ and hence in $V_{mat}(K)$. ∎

THEOREM 8.4.4

Let $\mathcal{V}$ be an operator space and let K be the corresponding dual matrix unit ball. Then

(a) $l^\infty_{mat}(K)$ is a von Neumann algebra;
(b) $V_{mat}(K)$ is an operator space which is completely isometric to $\mathcal{V}$; and
(c) $V_{mat}(K)$ generates $l^\infty_{mat}(K)$ as a von Neumann algebra.

PROOF

(a) Regard $l^\infty_{mat}(K)$ as a subalgebra of $B(\mathcal{H})$ where $\mathcal{H} = \bigoplus \mathbf{C}^m$, the sum being taken over all $m \in \mathbf{N}$ and all $X \in K_m$ as in the proof of Proposition 8.3.2. It is easy to see that $l^\infty_{mat}(K)$ is a unital $*$-subalgebra of $B(\mathcal{H})$, and the conditions (b) and (c) of Definition 8.4.1 are clearly preserved by bounded weak operator limits, so $l^\infty_{mat}(K)$ is a von Neumann algebra.

(b) This follows from Proposition 8.3.3.

(c) Call an element $X \in K_n \subset M_n(\mathcal{V}^*)$ reducible if there is a nontrivial projection $P \in M_n$ such that $PX = XP$; also, say that $X, Y \in K_n$ are unitarily equivalent if there exists a unitary matrix $U \in M_n$ such that $Y = U^*XU$.

Let $\mathcal{M}$ be the W*-subalgebra of $l^\infty_{mat}(K)$ generated by $V_{mat}(K)$. We first claim that for any irreducible $X \in K_n$, the C*-algebra $\{f_n(X) : f \in \mathcal{M}\}$ equals M_n. Suppose not; then by the double commutant theorem (Theorem 6.5.7) there is a nontrivial projection $P \in M_n$ such that $Pf_n(X) = f_n(X)P$ for all $f \in \mathcal{M}$, and hence for all $f \in V_{mat}(K)$. By linearity this implies $f_n(PX - XP) = 0$ for all $f \in V_{mat}(K)$, and the lemma then yields $PX - XP = 0$, contradicting irreducibility. This proves the first claim.

We next claim that if $X \in K_m$, $Y \in K_n$ are irreducible and (in case $m = n$) unitarily inequivalent then the C*-algebra $\mathcal{A} = \{f_m(X) \oplus f_n(Y) : f \in \mathcal{M}\}$ equals $M_m \oplus M_n$. To see this, consider the natural $*$-homomorphisms $\pi_1 : \mathcal{A} \to M_m$ and $\pi_2 : \mathcal{A} \to M_n$; note that they are surjective by the first claim, and the intersection of their kernels is zero. If $m \neq n$, this implies that $\mathcal{A} = M_m \oplus M_n$ (cf. Theorem 11.1.2 and Proposition 6.4.1). If $m = n$ then it is also possible that $\mathcal{A} = M_m$. By Proposition 6.4.7, there then exists a unitary matrix $U \in M_m$ such that $\pi_2 = U^*\pi_1 U$, i.e. $f_m(Y) = U^* f_m(X)U$ for all $f \in \mathcal{M}$. But this implies $f_m(Y - U^*XU) = 0$ for all $f \in V_{mat}(K)$, and hence $Y - U^*XU = 0$ by the lemma. That is, X and Y are unitarily equivalent, contradicting our assumption. This proves the second claim.

Now let f be any self-adjoint element of $l^\infty_{mat}(K)$; we will show that $f \in \mathcal{M}$. This is sufficient to verify that $l^\infty_{mat}(K) = \mathcal{M}$. For any finite set Λ of irreducible, unitarily inequivalent elements X_i,

$$\Lambda = \{X_i \in K_{n_i} : 1 \leq i \leq r\},$$

it is a consequence of the second claim that there exists $f^\Lambda \in \mathcal{M}$ such that $f^\Lambda_{n_i}(X_i) = f_{n_i}(X_i)$ for $1 \leq i \leq r$. By taking its real part we may assume f^Λ is self-adjoint, and by truncation using functional calculus we may then assume $\|f^\Lambda\| \leq \|f\|$.

It follows from conditions (b) and (c) in the definition of $l^\infty_{mat}(K)$ that f is determined by its values on irreducible, unitarily inequivalent elements of K. Thus the net (f^Λ) eventually agrees with f on any finite subset of K, where this net is ordered by setting $\Lambda \leq \Lambda'$ if every element of Λ is unitarily equivalent to some element of Λ'. Now consider the natural action of $l^\infty_{mat}(K)$ on the Hilbert space $\mathcal{H} = \bigoplus \mathbf{C}^m$ which appeared in the proof of part (a) of this theorem. We have just shown that $\langle f^\Lambda v, w\rangle \to \langle fv, w\rangle$ for any elements v and w of the algebraic direct sum of the $\mathbf{C}^m$, and since this is a dense subspace of $\mathcal{H}$ and the net (f^Λ) is bounded, we therefore have $f^\Lambda \to f$ weak*. Thus f belongs to the von Neumann algebra generated by $V_{mat}(K)$, and this completes the proof. ∎

Next, we define a matrix version of $C(K)$.

DEFINITION 8.4.5 *For any dual matrix unit ball K, let $C_{mat}(K)$ be the unital C*-algebra generated by $V_{mat}(K)$ in $l^\infty_{mat}(K)$.*

Note that this is *not* the same as the set of $f \in l^\infty_{mat}(K)$ such that each f_n is continuous. Indeed, it is easy to see that for any $f \in C_{mat}(K)$ the sequence (f_n) must be uniformly continuous. Even this condition is not sufficient to imply $f \in C_{mat}(K)$, however. But the next theorem does give us an abstract characterization of $C_{mat}(K)$.

If $\mathcal{A}$ and $\mathcal{B}$ are unital C*-algebras, we use the notation $\hom(\mathcal{A}, \mathcal{B})$ to denote the set of unital $*$-homomorphisms from $\mathcal{A}$ into $\mathcal{B}$. Note that if

$\mathcal{B} = \mathbf{C}$ then this is just the spectrum of $\mathcal{A}$.

THEOREM 8.4.6

Let $\mathcal{V}$ be an operator space and let K be the corresponding dual matrix unit ball. Then

(a) $C_{mat}(K)$ is the universal unital C-algebra which completely isometrically contains $\mathcal{V}$, and*
(b) $\hom(C_{mat}(K), M_n)$ is canonically homeomorphic to K_n for all n.

PROOF

(a) By definition, K_n consists of all completely contractive linear maps from $\mathcal{V}$ into M_n. Recalling the identification of $\mathcal{V}$ with $V_{mat}(K)$ in Theorem 8.4.4 (b), we see that $C_{mat}(K)$ is just the C*-algebra generated by the image of $\mathcal{V}$ in the direct sum of all of its finite dimensional completely contractive representations.

Thus, we need to know that any C*-algebra which completely isometrically contains $\mathcal{V}$ is residually finite in the following sense. Given any completely contractive representation $T : \mathcal{V} \to B(\mathcal{H})$ of $\mathcal{V}$, the norm of any element of the $*$-algebra generated by $T(\mathcal{V})$ must be approximated by the norms of the corresponding operators in finite dimensional representations of $\mathcal{V}$. In other words, if p is a $*$-polynomial in elements of $\mathcal{V}$ then $\|Tp\|$ (interpreting this notation in the obvious way) must be approximated by $\|T'p\|$ for completely contractive representations $T' : \mathcal{V} \to M_n$.

Fix a completely contractive representation $T : \mathcal{V} \to B(\mathcal{H})$ and a $*$-polynomial p in the elements of $\mathcal{V}$. Given $\epsilon > 0$, choose a unit vector $v \in \mathcal{H}$ such that $\|Tp(v)\| \geq \|Tp\| - \epsilon$. Let $\mathcal{K}$ be the finite dimensional subspace of $\mathcal{H}$ spanned by v and the vectors $Tp'(v)$ as p' ranges over all subpolynomials of p; let P be the orthogonal projection onto $\mathcal{K}$; and let $T' = PT|_{\mathcal{K}}$ be the induced completely contractive representation of $\mathcal{V}$ on $\mathcal{K}$. Then $T'p(v) = Tp(v)$, and hence $\|T'p\| \geq \|Tp\| - \epsilon$. Thus the norm of p in the arbitrary representation T is approximated by its norm in the finite dimensional representation T', as we needed to show.

(b) For any $X \in K_n$ let $\hat{X} : C_{mat}(K) \to M_n$ be evaluation at X, i.e., $\hat{X}(f) = f_n(X)$. Then define $\phi : K_n \to \hom(C_{mat}(K), M_n)$ by $\phi(X) = \hat{X}$. It is easy to see that this map is one-to-one and continuous. As $\hom(C_{mat}(K), M_n)$ is clearly Hausdorff, ϕ is a homeomorphism, so we only need to show surjectivity.

Let $\pi : C_{mat}(K) \to M_n$ be any $*$-homomorphism. Then the restriction of π to $V_{mat}(K)$ is a completely contractive linear map. Hence there exists $X \in K_n$ such that π agrees with $\hat{X}$ on $V_{mat}(K)$. But $V_{mat}(K)$ generates $C_{mat}(K)$, so $\pi = \hat{X}$. This shows that ϕ is surjective. ∎

COROLLARY 8.4.7
Let $\mathcal{V}$ and $\mathcal{W}$ be operator spaces and let K and L be the corresponding dual matrix unit balls. Then any completely contractive linear map from $\mathcal{V}$ to $\mathcal{W}$ extends uniquely to a unital $*$-homomorphism from $C_{mat}(K)$ to $C_{mat}(L)$ and to a weak continuous unital $*$-homomorphism from $l^\infty_{mat}(K)$ to $l^\infty_{mat}(L)$.*

PROOF The first statement follows immediately from the universality of $C_{mat}(K)$ proven in part (a) of the theorem. (It can also be shown by the following argument.)

Let $T : \mathcal{V} \to \mathcal{W}$ be a completely contractive linear map and let $T^\# : L \to K$ be the adjoint continuous completely linear map given by composition with T. Then define $\tilde{T} : l^\infty_{mat}(K) \to l^\infty_{mat}(L)$ by $\tilde{T}f(X) = f(T^\# X)$ for $f \in l^\infty_{mat}(K)$ and $X \in L$. It is now straightforward to check that $\tilde{T}$ is a weak* continuous unital $*$-homomorphism which extends T. Uniqueness follows from the fact that $V_{mat}(K)$ generates $l^\infty_{mat}(K)$. ∎

8.5 Operator systems

As we have seen, operator spaces are quantum versions of dual unit balls, i.e., compact balanced convex sets. In this section we will discuss the quantum version of a compact convex set, which is called an operator system. The classical analogs of operator systems are function systems, which are defined as follows.

DEFINITION 8.5.1 *A real ordered vector space is a real vector space $\mathcal{V}$ equipped with a partial ordering for which $\xi \le \eta$ implies $\xi + \zeta \le \eta + \zeta$ and $a\xi \le a\eta$ for all $\zeta \in \mathcal{V}$ and $a \ge 0$.*

Let $\mathcal{V}$ be a real ordered vector space. An Archemedian order unit for $\mathcal{V}$ is an element $\varepsilon \in \mathcal{V}$ such that

(a) for each $\xi \in \mathcal{V}$ there is a real number $a > 0$ such that $a\xi \le \varepsilon$, and
(b) if $\xi \le a\varepsilon$ for all $a > 0$ then $\xi \le 0$.

A (real) function system is a real ordered vector space together with a distinguished Archimedian order unit ε, such that the order norm

$$\|\xi\| = \inf\{a \ge 0 : -a\varepsilon \le \xi \le a\varepsilon\}$$

is complete.

A state on a function system $\mathcal{V}$ is a positive (i.e., order preserving) linear functional $\omega : \mathcal{V} \to \mathbf{R}$ such that $\|\omega\| = 1$. The state space of $\mathcal{V}$ is the set $S(\mathcal{V})$ of all states on $\mathcal{V}$.

Observe that the Archimedian property implies

$$-\|\xi\|\varepsilon \leq \xi \leq \|\xi\|\varepsilon$$

for all $\xi \in \mathcal{V}$.

It is also possible to define complex function systems. To do this, we start with a complex vector space $\mathcal{V}$ equipped with an antilinear involution $\xi \mapsto \xi^*$. Then we require that the real part of $\mathcal{V}$, that is, the set $\operatorname{Re}\mathcal{V}$ of self-adjoint elements of $\mathcal{V}$, satisfy the previous definition. We norm $\mathcal{V}$ by setting

$$\|\xi\| = \sup_{|a|=1} \|\operatorname{Re} a\xi\|,$$

where the norm on the right side is the order norm on $\operatorname{Re}\mathcal{V}$. It is easy to see that the complexification of any real function system is a complex function system, and there is a perfect equivalence between the two definitions. We will work with real function systems because in the present setting the complex part is superfluous; however, when we pass to operator systems there is a very good reason for using complex scalars (see the comment following Definition 8.5.7).

Example 8.5.2

Let S be a compact convex subset of a locally convex TVS. Then the space $A(S)$ of real continuous affine functions on S is a function system.

Example 8.5.3

The self-adjoint part of any unital C*-algebra $\mathcal{A}$ is a function system, and the states on this function system are precisely the restrictions of the states on $\mathcal{A}$ in the sense of Definition 5.6.3.

As per the comment following Definition 5.6.3, states can equivalently be defined as positive linear functionals ω such that $\omega(\varepsilon) = 1$. Moreover, any linear functional $\omega : \mathcal{V} \to \mathbf{R}$ which satisfies $\|\omega\| = \omega(\varepsilon) = 1$ must be positive and hence a state, since $\xi \geq 0$ implies

$$\big|\|\xi\| - \omega(\xi)\big| = \omega(\|\xi\|\varepsilon - \xi) \leq \omega(\|\xi\|\varepsilon) = \|\xi\|$$

and therefore $\omega(\xi) \geq 0$. Thus, of the three conditions that ω be positive, that $\|\omega\| = 1$, and that $\omega(\varepsilon) = 1$, any two imply the third.

The next result justifies the term "function system," and together with Example 8.5.2 it also establishes a correspondence between function systems and compact convex sets.

PROPOSITION 8.5.4

Let $\mathcal{V}$ be a function system. Then $S = S(\mathcal{V})$ is a compact convex subset of $\mathcal{V}^*$, and the natural map $\xi \mapsto \hat{\xi}$ is an isometric order isomorphism of $\mathcal{V}$ onto the space $A(S)$ of real continuous affine functions on S.

PROOF It is straightforward to check that $S(\mathcal{V})$ is compact and convex. Also, the map $\xi \mapsto \hat{\xi}$ is clearly contractive and preserves order.

For any $\xi \in \mathcal{V}$ define

$$|\xi|^- = \sup\{a \in \mathbf{R} : a\varepsilon \leq \xi\}$$

and

$$|\xi|^+ = \inf\{a \in \mathbf{R} : \xi \leq a\varepsilon\}.$$

Then for any $|\xi|^- \leq a \leq |\xi|^+$, a short computation shows that the linear functional ω_0 defined on $\mathrm{span}(\xi, \varepsilon)$ by setting $\omega_0(\varepsilon) = 1$ and $\omega_0(\xi) = a$ satisfies $\|\omega_0\| = 1$. So ω_0 extends to a state ω on $\mathcal{V}$ by the Hahn-Banach theorem and the comment which preceded this proposition. In particular, taking a to be either $\|\xi\|$ or $-\|\xi\|$ (one of them must lie between $|\xi|^-$ and $|\xi|^+$), this shows that $\|\hat{\xi}\| \geq \|\xi\|$. So the map $\xi \mapsto \hat{\xi}$ is isometric. Also, if $\xi \in \mathcal{V}$ and $\xi \not\geq 0$, then we can take $a < 0$ and obtain $\hat{\xi} \not\geq 0$, and this shows that $\xi \mapsto \hat{\xi}$ is an order isomorphism.

We must show that this map is onto. Let $f \in A(S)$. Then define $T : \mathcal{V}^* \to \mathbf{R}$ by setting

$$T(a\omega - b\rho) = af(\omega) - bf(\rho)$$

for all $a, b \geq 0$ and $\omega, \rho \in S$. This map is well-defined by the following argument. Suppose $a\omega - b\rho = a'\omega' - b'\rho'$. Applying both sides to ε yields $a - b = a' - b'$, and we also have

$$\frac{a'\omega' + b\rho}{a' + b} \in S.$$

Thus

$$\begin{aligned} af(\omega) + b'f(\rho') &= (a + b')f\left(\frac{a\omega + b'\rho'}{a + b'}\right) \\ &= (a' + b)f\left(\frac{a'\omega' + b\rho}{a' + b}\right) \\ &= a'f(\omega') + bf(\rho), \end{aligned}$$

and we conclude that $af(\omega) - bf(\rho) = a'f(\omega') - b'f(\rho')$. So T is well-defined.

Next, we claim that T is defined on all of $\mathcal{V}^*$. To see this let

$$S' = \{t\omega - (1 - t)\rho : \omega, \rho \in S \text{ and } t \in [0, 1]\}.$$

This set is weak* compact because it is a continuous image of the compact set $S \times S \times [0,1]$, and it is also convex. So for any $\omega \in \mathcal{V}^*$ not belonging to S', there exists $\xi \in \mathcal{V}$ such that $\rho(\xi) \leq 1$ for all $\rho \in S'$ but $\omega(\xi) > 1$. Using the fact that the map $\xi \mapsto \hat{\xi}$ is isometric, and since $S \subset S' = -S'$, we obtain that $\|\xi\| \leq 1$, so we must have $\|\omega\| > 1$. Thus S' contains the unit ball of $\mathcal{V}^*$ (in fact the two are equal), and so T is defined everywhere on $\mathcal{V}^*$.

Now T is weak* continuous on S, so it is weak* continuous on S' and therefore on all of $\mathcal{V}^*$ by the Krein-Smulian theorem. Thus there exists $\xi \in \mathcal{V}$ such that $T(\omega) = \omega(\xi)$ for all $\omega \in \mathcal{V}^*$. In particular, for any $\omega \in S$ we have

$$\hat{\xi}(\omega) = \omega(\xi) = T(\omega) = f(\omega).$$

So $f = \hat{\xi}$, and we have shown that the map $\xi \mapsto \hat{\xi}$ takes $\mathcal{V}$ onto $A(S)$. ∎

There is also a correspondence between affine maps of compact convex sets and positive unital maps between function systems.

Example 8.5.5

Let $\mathcal{V}$ and $\mathcal{W}$ be function systems and let $T : \mathcal{V} \to \mathcal{W}$ be a positive unital map. Then composition with T defines a weak* continuous affine map from $S(\mathcal{W})$ into $S(\mathcal{V})$.

Example 8.5.6

Let S and S' be compact convex sets in locally convex TVSs and let $\phi : S \to S'$ be a continuous affine map. Then composition with ϕ defines a positive unital map from $A(S')$ into $A(S)$.

Now we define the Hilbert space analog of classical function systems.

DEFINITION 8.5.7 *If E is a vector space equipped with an involution, we say that a subspace of E is* <u>self-adjoint</u> *if it is stable under the involution.*

An <u>operator system</u> *is a self-adjoint, unital operator space $\mathcal{V}$. In other words, it is a closed subspace $\mathcal{V} \subset B(\mathcal{H})$ such that $I \in \mathcal{V}$, and $\xi \in \mathcal{V}$ implies $\xi^* \in \mathcal{V}$.*

A linear map $T : \mathcal{V} \to \mathcal{W}$ between operator systems is <u>completely positive</u> *if $T^{(n)} : M_n(\mathcal{V}) \to M_n(\mathcal{W})$ is positive for all $n \in \mathbf{N}$.*

Unlike the case of classical function systems, here there is no equivalent real version of the definition because positive elements of $M_n(\mathcal{V})$ may contain non self-adjoint entries for $n > 1$, and so the notion of complete positivity involves non self-adjoint elements of $\mathcal{V}$ in an essential way.

The same example which distinguishes boundedness from complete boundedness also distinguishes positivity from complete positivity: the transpose map on M_2 is positive but not completely positive.

We conclude this section with an abstract characterization of operator systems. If the vector space $\mathcal{V}$ is equipped with an antilinear involution * (i.e., a map $^* : \mathcal{V} \to \mathcal{V}$ such that $(a\xi)^* = \bar{a}\xi^*$ and $\xi^{**} = \xi$ for all $\xi \in \mathcal{V}$), then we define a corresponding involution on $M_n(\mathcal{V})$ by taking the transpose of a matrix and applying * to each entry.

DEFINITION 8.5.8 *A* matrix ordered space *is a complex vector space $\mathcal{V}$ equipped with an antilinear involution $\xi \mapsto \xi^*$ together with a vector space ordering on the self-adjoint part of each $M_n(\mathcal{V})$, such that if $\Xi \in M_n(\mathcal{V})$ and $A \in M_{n,m}$ then $\Xi \geq 0$ implies $A^*\Xi A \geq 0$. It is* unital *if there is a distinguished Archimedean order unit $\varepsilon \in \mathcal{V}$ such that*

$$E_n = \begin{bmatrix} \varepsilon & \cdots & 0 \\ \vdots & \ddots & \vdots \\ 0 & \cdots & \varepsilon \end{bmatrix}$$

is an Archimedean order unit for $M_n(\mathcal{V})$, for all n.

The matrix order norm *on a unital matrix ordered space $\mathcal{V}$ is defined by*

$$\|\Xi\| = \inf\left\{a \geq 0 : \begin{bmatrix} aE_n & \Xi \\ \Xi^* & aE_n \end{bmatrix} \geq 0\right\}$$

for $\Xi \in M_n(\mathcal{V})$.

Complete positivity for maps between matrix ordered spaces is defined just as for maps between operator systems.

The above definition of the matrix order norm is based on the fact that it gives the correct norm in $B(\mathcal{H})$ (and hence in any operator system), as a short computation shows. Thus:

Example 8.5.9

Every operator system is a unital matrix ordered space.

For self-adjoint elements of $\mathcal{V}$, the matrix order norm of Definition 8.5.8 also agrees with the order norm on $M_n(\mathcal{V})$ given in Definition 8.5.1. For if $\begin{bmatrix} E_n & \Xi \\ \Xi & E_n \end{bmatrix} \geq 0$ then

$$2(E_n - \Xi) = \begin{bmatrix} I_n & -I_n \end{bmatrix} \begin{bmatrix} E_n & \Xi \\ \Xi & E_n \end{bmatrix} \begin{bmatrix} I_n \\ -I_n \end{bmatrix} \geq 0$$

and

$$2(E_n + \Xi) = [\, I_n \quad I_n \,] \begin{bmatrix} E_n & \Xi \\ \Xi & E_n \end{bmatrix} \begin{bmatrix} I_n \\ I_n \end{bmatrix} \geq 0,$$

and hence $-E_n \leq \Xi \leq E_n$; while conversely, if $-E_n \leq \Xi \leq E_n$ then

$$2 \begin{bmatrix} E_n & \Xi \\ \Xi & E_n \end{bmatrix} = \begin{bmatrix} I_n & I_n \\ I_n & -I_n \end{bmatrix} \begin{bmatrix} E_n + \Xi & 0 \\ 0 & E_n - \Xi \end{bmatrix} \begin{bmatrix} I_n & I_n \\ I_n & -I_n \end{bmatrix} \geq 0.$$

Also, observe that if $\mathcal{V}$ is a matrix ordered space and $\omega : \mathcal{V} \to \mathbf{C}$ is positive, then ω is completely positive. To see this let $\Xi \in M_n(\mathcal{V})$ and suppose $\Xi \geq 0$. Then for any $v \in \mathbf{C}^n \cong M_{n,1}$ we have

$$\langle \omega^{(n)}(\Xi)v, v \rangle = \omega(v^* \Xi v) \geq 0,$$

and therefore $\omega^{(n)}(\Xi) \geq 0$.

THEOREM 8.5.10

Let $\mathcal{V}$ be a unital matrix ordered space which is complete for the matrix order norm. Then $\mathcal{V}$ is completely order isomorphic to an operator system via a completely isometric unital map.

PROOF Let T be the direct sum of all completely positive unital maps from $\mathcal{V}$ into M_n for $n \in \mathbf{N}$. It is clear that T is unital and that it preserves order. We must show that it is a complete order isomorphism; the fact that it is completely isometric will then follow because order and the unit determine the norm. Thus, let $\Xi \in M_n(\mathcal{V})$ and suppose $\Xi \not\leq 0$. We must find a completely positive unital map $T_0 : \mathcal{V} \to M_m$ such that $T_0^{(n)}(\Xi) \not\leq 0$.

$M_n(\mathcal{V})$ is a complex function system, so Proposition 8.5.4 (together with the first comment preceding this theorem) implies the existence of a state $\Omega : M_n(\mathcal{V}) \to \mathbf{C}$ such that $\Omega(\Xi) \not\leq 0$. By the second comment preceding this theorem, Ω is completely positive. Say $\Omega = [\omega_{ij}]$ with $\omega_{ij} \in \mathcal{V}^*$ and define $\tilde{\Omega} : \mathcal{V} \to M_n$ by $\tilde{\Omega}(\xi) = [\omega_{ij}(\xi)]$. We claim that $\tilde{\Omega}$ is also completely positive. To see this, let $H = [\eta_{ij}] \in M_m(\mathcal{V})$ be positive and define $\tilde{H} = [\tilde{\eta}_{ikp,jlq}] \in M_{mn^2}(\mathcal{V})$ by

$$\tilde{\eta}_{ikp,jlq} = \begin{cases} \eta_{i,j} & \text{if } k = p \text{ and } l = q \\ 0 & \text{otherwise;} \end{cases}$$

then $\tilde{H}$ is positive because we have $\tilde{H} = A^* H A$ where $A = [a_{i,jlq}] \in M_{m,mn}$ is the matrix

$$a_{i,jlq} = \begin{cases} 1 & \text{if } i = j \text{ and } l = q \\ 0 & \text{otherwise.} \end{cases}$$

So $\tilde{\Omega}^{(m)}(H) = \Omega^{(mn)}(\tilde{H}) \geq 0$ in M_{mn}, and this shows that $\tilde{\Omega}$ is completely positive. Moreover, in the case $m = n$ we have

$$\tilde{\Omega}^{(n)}(\Xi) = \Omega^{(n^2)}(A^*\Xi A) = \Omega(\Xi)A^*A \not\leq 0.$$

Thus $\tilde{\Omega}$ is a completely positive map from $\mathcal{V}$ to M_n and $\tilde{\Omega}^{(n)}(\Xi) \not\leq 0$.

Finally, $\tilde{\Omega}(\varepsilon)$ is a positive matrix in M_n, so there is a matrix $B \in M_{n,m}$ such that $B^*\tilde{\Omega}(\varepsilon)B$ is the identity in M_m, where m is the rank of $\tilde{\Omega}(\varepsilon)$. Then $T_0 = B^*\tilde{\Omega}B$ is the desired completely positive unital map. ∎

8.6 Notes

See [22] for more on operator spaces generally.

Ruan's theorem (Theorem 8.2.11) was proven in [62]. The separation theorem presented in Theorem 8.2.10 is taken from [23]. There is also a Hahn-Banach theorem for completely bounded maps of operator spaces into $B(\mathcal{H})$, known as the Arveson-Wittstock extension theorem; see [77].

For more on the dual of an operator space see [9]. Proposition 8.3.4 is from [44]; that paper also contains an example of an operator space which is a dual Banach space but not a dual operator space. Section 8.4 is based on [70].

Our treatment of function systems follows [21]. The abstract characterization of operator systems is from [11].

Chapter 9

Hilbert Modules

9.1 Continuous Hilbert bundles

Measurable Hilbert bundles were introduced in Chapter 2 (Definition 2.4.8), and they played an important role in Chapters 3, 5, and 6. We begin this chapter by describing a continuous version of the construction; next we revisit the measurable version; and then we introduce Hilbert modules over C*- and von Neumann algebras, which are the quantum analogs of Hilbert bundles.

DEFINITION 9.1.1 *Let X be a compact Hausdorff space. A covering space of X is a topological space Y together with a continuous open surjection $p : Y \to X$. A continuous Hilbert bundle over X is then a covering space $\mathcal{X}$ such that $\mathcal{H}_x = p^{-1}(x)$ is equipped with a Hilbert space structure for each $x \in X$, and satisfying the following conditions:*

(a) the map $y \mapsto \|y\|$ is continuous from $\mathcal{X}$ to $\mathbf{R}$;
(b) the map $(y, z) \mapsto y + z$ is continuous from $\mathcal{X} \times \mathcal{X}$ to $\mathcal{X}$;
(c) for each $a \in \mathbf{C}$, the map $y \mapsto ay$ is continuous from $\mathcal{X}$ to $\mathcal{X}$; and
(d) for any neighborhood $\mathcal{O}$ of the origin of $\mathcal{H}_x$ in $\mathcal{X}$ there exists a neighborhood $\mathcal{O}'$ of x in X and an $\epsilon > 0$ such that

$$\{y \in \mathcal{X} : p(y) \in \mathcal{O}' \text{ and } \|y\| < \epsilon\} \subset \mathcal{O}.$$

A section of $\mathcal{X}$ is a function $\phi : X \to \mathcal{X}$ such that $\phi(x) \in \mathcal{H}_x$ for all $x \in X$. The set of all continuous sections of $\mathcal{X}$ is denoted $\mathcal{S}(\mathcal{X})$.

The prototypical examples of continuous Hilbert bundles are geometric. Let X be a smooth manifold; then at each point x of X we have a real vector space T_x which consists of all tangent vectors at x. This is the tangent space at x, and the disjoint union of all of the tangent spaces is the tangent bundle $TX = \bigcup_x T_x$. A smooth manifold is Riemannian if each tangent space is equipped with an inner product and these vary in an appropriately smooth manner as x varies. Now each tangent space is a real Hilbert space, so complexifying TX gives rise to a continuous Hilbert bundle over X.

Example 9.1.2

Let X be a compact Riemannian manifold and let $\mathcal{X} = TX + iTX$ be the complexification of the tangent bundle of X. Then $\mathcal{X}$ is a continuous Hilbert bundle.

The space of continuous sections of a continuous Hilbert bundle has a special structure. It is a vector space, as sections can be multiplied by scalars and added pointwise. But in fact a continuous section $\phi \in \mathcal{S}(\mathcal{X})$ can be multiplied pointwise by any continuous function $f \in C(X)$, and the product is again a continuous section. Thus $\mathcal{S}(\mathcal{X})$ is a module over $C(X)$.

$\mathcal{S}(\mathcal{X})$ has even more structure. For any continuous sections ϕ and ψ of $\mathcal{X}$, we can take their inner product pointwise, and the result will be a continuous function on X. That is, we define

$$\langle \phi, \psi \rangle (x) = \langle \phi(x), \psi(x) \rangle;$$

then $\langle \phi, \psi \rangle \in C(X)$ for any $\phi, \psi \in \mathcal{S}(\mathcal{X})$. Continuity follows from the fact that the map $(y, z) \mapsto \langle y, z \rangle$ is continuous on $\mathcal{X} \times \mathcal{X}$, which follows from property (a) of continuous Hilbert bundles by polarization.

This motivates the following definition.

DEFINITION 9.1.3 *Let X be a compact Hausdorff space and let $\mathcal{E}$ be a $C(X)$-module. A $C(X)$-valued pseudo inner product on $\mathcal{E}$ is a map $\langle \cdot, \cdot \rangle : \mathcal{E} \times \mathcal{E} \to C(X)$ satisfying*

(a) $\langle f\phi_1 + g\phi_2, \psi \rangle = f\langle \phi_1, \psi \rangle + g\langle \phi_2, \psi \rangle$;
(b) $\overline{\langle \phi, \psi \rangle} = \langle \psi, \phi \rangle$; *and*
(c) $\langle \phi, \phi \rangle \geq 0$

for all $f, g \in C(X)$ and $\phi_1, \phi_2, \phi, \psi \in \mathcal{E}$. We write $|\phi| = \langle \phi, \phi \rangle^{1/2}$ and define $\|\phi\|$ to be the supremum norm of $|\phi|$ in $C(X)$.

If $\|\phi\| = 0$ implies $\phi = 0$ then $\langle \cdot, \cdot \rangle$ is a C(X)-valued inner product, and a Hilbert module over $C(X)$ is a $C(X)$-module equipped with a $C(X)$-valued inner product whose corresponding norm $\| \cdot \|$ is complete.

Observe that if $\langle \cdot, \cdot \rangle$ is a $C(X)$-valued pseudo inner product then the map $(\phi, \psi) \mapsto \langle \phi, \psi \rangle (x)$ is a pseudo inner product for each $x \in X$. It follows that $\|\phi\| = \sup \langle \phi, \phi \rangle^{1/2}(x)$ is a pseudonorm. Thus $\| \cdot \|$ really is a norm if $\langle \cdot, \cdot \rangle$ is a $C(X)$-valued inner product.

Example 9.1.4

Let $\mathcal{X}$ be a continuous Hilbert bundle over a compact Hausdorff space X. Then $\mathcal{S}(X)$ is a Hilbert module over $C(X)$.

This example has a converse: every Hilbert module is the module of continuous sections of a continuous Hilbert bundle. The construction goes as follows. Given a Hilbert module $\mathcal{E}$ over $C(X)$ and an element $x \in X$, let $\mathcal{I}_x = \{f \in C(X) : f(x) = 0\}$ and define $\mathcal{H}_x = \mathcal{E}/\mathcal{I}_x\mathcal{E}$. Here $\mathcal{I}_x\mathcal{E}$ is the closed span of the set of elements $f\phi$ with $f \in \mathcal{I}_x$ and $\phi \in \mathcal{E}$. The $C(X)$-valued inner product on $\mathcal{E}$ then passes to an ordinary inner product on $\mathcal{H}_x$, and $\mathcal{H}_x$ is the quotient of a Banach space by a closed subspace, so it is complete. Thus each $\mathcal{H}_x$ is a Hilbert space. Also, for each $\phi \in \mathcal{E}$ let $\phi^x = \phi + \mathcal{I}_x$ be its projection in $\mathcal{H}_x$. Now we need the following proposition.

PROPOSITION 9.1.5
Let X be a compact Hausdorff space, let $\mathcal{E}$ be a Hilbert module over $C(X)$, and let $\mathcal{X} = \bigcup \mathcal{H}_x$ be the disjoint union of the Hilbert spaces $\mathcal{H}_x$ defined above. Then there is a unique topology on $\mathcal{X}$ which makes it a continuous Hilbert bundle over X and such that the map $x \mapsto \phi^x$ is a continuous section for all $\phi \in \mathcal{E}$.

PROOF The desired topology is characterized by the condition that $\phi_\kappa^{x_\kappa} \to \phi^x$ if and only if $x_\kappa \to x$ and $|\phi_\kappa - \phi|(x) \to 0$. A basis for this topology is given by the sets

$$\{\phi^x \in \mathcal{X} : \phi \in \mathcal{E},\ x \in \mathcal{O},\ \text{ and } \|\phi^x - \psi^x\| \leq \epsilon\},$$

where $\mathcal{O}$ is an open subset of X, $\psi \in \mathcal{E}$, and $\epsilon > 0$. It is routine but tedious to verify that this is a basis for a topology and that it makes $\mathcal{X}$ into a continuous Hilbert bundle with the stated property. ∎

We write $\mathcal{B}(\mathcal{E})$ for the continuous Hilbert bundle provided by Proposition 9.1.5. The following result shows that there is a perfect correspondence between continuous Hilbert bundles and Hilbert modules. The proof is rather long and we therefore omit it.

THEOREM 9.1.6
Let X be a compact Hausdorff space.

(a) $\mathcal{S}(\mathcal{B}(\mathcal{E}))$ is canonically isomorphic to $\mathcal{E}$, for any Hilbert module $\mathcal{E}$ over $C(X)$.

(b) $\mathcal{B}(\mathcal{S}(\mathcal{X}))$ is canonically isomorphic to $\mathcal{X}$, for any continuous Hilbert bundle $\mathcal{X}$ over X.

9.2 Hilbert L^∞-modules

It is typically the case that the natural L^∞ (or von Neumann algebra) version of a $C(X)$ (or C*-algebra) construction involves dual spaces and weak* topologies in place of Banach spaces and norm topologies. In the case of Hilbert modules, the appropriate requirement is self-duality in the following sense.

DEFINITION 9.2.1 *Let X be a σ-finite measure space and let $\mathcal{E}$ be a Hilbert module over $L^\infty(X)$. Then the dual module $\mathcal{E}'$ is the set of bounded $L^\infty(X)$-linear maps from $\mathcal{E}$ into $L^\infty(X)$. The module structure of $\mathcal{E}'$ is defined by*

$$(f \cdot \Phi)(\phi) = \bar{f}\Phi(\phi),$$

for $f \in L^\infty(X)$, $\Phi \in \mathcal{E}'$, and $\phi \in E$.

For any $\phi \in \mathcal{E}$, let $\hat{\phi} \in \mathcal{E}'$ be the map $\hat{\phi}(\psi) = \langle \psi, \phi \rangle$. If every element of $\mathcal{E}'$ is of the form $\hat{\phi}$ for some $\phi \in \mathcal{E}$, then we say that $\mathcal{E}$ is self-dual.

It is fairly easy to show that any self-dual Hilbert module over $L^\infty(X)$ is a dual Banach space. We will prove this in greater generality in Proposition 9.4.2. Anyway under suitable separability assumptions we can completely characterize the structure of self-dual Hilbert modules over $L^\infty(X)$, and duality follows easily from this characterization.

The measurable Hilbert bundles used in earlier chapters give rise to examples of self-dual Hilbert L^∞-modules. First we present the construction, and then we prove that it satisfies the condition of self-duality.

DEFINITION 9.2.2 *Let X be a σ-finite measure space and let $\mathcal{X} = \bigcup(X_n \times \mathcal{H}_n)$ be a (separable) measurable Hilbert bundle over X, as in Definition 2.4.8. The Hilbert module of L^∞ sections of $\mathcal{X}$ is the set $L^\infty(X;\mathcal{X})$ of weakly measurable essentially bounded functions $\phi : X \to \bigcup \mathcal{H}_n$ with the property that $\phi(x) \in \mathcal{H}_n$ for all $x \in X_n$.*

Recall that "weakly measurable" means that for each $n \in \mathbf{N}$ and $v \in \mathcal{H}_n$ the function $x \mapsto \langle \phi(x), v \rangle$ is measurable on X_n.

PROPOSITION 9.2.3
Let X be a σ-finite measure space and let $\mathcal{X} = \bigcup(X_n \times \mathcal{H}_n)$ be a measurable Hilbert bundle over X. Then $L^\infty(X;\mathcal{X})$ is a self-dual Hilbert module over $L^\infty(X)$.

PROOF It is straightforward to check that $L^\infty(X;\mathcal{X})$ is a Hilbert module over $L^\infty(X)$. To verify self-duality, let $\Phi \in L^\infty(X;\mathcal{X})'$. Assume $\mathcal{X} = X \times \mathcal{H}$ where $\mathcal{H}$ is separable and infinite dimensional; the finite dimensional case is similar, and passing to a disjoint union of such bundles is easy.

Let (e_n) be an orthonormal basis of $\mathcal{H}$. Define $f_n = \Phi(1_X e_n)$. Then $f_n \in L^\infty(X)$, and for any $g_1, \ldots, g_n \in L^\infty(X)$ we have

$$\Phi\left(\sum_{i=1}^n g_i e_i\right) = \sum_{i=1}^n g_i f_i = \left\langle \sum_{i=1}^n g_i e_i, \phi_n \right\rangle$$

where $\phi_n = \sum_1^n \bar{f}_i e_i$. In particular, we have $\Phi(\phi_n) = |\phi_n|^2$, so boundedness of Φ implies that the sequence (ϕ_n) is bounded in norm. But $|\phi_n|$ is increasing, so it follows that $|\phi_n|(x)$ is Cauchy for almost every $x \in X$.

Thus $\phi = \lim \phi_n$ exists almost everywhere, is weakly measurable, and is essentially bounded, and we have $\Phi(\psi) = \langle \psi, \phi \rangle$ for any ψ which can be written as a finite $L^\infty(X)$-linear combination of the sections $1_X e_n$. It is not yet clear that this equality holds for all $\psi \in L^\infty(X;\mathcal{X})$, because finite linear combinations of the sections $1_X e_n$ are in general not dense. (For example, they cannot approximate in norm a section which for all n constantly takes the value e_n on some positive measure set.) However, for any $\psi \in L^\infty(X;\mathcal{X})$ and any $\epsilon > 0$ we can find a measurable partition (X_m) of X and a sequence of sections ψ_m which are finite $L^\infty(X)$-linear combinations of the sections $1_X e_n$, such that ψ_m is supported on X_m and $\|\psi - \sum \psi_m\| \leq \epsilon$. Then

$$\chi_{X_m} \cdot \Phi\left(\sum \psi_i\right) = \Phi(\psi_m) = \langle \psi_m, \phi \rangle = \chi_{X_m} \cdot \left\langle \sum \psi_i, \phi \right\rangle$$

for all m, and hence $\Phi(\sum \psi_i) = \langle \sum \psi_i, \phi \rangle$. It now follows by continuity that $\Phi(\psi) = \langle \psi, \phi \rangle$ for all $\psi \in L^\infty(X;\mathcal{X})$, and we conclude that $L^\infty(X;\mathcal{X})$ is self-dual. ∎

Now we want to show that under reasonable separability assumptions all self-dual Hilbert L^∞-modules arise in the above fashion from measurable Hilbert bundles. The appropriate condition is that the module should be separable for the weak topology induced by the inner product, that is, the weakest topology on $\mathcal{E}$ such that for all $\psi \in \mathcal{E}$ the map $\phi \mapsto \langle \phi, \psi \rangle$ is continuous as a map into $L^\infty(X)$ with the weak* topology. Thus, if (ϕ_κ) is a bounded net in $\mathcal{E}$ then $\phi_\kappa \to \phi$ weakly if and only if $\langle \phi_\kappa, \psi \rangle \to \langle \phi, \psi \rangle$ weak* in $L^\infty(X)$ for all $\psi \in \mathcal{E}$. We will show in Proposition 9.4.2 that on bounded sets this is actually a weak* topology.

THEOREM 9.2.4

Let X be a σ-finite measure space and let $\mathcal{E}$ be a self-dual Hilbert module *over $L^\infty(X)$ which is weakly separable. Then there is a measurable Hilbert bundle $\mathcal{X}$ over X such that $\mathcal{E} \cong L^\infty(X;\mathcal{X})$.*

PROOF Any σ-finite measure can be replaced by a probability measure with which it is mutually absolutely continuous, so without loss of generality suppose X is a probability measure space and let μ be the measure on X. Define an inner product $[\cdot,\cdot]$ on $\mathcal{E}$ by setting $[\phi,\psi] = \int \langle\phi,\psi\rangle\, d\mu$ and let $\mathcal{H}$ be the Hilbert space completion of $\mathcal{E}$ for this inner product. If $\phi_n \to \phi$ weakly then $[\phi_n,\psi] \to [\phi,\psi]$ for all $\psi \in \mathcal{E}$, which implies that $\mathcal{H}$ is weakly separable and hence separable. Let $\|\cdot\|_2$ denote the norm on $\mathcal{H}$.

For each measurable subset S of X, let $E(S)$ be the closure in $\mathcal{H}$ of $\chi_S \cdot \mathcal{E}$. It is straightforward to verify that E is an $\mathcal{H}$-valued spectral measure on X.

By Corollary 3.4.3 there is a probability measure μ' on X, a measurable Hilbert bundle $\mathcal{X}$ over X, and an isometric isomorphism U : $L^2(X;\mathcal{X}) \cong \mathcal{H}$ such that $U(L^2(S;\mathcal{X}|_S)) = E(S)$ for every measurable set $S \subset X$. If S is null with respect to μ then $E(S) = 0$, so μ' is absolutely continuous with respect to μ. Possibly letting $\mathcal{X}$ be zero on a positive measure set, we can therefore assume that the two measures are mutually absolutely continuous. Having done this, it is not hard to see that by modifying U we can actually take $\mu' = \mu$.

Next we show that U^{-1} takes $\mathcal{E}$ isometrically into $L^\infty(X;\mathcal{X})$. First, the Cauchy-Schwarz inequality applied pointwise shows that

$$|\langle\phi,\psi\rangle|(x) \le |\phi|(x)|\psi|(x) \le |\phi|(x)\|\psi\|$$

almost everywhere, and hence $\|\langle\phi,\psi\rangle\|_2 \le \|\psi\|\|\phi\|_2$ for any $\phi,\psi \in \mathcal{E}$. It follows that for all $\psi \in \mathcal{E}$ the map $\phi \mapsto \langle\phi,\psi\rangle$ from $\mathcal{H}$ to $L^2(X)$ has norm at most $\|\psi\|$. Conversely, for any $\epsilon > 0$ let S be a positive measure set on which $|\psi| \ge \|\psi\| - \epsilon$; then taking $\phi = \chi_S\psi$ yields

$$\|\phi\|_2^2 = \int_S \langle\psi,\psi\rangle \le \|\psi\|^2\mu(S),$$

while

$$\|\langle\phi,\psi\rangle\|_2^2 = \int_S |\langle\psi,\psi\rangle|^2 \ge (\|\psi\| - \epsilon)^4\mu(S),$$

and we conclude that the map $\phi \mapsto \langle\phi,\psi\rangle$ from $\mathcal{H}$ to $L^2(X)$ has norm exactly $\|\psi\|$. The same argument shows that the L^∞ norm of $U^{-1}\psi$ equals the norm of the map $U^{-1}\phi \mapsto \langle U^{-1}\phi, U^{-1}\psi\rangle$. So the fact that U is isometric for L^2 norms implies that U^{-1} takes $\mathcal{E}$ isometrically into $L^\infty(X;\mathcal{X})$.

It is clear that $U^{-1} : \mathcal{E} \to L^\infty(X;\mathcal{X})$ respects $L^\infty(X)$-module structure. For any $\phi \in \mathcal{E}$ let $f = \langle \phi, \phi \rangle$ and $g = \langle U^{-1}\phi, U^{-1}\phi \rangle$. By $L^\infty(X)$-linearity and the fact that U^{-1} is an isometry for L^∞ norms, we have $\|f|_S\|_\infty = \|g|_S\|_\infty$ for any $S \subset X$. Since f and g are both positive, this implies $f = g$ almost everywhere; polarization then implies that $\langle \phi, \psi \rangle = \langle U^{-1}\phi, U^{-1}\psi \rangle$ for all $\phi, \psi \in \mathcal{E}$. Thus U^{-1} is compatible with the $L^\infty(X)$-valued inner product.

Finally, to show U^{-1} maps $\mathcal{E}$ onto $L^\infty(X;\mathcal{X})$, let $\psi \in L^\infty(X;\mathcal{X})$. Then the map $\phi \mapsto \langle U^{-1}\phi, \psi \rangle$ is a bounded $L^\infty(X)$-linear map from $\mathcal{E}$ into $L^\infty(X)$, so by self-duality there exists $\psi' \in \mathcal{E}$ such that $\langle \phi, U\psi \rangle = \langle U^{-1}\phi, \psi \rangle = \langle \phi, \psi' \rangle$ for all $\phi \in \mathcal{E}$. Then $U\psi - \psi'$ is orthogonal (in $\mathcal{H}$) to every element of $\mathcal{E}$, and since $\mathcal{E}$ is dense in $\mathcal{H}$ it follows that $U\psi - \psi' = 0$. Thus $\psi = U^{-1}\psi'$, and we have shown that U^{-1} maps $\mathcal{E}$ onto $L^\infty(X;\mathcal{X})$. ∎

Conversely, every Hilbert module constructed as in Definition 9.2.2 is weakly separable. So Theorem 9.2.4 exactly characterizes weakly separable self-dual Hilbert L^∞-modules.

9.3 Hilbert C*-modules

There is a C*-algebraic version of the Hilbert $C(X)$-modules introduced in Section 9.1; in fact, due to noncommutativity there are two versions, left Hilbert modules and right Hilbert modules. There is no essential difference between the two cases, but in various circumstances one or the other may be more convenient. The basic conflict lies in the fact that Hilbert space inner products are antilinear in the second variable, but if operators act from the left then it is natural to have scalars act from the right, which accords better with linearity in the second variable.

We use the symbols α and β to denote elements of an abstract C*-algebra.

DEFINITION 9.3.1 *Let $\mathcal{A}$ be a C*-algebra and let $\mathcal{E}$ be a left $\mathcal{A}$-module. An* <u>*$\mathcal{A}$-valued pseudo inner product*</u> *on $\mathcal{E}$ is a map $\langle \cdot, \cdot \rangle : \mathcal{E} \times \mathcal{E} \to \mathcal{A}$ satisfying*

(a) $\langle \alpha\xi_1 + \beta\xi_2, \eta \rangle = \alpha\langle \xi_1, \eta \rangle + \beta\langle \xi_2, \eta \rangle$;
(b) $\langle \xi, \eta \rangle^* = \langle \eta, \xi \rangle$; *and*
(c) $\langle \xi, \xi \rangle \geq 0$

for all $\alpha, \beta \in \mathcal{A}$ and $\xi_1, \xi_2, \xi, \eta \in \mathcal{E}$. We write $|\xi| = \langle \xi, \xi \rangle^{1/2}$ and define $\|\xi\|$ to be the norm of $|\xi|$ in $\mathcal{A}$.

If $\|\xi\| = 0$ implies $\xi = 0$ then $\langle \cdot, \cdot \rangle$ is an <u>*$\mathcal{A}$-valued inner product*</u> *and a* <u>*left Hilbert module*</u> *over $\mathcal{A}$ is a left $\mathcal{A}$-module equipped with an $\mathcal{A}$-valued* *inner product whose corresponding norm* $\|\cdot\|$ *is complete.*

Right Hilbert modules *are defined similarly, with the axiom*

(a') $\langle \xi, \eta_1\alpha + \eta_2\beta\rangle = \langle \xi, \eta_1\rangle\alpha + \langle \xi, \eta_2\rangle\beta$

in place of (a).

This definition is contingent on our verifying that $\|\cdot\|$ really is a pseudonorm, which we will do momentarily.

It is easy to check that if $\mathcal{A}^+$ is the unitization of $\mathcal{A}$ (as in Definition 5.3.1) and $\mathcal{E}$ is a Hilbert module over $\mathcal{A}$ then $\mathcal{E}$ is also a Hilbert module over $\mathcal{A}^+$ via the obvious action of $\mathcal{A}^+$ on $\mathcal{E}$.

Left modules can be converted into right modules in the following way. Let $\mathcal{E}$ be a left $\mathcal{A}$-module. Then define $\mathcal{E}^{op}$ to be the set $\mathcal{E}$, denoting by $\bar{\xi}$ the element of $\mathcal{E}^{op}$ corresponding to $\xi \in \mathcal{E}$, with addition as defined in $\mathcal{E}$ but with module multiplication given by $\bar{\xi}\alpha = \overline{\alpha^*\xi}$. It is easy to see that $\mathcal{E}^{op}$ is a right $\mathcal{A}$-module, and an $\mathcal{A}$-valued pseudo inner product on $\mathcal{E}$ is converted to one on $\mathcal{E}^{op}$ by setting $\langle\bar{\xi},\bar{\eta}\rangle = \langle \eta, \xi\rangle^*$. Similarly, right modules can be converted into left modules, so there is no essential difference between left and right Hilbert modules. Thus we will feel free to use whichever version is more convenient at any given moment.

Our immediate burden is to prove that $\|\cdot\|$ is a pseudonorm on $\mathcal{E}$; this amounts to proving the triangle inequality, which is a consequence of the following version of the Cauchy-Schwarz inequality.

PROPOSITION 9.3.2

Let $\mathcal{A}$ *be a C*-algebra and let* $\mathcal{E}$ *be a left* $\mathcal{A}$*-module equipped with an* $\mathcal{A}$*-valued pseudo inner product* $\langle\cdot,\cdot\rangle$*. Then*

$$\langle \xi, \eta\rangle\langle \eta, \xi\rangle \leq |\xi|^2\|\eta\|^2$$

for all $\xi, \eta \in \mathcal{E}$.

PROOF Let ω be a state on $\mathcal{A}$. Then the map $(\xi,\eta) \mapsto \omega(\langle\xi,\eta\rangle)$ is a pseudo inner product on $\mathcal{E}$ and it therefore satisfies the ordinary Cauchy-Schwarz inequality. Thus, letting $\alpha = \langle \xi, \eta\rangle$, we have

$$\begin{aligned}\omega(\alpha\alpha^*) &= \omega(\langle \alpha\eta, \xi\rangle)\\ &\leq \omega(\langle\alpha\eta,\alpha\eta\rangle)^{1/2}\omega(\langle\xi,\xi\rangle)^{1/2}\\ &= \omega(\alpha|\eta|^2\alpha^*)^{1/2}\omega(|\xi|^2)^{1/2}\\ &\leq \|\eta\|\omega(\alpha\alpha^*)^{1/2}\omega(|\xi|^2)^{1/2},\end{aligned}$$

which implies $\omega(\alpha\alpha^*) \leq \|\eta\|^2\omega(|\xi|^2)$. Since this is true for any state, it follows that $\langle\xi,\eta\rangle\langle\eta,\xi\rangle = \alpha\alpha^* \leq |\xi|^2\|\eta\|^2$. ∎

In particular, we have $\|\langle\xi,\eta\rangle\| \leq \|\xi\|\|\eta\|$, and this implies the triangle inequality for $\|\cdot\|$ in the usual way. So $\|\xi\| = \|\langle\xi,\xi\rangle\|^{1/2}$ is indeed a pseudonorm. It also follows, just as in the scalar case, that factoring out null vectors converts an $\mathcal{A}$-valued pseudo inner product into an $\mathcal{A}$-valued inner product, and completing an $\mathcal{A}$-valued inner product yields a Hilbert module. So there is no obstruction to making a Hilbert module out of any $\mathcal{A}$-module equipped with an $\mathcal{A}$-valued pseudo inner product.

Example 9.3.3

(a) Let $\mathcal{A}$ be a C*-algebra and X a set. Define $l^2(X;\mathcal{A})$ to be the set of functions $f : X \to \mathcal{A}$ such that $\sum_x f(x)f(x)^*$ converges in $\mathcal{A}$. Then $l^2(X;\mathcal{A})$ is a left Hilbert $\mathcal{A}$-module with the obvious left action of $\mathcal{A}$ and the inner product

$$\langle f,g\rangle = \sum_{x\in X} f(x)g(x)^*.$$

The proof that this expression is well-defined is similar to the scalar case (Example 2.1.7), as is the proof that $l^2(X;\mathcal{A})$ is complete.
(b) Now let X be a σ-finite measure space. The simplest way to define $L^2(X;\mathcal{A})$ is to take a completion of the set of all simple functions from X to $\mathcal{A}$ whose support has finite measure. This set has an $\mathcal{A}$-valued pseudo inner product defined by linear extension of the equation

$$\langle\alpha\chi_S,\beta\chi_T\rangle = \mu(S\cap T)\alpha\beta^*,$$

for $\alpha,\beta\in\mathcal{A}$ and S and T finite measure subsets of X.

Next we describe the Hilbert module analog of $B(\mathcal{H})$.

DEFINITION 9.3.4 *Let $\mathcal{E}$ be a Hilbert module over a C*-algebra $\mathcal{A}$. A bounded $\mathcal{A}$-linear map $A : \mathcal{E}\to\mathcal{E}$ is* <u>adjointable</u> *if there exists a bounded $\mathcal{A}$-linear map $A^* : \mathcal{E}\to\mathcal{E}$ such that $\langle A\xi,\eta\rangle = \langle\xi,A^*\eta\rangle$ for all $\xi,\eta\in\mathcal{E}$. We denote the set of all bounded adjointable $\mathcal{A}$-linear maps from $\mathcal{E}$ to itself by $B(\mathcal{E})$.*

Adjointability is not automatic, even in the commutative case. For example, let $\mathcal{E} = C[0,1]\oplus C_0(0,1]$, where we identify $C_0(0,1]$ with the continuous functions on $C[0,1]$ which vanish at 0. The action $h(f\oplus g) = hf\oplus hg$ and the inner product $\langle f_1\oplus g_1, f_2\oplus g_2\rangle = f_1\bar{f}_2 + g_1\bar{g}_2$ make $\mathcal{E}$ into a left Hilbert $C[0,1]$-module. Now consider the map $A : \mathcal{E}\to\mathcal{E}$ defined by $A(f\oplus g) = g\oplus 0$. This is clearly a bounded $\mathcal{A}$-linear map, but

$$\langle A(f\oplus g), 1\oplus 0\rangle = g,$$

so that if A were adjointable we would have

$$\langle f \oplus g, A^*(1 \oplus 0)\rangle = g$$

for all f and g. But there is no possible value of $A^*(1 \oplus 0)$ for which this can hold.

Adjointability is automatic in the von Neumann algebra setting, however: see the comment following Definition 9.4.1.

LEMMA 9.3.5
Let $\mathcal{E}_1$ and $\mathcal{E}_2$ be left Hilbert modules over a C-algebra $\mathcal{A}$ and let $A : \mathcal{E}_1 \to \mathcal{E}_2$ be a bounded $\mathcal{A}$-linear map. Then $|A\xi|^2 \leq \|A\|^2|\xi|^2$ for all $\xi \in \mathcal{E}_1$.*

PROOF Fix $\epsilon > 0$ and define $f \in C_0(\mathbf{R})$ by $f(t) = 0$ for $t \leq \epsilon$, $f(t) = 1$ for $t \geq 2\epsilon$, and $f(t) = (t - \epsilon)/\epsilon$ for $\epsilon \leq t \leq 2\epsilon$. Let $\eta = f(|\xi|)\xi$. Then, working momentarily in the unitization of $\mathcal{A}$ (which we can do by the comment following Definition 9.3.1), we have

$$|\xi - \eta| = (I_{\mathcal{A}} - f(|\xi|))|\xi| \to 0$$

as $\epsilon \to 0$. It follows that $\eta \to \xi$, and hence $A\eta \to A\xi$, as $\epsilon \to 0$, so it will suffice to show that $|A\eta|^2 \leq \|A\|^2|\xi|^2$. (This is because

$$\begin{aligned} |A\xi|^2 &= |A\eta + A(\xi - \eta)|^2 \\ &= |A\eta|^2 + 2\mathrm{Re}\langle A\eta, A(\xi - \eta)\rangle + |A(\xi - \eta)|^2, \end{aligned}$$

and the last two terms go to zero in norm as $\eta \to \xi$.)

Now fix $g \in C_0(\mathbf{R})$ such that $g(0) = 0$ and $g(t) = 1/t$ for $t \geq \epsilon$, and define $\zeta = fg(|\xi|)\xi$. We have $|\zeta| = f(|\xi|) \leq I_{\mathcal{A}}$, so $\|\zeta\| \leq 1$; therefore $\|A\zeta\| \leq \|A\|$, which implies

$$f(|\xi|)|A\xi|^2 f(|\xi|) = |\xi||A\zeta|^2|\xi| \leq \|A\||\xi|^2.$$

Hence $|A\eta|^2 \leq \|A\|^2|\xi|^2$, as desired. ∎

PROPOSITION 9.3.6
Let $\mathcal{E}$ be a left Hilbert module over a C-algebra $\mathcal{A}$. Then $B(\mathcal{E})$ is an abstract C*-algebra.*

PROOF Without loss of generality assume $\mathcal{A}$ is unital. For each state ω on $\mathcal{A}$ let $\mathcal{H}_\omega$ be the Hilbert space formed from the pseudo inner product $\omega(\langle\cdot,\cdot\rangle)$ on $\mathcal{E}$ by factoring out null vectors and completing. Let

$\mathcal{H} = \bigoplus \mathcal{H}_\omega$ be the direct sum of all of these Hilbert spaces. Then define a map $\pi : B(\mathcal{E}) \to B(\mathcal{H})$ by setting $\pi(A)(\xi_\omega) = (A\xi)_\omega$ for $A \in B(\mathcal{E})$, where ξ_ω is the element of $\mathcal{H}_\omega$ corresponding to $\xi \in \mathcal{E}$.

Applying an arbitrary state ω to both sides of the inequality $|A\xi|^2 \leq \|A\|^2|\xi|^2$ proven in the lemma, we find that π is well-defined, $\pi(A) \in B(\mathcal{H})$, and $\|\pi(A)\| \leq \|A\|$. Conversely, for any $\xi \in \mathcal{E}$ and $\epsilon > 0$ we can find a state ω on $\mathcal{A}$ such that $\omega(|A\xi|^2) \geq \|A\xi\|^2 - \epsilon$; then

$$\|\pi(A)\xi_\omega\|^2 = \|(A\xi)_\omega\|^2 = \omega(|A\xi|^2) \geq \|A\xi\|^2 - \epsilon.$$

But $\|\xi_\omega\|^2 = \omega(|\xi|^2) \leq \|\xi\|^2$, so this shows that $\|\pi(A)\| \geq \|A\xi\|/\|\xi\|$. As this is true for all ξ, we conclude that π is an isometry.

It is clear that $B(\mathcal{E})$ is complete in norm and π is a $*$-homomorphism. So $B(\mathcal{E})$ is $*$-isomorphic to a C*-subalgebra of $B(\mathcal{H})$. ∎

As a first illustration of the way Proposition 9.3.6 can be used, we now present a version of the GNS construction for completely positive maps between C*-algebras. Let $\mathcal{A}$ and $\mathcal{B}$ be C*-algebras and let $T : \mathcal{A} \to \mathcal{B}$ be completely positive (Definition 8.5.7). Regard the algebraic tensor product $\mathcal{A} \otimes \mathcal{B}$ as a right $\mathcal{B}$-module via the action $(\alpha \otimes \beta)\gamma = \alpha \otimes \beta\gamma$. We define a $\mathcal{B}$-valued pseudo inner product on $\mathcal{A} \otimes \mathcal{B}$ by setting

$$\langle \alpha_1 \otimes \beta_1, \alpha_2 \otimes \beta_2 \rangle = \beta_1^* T(\alpha_1^* \alpha_2)\beta_2.$$

Complete positivity of T comes into the verification of positivity of the inner product. Namely, if $\sum_1^n \alpha_i \otimes \beta_i$ is an arbitrary element of $\mathcal{A} \otimes \mathcal{B}$ then we have

$$\begin{aligned} \left\langle \sum_i \alpha_i \otimes \beta_i, \sum_i \alpha_i \otimes \beta_i \right\rangle &= \sum_{i,j} \beta_i^* T(\alpha_i^* \alpha_j)\beta_j \\ &= B^* T^{(n)}(A^*A)B \end{aligned}$$

where

$$A = [\,\alpha_1 \quad \cdots \quad \alpha_n\,] \qquad \text{and} \qquad B = \begin{bmatrix} \beta_1 \\ \vdots \\ \beta_n \end{bmatrix}.$$

The matrix $A^*A \in M_n(\mathcal{A})$ is positive, so $T^{(n)}(A^*A) \in M_n(\mathcal{B})$ is also positive, and hence $B^*T^{(n)}(A^*A)B$ is positive in $\mathcal{B}$, as desired.

Let $\mathcal{E}_T$ be the Hilbert $\mathcal{B}$-module formed from $\mathcal{A} \otimes \mathcal{B}$ by factoring out null vectors and completing. The following proposition, which generalizes Theorem 5.6.6 (cf. the comment preceding Theorem 8.5.10), is now a short calculation.

PROPOSITION 9.3.7
Let $\mathcal{A}$ and $\mathcal{B}$ be C-algebras and let $T : \mathcal{A} \to \mathcal{B}$ be completely positive. Then there is a unique $*$-homomorphism $\pi : \mathcal{A} \to B(\mathcal{E}_T)$ such that $\pi(\alpha)(\beta \otimes \gamma) = \alpha\beta \otimes \gamma$ for all $\alpha, \beta \in \mathcal{A}$ and $\gamma \in \mathcal{B}$. If $\mathcal{A}$ and $\mathcal{B}$ are unital and $T(I_{\mathcal{A}}) = I_{\mathcal{B}}$, then $\|\bar{I}_{\mathcal{A}} \otimes \bar{I}_{\mathcal{B}}\| = 1$ in $\mathcal{E}_T$ and*

$$T(\alpha) = \langle \pi(\alpha)(\bar{I}_{\mathcal{A}} \otimes \bar{I}_{\mathcal{B}}), \bar{I}_{\mathcal{A}} \otimes \bar{I}_{\mathcal{B}} \rangle$$

for all $\alpha \in \mathcal{A}$.

9.4 Hilbert W*-modules

In the von Neumann algebra setting it is natural to consider self-dual modules, as we did in Section 9.2. The following is the general definition.

DEFINITION 9.4.1 *Let $\mathcal{E}$ be a left Hilbert module over a C*-algebra $\mathcal{A}$. Then the dual module $\mathcal{E}'$ is the set of bounded $\mathcal{A}$-linear maps from $\mathcal{E}$ into $\mathcal{A}$. The module structure of $\mathcal{E}'$ is defined by*

$$(\alpha \cdot \Xi)(\xi) = \Xi(\xi)\alpha^*$$

for $\alpha \in \mathcal{A}$, $\Xi \in \mathcal{E}'$, and $\xi \in \mathcal{E}$.

For any $\eta \in \mathcal{E}$, let $\hat{\eta} \in \mathcal{E}'$ be the map $\hat{\eta}(\xi) = \langle \xi, \eta \rangle$. If every element of $\mathcal{E}'$ is of the form $\hat{\eta}$ for some $\eta \in \mathcal{E}$, then we say that $\mathcal{E}$ is self-dual.

For right Hilbert modules, we define $(\Xi \cdot \alpha)(\xi) = \alpha^*\Xi(x)$ and $\hat{\eta}(\xi) = \langle \eta, \xi \rangle$.

Self-duality has many nice consequences. For example, in contrast to the general case (see the comment following Definition 9.3.4), here any bounded module map $A : \mathcal{E}_1 \to \mathcal{E}_2$ has a bounded adjoint A^*. This is easily seen, by essentially the same proof as in the Hilbert space case.

Before giving examples of self-dual Hilbert modules we need to develop some machinery. First we prove that every self-dual Hilbert module over a von Neumann algebra is a dual Banach space. In proving this result we need the fact that weak* continuous linear functionals are abundant on any von Neumann algebra $\mathcal{M}$. For this it is sufficient to note that for any $v, w \in \mathcal{H}$ the map $A \mapsto \langle Av, w \rangle$ is weak* continuous on $B(\mathcal{H})$, and hence on any von Neumann algebra $\mathcal{M} \subset B(H)$.

PROPOSITION 9.4.2
Let $\mathcal{E}$ be a self-dual left Hilbert module over a von Neumann algebra $\mathcal{M}$. Then $\mathcal{E}$ can be identified with the dual of a Banach space in such a way that a bounded net (ξ_κ) in $\mathcal{E}$ converges weak if and only if the net $\langle \xi_\kappa, \eta \rangle$ converges weak* in $\mathcal{M}$ for every $\eta \in \mathcal{E}$.*

PROOF Define a topology on the unit ball of $\mathcal{E}$ by saying that $\xi_\kappa \to \xi$ if and only if $\langle \xi_\kappa, \eta \rangle \to \langle \xi, \eta \rangle$ weak* in $\mathcal{M}$ for all $\eta \in \mathcal{E}$. In other words, this is the weakest topology which makes the maps $\xi \mapsto \langle \xi, \eta \rangle$ continuous with respect to the weak* topology on $\mathcal{M}$.

Let $\mathcal{V} \subset \mathcal{E}^*$ be the set of bounded linear functionals whose restriction to the unit ball of $\mathcal{E}$ is continuous for the above topology. It is routine to verify that $\mathcal{V}$ is a closed subspace of $\mathcal{E}^*$, so $\mathcal{V}$ is a Banach space. Let $T : \mathcal{E} \to \mathcal{V}^*$ be the natural map, i.e., $T\xi(\omega) = \omega(\xi)$.

For every $\eta \in \mathcal{E}$ and every weak* continuous linear functional ρ on $\mathcal{M}$, the map $\xi \mapsto \rho(\langle \xi, \eta \rangle)$ belongs to $\mathcal{V}$. For any $\xi \in \mathcal{E}$ and any $\epsilon > 0$ we can choose ρ so that $\|\rho\| = 1$ and $|\rho(\langle \xi, \xi \rangle)| \geq \|\xi\|^2 - \epsilon$. This implies that $\|T\xi\| \geq \|\xi\|$, and as T is clearly nonexpansive we conclude that it is an isometry.

To show that T is surjective, let ω belong to the unit ball of $\mathcal{V}^*$. We can extend ω to a linear functional $\tilde{\omega}$ on $\mathcal{E}^*$ with norm at most 1, and then we can find a net (ξ_κ) in the unit ball of $\mathcal{E}$ which converges weak* to $\tilde{\omega}$ in $\mathcal{E}^{**}$. Passing to a subnet, we may suppose that $\langle \eta, \xi_\kappa \rangle$ converges weak* in $\mathcal{M}$ for all $\eta \in \mathcal{E}$; then the map $\eta \mapsto \lim \langle \eta, \xi_\kappa \rangle$ belongs to $\mathcal{E}'$, so by self-duality there exists $\xi \in \mathcal{E}$ such that $\langle \eta, \xi \rangle = \lim \langle \eta, \xi_\kappa \rangle$ for all $\eta \in \mathcal{E}$. From the original definition of $\mathcal{V}$ it is clear that $\omega(\xi_\kappa) \to \omega(\xi)$ for all $\omega \in \mathcal{V}$, so therefore $T\xi_\kappa \to T\xi$ weak* in $\mathcal{V}^*$. But $T\xi_\kappa \to \omega$ weak* in $\mathcal{V}^*$ as well, so we must have $\omega = T\xi$. Thus T is a surjective isometric isomorphism from $\mathcal{E}$ onto $\mathcal{V}^*$.

If $\rho(\langle \xi_\kappa, \eta \rangle) \to \rho(\langle \xi, \eta \rangle)$ for every weak* continuous linear functional ρ on $\mathcal{M}$ then $\langle \xi_\kappa, \eta \rangle \to \langle \xi, \eta \rangle$ weak*. This implies that the restriction of T^{-1} to the unit ball of $\mathcal{V}^*$ is continuous for the topology on $\mathcal{E}$ described in the statement of the theorem. But the unit ball of $\mathcal{V}^*$ is compact and the unit ball of $\mathcal{E}$ is Hausdorff, so $T^{-1} : (\mathcal{V}^*)_1 \to (\mathcal{E})_1$ must be a homeomorphism. This shows that the weak* topology on the unit ball of $\mathcal{E}$ is as stated. ∎

As a consequence of the preceding theorem, we get the von Neumann algebra version of Proposition 9.3.6.

COROLLARY 9.4.3

Let $\mathcal{E}$ be a self-dual left Hilbert module over a von Neumann algebra $\mathcal{M}$. Then $B(\mathcal{E})$ is an abstract von Neumann algebra.

PROOF The proof is similar to the proof of Proposition 9.3.6. For each weak* continuous state ω on $\mathcal{M}$, form the corresponding Hilbert space $\mathcal{H}_\omega$; let $\mathcal{H}$ be their direct sum; and let $\pi : B(\mathcal{E}) \to B(\mathcal{H})$ be the natural map. It is clear that π is a $*$-homomorphism. Existence of sufficiently many weak* continuous states to ensure that π is an isomorphism follows from the observation made just prior to Proposition 9.4.2.

We must show that $\pi(B(\mathcal{E}))$ is weak* closed in $B(\mathcal{H})$. Thus, let (A_κ) be a bounded net in $\pi(B(\mathcal{E}))$, and suppose $A_\kappa \to A$ weak*. We must show $A \in \pi(B(\mathcal{E}))$.

Let $B_\kappa = \pi^{-1}(A_\kappa)$; then (B_κ) is a bounded net in $B(\mathcal{E})$, and by passing to a subnet we may suppose that $B_\kappa \xi$ converges weak* in $\mathcal{E}$ for every $\xi \in \mathcal{E}$. Then $B\xi = \lim B_\kappa \xi$ defines an operator $B \in B(\mathcal{E})$ which satisfies $\pi(B) = A$. So $\pi(B(\mathcal{E}))$ is weak* closed in $B(\mathcal{H})$, and hence $B(\mathcal{E}) \cong \pi(B(\mathcal{E}))$ is a von Neumann algebra. ∎

Every Hilbert module $\mathcal{E}$ over a von Neumann algebra $\mathcal{M}$ can be completed to a self-dual module. The simplest way to do this is by embedding $\mathcal{E}$ in its dual module $\mathcal{E}'$, giving $\mathcal{E}'$ a Hilbert module structure, and then showing that $\mathcal{E}'$ is self-dual. The natural embedding is the map $\xi \mapsto \hat{\xi}$. We will define an inner product on $\mathcal{E}'$ by continuous extension with respect to the following topology. We say that a net (Ξ_κ) in $\mathcal{E}'$ converges <u>weakly</u> to Ξ if $\Xi_\kappa(\xi) \to \Xi(\xi)$ weak* in $\mathcal{M}$, for every $\xi \in \mathcal{E}$. It will turn out that $\mathcal{E}'$ is a dual space (via Proposition 9.4.2), and by the following lemma this topology therefore agrees with the weak* topology on bounded sets.

The main step in the construction of an $\mathcal{M}$-valued inner product on $\mathcal{E}'$ is showing density of $\mathcal{E}$ in $\mathcal{E}'$. We address this first.

LEMMA 9.4.4

Let $\mathcal{E}$ be a left Hilbert module over a von Neumann algebra $\mathcal{M}$. Then the map $\xi \mapsto \hat{\xi}$ takes the unit ball of $\mathcal{E}$ into a weakly dense subset of the unit ball of $\mathcal{E}'$.

PROOF Let $\Xi \in \mathcal{E}'$ and suppose $\|\Xi\| \leq 1$. Fixing $\xi_1, \ldots, \xi_n \in \mathcal{E}$, it will suffice to find a bounded sequence $(\eta_k) \subset \mathcal{E}$ such that $\hat{\eta}_k(\xi_i) \to \Xi(\xi_i)$ weak* in $\mathcal{M}$ for $1 \leq i \leq n$.

Fix k, let $\epsilon = 1/k$, and for $\xi \in \mathcal{E}$ write $|\xi|_\epsilon = (\langle \xi, \xi \rangle + \epsilon I_{\mathcal{M}})^{1/2}$. Observe that $|\xi|_\epsilon$ is invertible and $|\xi| \cdot |\xi|_\epsilon^{-1} \leq I_{\mathcal{M}}$. Let $\xi_1' = |\xi_1|_\epsilon^{-1} \xi_1$ and inductively define ξ_i' by an approximate Gramm-Schmidt process, setting $\xi_{i+1}' = |\zeta_{i+1}|_\epsilon^{-1} \zeta_{i+1}$ where

$$\zeta_{i+1} = \xi_{i+1} - \langle \xi_{i+1}, \xi_1' \rangle \xi_1' - \cdots - \langle \xi_{i+1}, \xi_i' \rangle \xi_i'.$$

Finally we define

$$\eta_k = \Xi(\xi_1')\xi_1' + \cdots + \Xi(\xi_n')\xi_n'.$$

One can prove by induction on i that the $\mathcal{M}$-linear span E of $\{\xi_1', \ldots, \xi_n'\}$

is independent of ϵ; that $\xi_1, \ldots, \xi_n \in E$; that $\langle \xi_i', \xi_j' \rangle$ weak* converges as $k \to \infty$ to a projection in $\mathcal{M}$ if $i = j$ and to 0 if $i \neq j$; and that $\langle \xi, \eta_k \rangle \to \Xi(\xi)$ weak* in $\mathcal{M}$ as $k \to \infty$, for all $\xi \in E$. Thus (η_κ) has the desired property. ∎

LEMMA 9.4.5

Let $\mathcal{E}_1$ and $\mathcal{E}_2$ be left Hilbert modules over a von Neumann algebra $\mathcal{M}$ and let $A : \mathcal{E}_1 \to \mathcal{E}_2$ be a bounded $\mathcal{M}$-linear map. Let (ξ_κ) be a bounded net in $\mathcal{E}_1$. Then $\langle \xi_\kappa, \eta \rangle \to \langle \xi, \eta \rangle$ weak in $\mathcal{M}$ for all $\eta \in \mathcal{E}_1$ implies $\langle A\xi_\kappa, \zeta \rangle \to \langle A\xi, \zeta \rangle$ weak* in $\mathcal{M}$ for all $\zeta \in \mathcal{E}_2$.*

PROOF We may suppose $\langle \xi_\kappa, \eta \rangle \to 0$ weak* in $\mathcal{M}$ for every $\eta \in \mathcal{E}_1$. Fix a weak* continuous state ω on $\mathcal{M}$ and let $\mathcal{H}_1$ and $\mathcal{H}_2$ be the corresponding Hilbert spaces constructed as in Proposition 9.3.6 for $\mathcal{E}_1$ and $\mathcal{E}_2$ respectively. Then $\omega(\langle \xi_\kappa, \eta \rangle) \to 0$ for all $\eta \in \mathcal{E}$, which implies that $\xi_\kappa \to 0$ weakly in $\mathcal{H}_1$. Applying ω to the result of Lemma 9.3.5 shows that A is bounded as a map from $\mathcal{H}_1$ to $\mathcal{H}_2$, and bounded maps between Hilbert spaces preserve weak continuity, so $A\xi_\kappa \to 0$ weakly in $\mathcal{H}_2$. Thus $\omega(\langle A\xi_\kappa, \zeta \rangle) \to 0$ for all $\zeta \in \mathcal{E}_2$. As this holds for any weak* continuous state ω, we conclude that $\langle A\xi_\kappa, \zeta \rangle \to 0$ weak* in $\mathcal{M}$ for all $\zeta \in \mathcal{E}_2$. ∎

THEOREM 9.4.6

Let $\mathcal{E}$ be a Hilbert module over a von Neumann algebra $\mathcal{M}$. Then the dual module $\mathcal{E}'$ carries a natural $\mathcal{M}$-valued inner product which makes it a self-dual Hilbert module, and $\mathcal{E}$ embeds as a weak dense submodule of $\mathcal{E}'$.*

PROOF Given $\Xi, H \in \mathcal{E}'$, let (ξ_κ) be a bounded net in $\mathcal{E}$ such that $\hat{\xi}_\kappa \to \Xi$ weakly. Such a sequence exists by Lemma 9.4.4. Then define an $\mathcal{M}$-valued inner product on $\mathcal{E}'$ by setting $\langle \Xi, H \rangle = \lim H(\xi_\kappa)$. Applying Lemma 9.4.5 to the map $H : \mathcal{E} \to \mathcal{M}$ shows that this limit exists and is independent of the choice of (ξ_κ).

We verify self-duality. Let $T : \mathcal{E}' \to \mathcal{M}$ be a bounded module homomorphism. Its restriction to $\mathcal{E}$ is an element Ξ of $\mathcal{E}'$; then $T = \hat{\Xi}$ on $\mathcal{E} \subset \mathcal{E}'$, and this implies $T = \hat{\Xi}$ by Lemmas 9.4.4 and 9.4.5.

The weak* topology on $\mathcal{E}'$ was characterized in Proposition 9.4.2, and that result, together with Lemma 9.4.4, implies that $\mathcal{E}$ is weak* dense in $\mathcal{E}'$. ∎

Lemma 9.4.4 also implies the following converse to Proposition 9.4.2.

PROPOSITION 9.4.7
Let $\mathcal{E}$ be a left Hilbert module over a von Neumann algebra $\mathcal{M}$. Suppose $\mathcal{E}$ can be identified with the dual of a Banach space in such a way that a bounded net (ξ_κ) in $\mathcal{E}$ converges weak if and only if the net $\langle \xi_\kappa, \eta \rangle$ converges weak* in $\mathcal{M}$ for every $\eta \in \mathcal{E}$. Then $\mathcal{E}$ is self-dual.*

PROOF Let $\Xi \in \mathcal{E}'$. By Lemma 9.4.4 there exists a bounded net (ξ_κ) in $\mathcal{E}$ such that $\hat{\xi}_\kappa \to \Xi$ weakly. By passing to a subnet, we may suppose that $\xi_\kappa \to \xi$ weak*; then for any $\eta \in \mathcal{E}$ we have

$$\langle \eta, \xi \rangle = \lim \langle \eta, \xi_\kappa \rangle = \Xi(\eta),$$

so $\hat{\xi} = \Xi$. Thus $\mathcal{E}$ is self-dual. ∎

Now we can describe the basic examples of self-dual Hilbert modules over von Neumann algebras.

Example 9.4.8

(a) Let $\mathcal{M}$ be a von Neumann algebra and let X be a set. Define $l^2(X; \mathcal{M})$ to be the set of functions $f : X \to \mathcal{M}$ such that $\sum_x f(x) f(x)^*$ is bounded. It is easy to see that boundedness of this sum implies that it converges weak* in $\mathcal{M}$. Note that this definition is *not* the same as the one given in Example 9.3.3 (a).

The proof that $l^2(X; \mathcal{M})$ is a left Hilbert module runs along familiar lines. Self-duality can be proven as follows. Given any F in the dual module, let $f(x) = F(I_{\mathcal{M}} \chi_x)^*$; then $\langle g, f \rangle = F(g)$ for all finitely supported $g : X \to \mathcal{M}$, and the fact that $F = \hat{f}$ then follows from Lemma 9.4.5.

(b) Now let X be a σ-finite measure space. We define $L^2(X; \mathcal{M})$ to be the dual of the completion of the simple functions with finite measure support (cf. Example 9.3.3 (b)). This is self-dual by Theorem 9.4.6. Again, this definition does not agree with Example 9.3.3 (b).

We now know that if $\mathcal{E}$ is a Hilbert module over a von Neumann algebra $\mathcal{M}$ then $\mathcal{E}'$ is a self-dual Hilbert $\mathcal{M}$-module and $\mathcal{E}$ embeds in $\mathcal{E}'$. Also, $B(\mathcal{E})$ is a C*-algebra and $B(\mathcal{E}')$ is a von Neumann algebra. The final result we need in this direction says that $B(\mathcal{E})$ embeds in $B(\mathcal{E}')$. This is useful because dualization is the usual method employed to construct self-dual Hilbert modules, and often one is interested in operators which are initially defined only on the original module.

PROPOSITION 9.4.9
Let $\mathcal{E}$ be a left Hilbert module over a von Neumann algebra $\mathcal{M}$ and let $A \in B(\mathcal{E})$. Then there is a unique extension $\tilde{A} \in B(\mathcal{E}')$ of $\mathcal{A}$. The map

$A \mapsto \tilde{A}$ is a $*$-isomorphism of $B(\mathcal{E})$ into $B(\mathcal{E}')$.

PROOF First, define $A^\# : \mathcal{E} \to \mathcal{E}'$ by $A^\#\xi(\eta) = \langle \xi, A\eta \rangle$; then define $\tilde{A} = A^{\#\#} : \mathcal{E}' \to \mathcal{E}'$ by $\tilde{A}\Xi(\xi) = \langle \Xi, A^\#\xi \rangle$. It is straightforward to check that $\tilde{A}$ is a left module map and that $\|\tilde{A}\| \leq \|A\|$. Moreover, $\tilde{A}$ extends A, so that $\|\tilde{A}\| = \|A\|$. Uniqueness follows from Lemmas 9.4.4 and 9.4.5.

This shows that the map $A \mapsto \tilde{A}$ from $B(\mathcal{E})$ into $B(\mathcal{E}')$ is well-defined and isometric. The fact that it preserves sums, products, and adjoints follows from uniqueness; for example, $\tilde{A}\tilde{B}$ is an extension of AB, and hence must equal $(AB)\tilde{}$. ∎

We conclude this section with a description of the structure of self-dual Hilbert modules over von Neumann algebras. The key tool is the following lemma, which generalizes Theorem 2.2.5. Here, given $\mathcal{E}_0 \subset \mathcal{E}$ we write $\mathcal{E}_0^\perp$ for the set of $\xi \in \mathcal{E}$ such that $\langle \xi, \eta \rangle = 0$ for all $\eta \in \mathcal{E}_0$.

LEMMA 9.4.10

Let $\mathcal{E}$ be a self-dual left Hilbert module over a von Neumann algebra $\mathcal{M}$ and let $\mathcal{E}_0$ be a weak closed submodule of $\mathcal{E}$. Then $\mathcal{E} = \mathcal{E}_0 \oplus \mathcal{E}_0^\perp$.*

PROOF Let $\xi \in \mathcal{E}$; we must find $\xi_1 \in \mathcal{E}_0$ and $\xi_2 \in \mathcal{E}_0^\perp$ such that $\xi = \xi_1 + \xi_2$.

By Proposition 9.4.7, $\mathcal{E}_0$ is self-dual for the Hilbert module structure it inherits from $\mathcal{E}$. Thus the identity map $A : \mathcal{E}_0 \to \mathcal{E}$ has an adjoint $A^* : \mathcal{E} \to \mathcal{E}_0$ by the comment following Definition 9.4.1. Let $P = AA^* : \mathcal{E} \to \mathcal{E}$, and define $\xi_1 = P\xi$ and $\xi_2 = \xi - P\xi$. It is routine to verify that $\xi_1 \in \mathcal{E}_0$ and $\xi_2 \in \mathcal{E}_0^\perp$. ∎

Let $\mathcal{M}$ be a von Neumann algebra. For any family $(\mathcal{E}_\kappa)$ of self-dual Hilbert modules, write $\bigoplus^\infty \mathcal{E}_\kappa$ for the set of sequences $\bigoplus \xi_\kappa$ such that $\xi_\kappa \in \mathcal{E}_\kappa$ for all κ and $\sum |\xi_\kappa|^2$ is bounded. Equivalently, $\bigoplus^\infty \mathcal{E}_\kappa$ is the dual of the norm closure of the algebraic direct sum of the modules $\mathcal{E}_\kappa$. Either way it is a self-dual Hilbert $\mathcal{M}$-module.

Observe also that if $\mathcal{I}$ is a weak* closed left ideal of $\mathcal{M}$ then the inner product $\langle \xi, \eta \rangle = \xi\eta^*$ and the natural left action of $\mathcal{M}$ make $\mathcal{I}$ into a left Hilbert module over $\mathcal{M}$. Moreover, $\mathcal{I}$ is self-dual by Proposition 9.4.7.

The following theorem characterizes the structure of self-dual Hilbert modules over von Neumann algebras.

THEOREM 9.4.11

Let $\mathcal{E}$ be a self-dual left Hilbert module over a von Neumann algebra

$\mathcal{M}$. Then there is a family $(\mathcal{I}_\kappa)$ of weak closed left ideals of $\mathcal{M}$ such that $\mathcal{E}$ is isometrically isomorphic to $\bigoplus^\infty \mathcal{I}_\kappa$.*

PROOF Suppose first that $\mathcal{E}$ is singly generated in the sense that there exists $\xi \in \mathcal{E}$ such that no proper weak* closed submodule of $\mathcal{E}$ contains ξ. We claim that $\mathcal{E}$ is isometrically isomorphic to a weak* closed left ideal of $\mathcal{M}$. To see this, let $\xi_n = (|\xi| + \frac{1}{n}I_{\mathcal{M}})^{-1}\xi$ and observe that $\|\xi_n\| \leq 1$. Thus the sequence (ξ_n) has a weak* cluster point η.

We have $|\xi|\eta = \xi$, so η also generates $\mathcal{E}$. Also

$$|\xi|\langle \eta, \eta\rangle^{1/2} = \langle \xi, \xi\rangle^{1/2} = |\xi|,$$

which implies that $|\eta|$ is the projection onto the closure of $\mathrm{ran}(|\xi|)$. Let $\mathcal{I} = \mathcal{M}|\eta|$. For any $\alpha, \beta \in \mathcal{I}$ we have $\langle \alpha\eta, \beta\eta\rangle = \alpha|\eta|^2\beta^* = \alpha\beta^*$; this shows that the map $\alpha \mapsto \alpha\eta$ from $\mathcal{I}$ into $\mathcal{E}$ is a Hilbert module isomorphism, and since η generates $\mathcal{E}$ it is surjective. This proves the claim.

The remainder of the proof is an easy application of Lemma 9.4.10 and Zorn's lemma. ∎

Incidentally, this proof shows that every weak* closed left ideal of a von Neumann algebra $\mathcal{M}$ is of the form $\mathcal{M}p$ for some projection $p \in \mathcal{M}$ (cf. Proposition 6.5.5).

9.5 Crossed products

We will now use the machinery of Hilbert modules to present an important general construction of C*- and von Neumann algebras, the reduced crossed product construction. This can be carried out without using Hilbert modules, but it then becomes substantially more technical.

We will need two basic facts about topological groups, that is, groups that are equipped with a topology which is compatible with the product and inverse operations in the natural way. First, every locally compact group has a unique (up to a positive scalar multiple) regular Borel measure h which is finite on compact sets, strictly positive on nonempty open sets, and invariant under left translation by any group element. Thus

$$\int_G f(xy)\, dh(y) = \int_G f(y)\, dh(y)$$

for any $f \in C_c(G)$ and $x \in G$. This is the <u>Haar measure</u>. Second, on every locally compact group G there is a homomorphism $\Delta : G \to \mathbf{R}^+$ into the multiplicative group of positive reals such that

$$\int_G f(x^{-1})\, dh(x) = \int_G f(x)\Delta(x)^{-1}\, dh(x)$$

for any $f \in C_c(G)$. This is the modular function.

For the sake of simplicity it is possible to restrict the following discussion to discrete groups, when Haar measure is just counting measure and the modular function is constantly 1. Outside the discrete case, the most important examples are the groups $\mathbf{R}^n$ and $\mathbf{T}^n$, for both of which Haar measure is Lebesgue measure and the modular function is again constantly 1.

DEFINITION 9.5.1

(a) A C-dynamical system is a triple $(\mathcal{A}, G, \theta)$ such that $\mathcal{A}$ is a C*-algebra, G is a topological group, and θ is a homomorphism from G into the group of $*$-automorphisms of $\mathcal{A}$ which is continuous for the point-norm topology, meaning that if $x_\kappa \to x$ in G then $\theta(x_\kappa)(\alpha) \to \theta(x)(\alpha)$ in norm, for every $\alpha \in \mathcal{A}$.*

(b) A W-dynamical system is a triple $(\mathcal{M}, G, \theta)$ such that $\mathcal{M}$ is a von Neumann algebra, G is a topological group, and θ is a homomorphism from G into the group of $*$-automorphisms of $\mathcal{M}$ which is continuous for the point-weak* topology, meaning that if $x_\kappa \to x$ in G then $\theta(x_\kappa)(\alpha) \to \theta(x)(\alpha)$ weak*, for every $\alpha \in \mathcal{M}$.*

We write θ_x for $\theta(x)$.

Intuitively, the reduced crossed product of a C*-dynamical system $(\mathcal{A}, G, \theta)$ is a C*-algebra generated by the elements of $\mathcal{A}$ and G in an appropriately "smoothed out" manner, and similarly for W*-dynamical systems. An important special case arises when $\mathcal{A} = \mathbf{C}$ and G acts trivially; then the crossed product construction produces the reduced group C*-algebra $C_r^*(G)$ and von Neumann algebra $W_r^*(G)$. The former is the C*-algebra generated by $C_c(G)$ (the continuous functions on G with compact support) acting by convolution on $L^2(G)$, and the latter is the weak* closure of the former. General reduced crossed products are defined as follows.

DEFINITION 9.5.2

(a) Let $(\mathcal{A}, G, \theta)$ be a C-dynamical system and let $C_c(G; \mathcal{A})$ be the set of all continuous functions from G into $\mathcal{A}$ with compact support. This is a right $\mathcal{A}$-module via pointwise right multiplication of elements of $\mathcal{A}$, and it has an $\mathcal{A}$-valued inner product defined by*

$$\langle f, g\rangle = \int_G f(x)^* g(x)\, dh(x)$$

(cf. Example 9.3.3). Let $\mathcal{E}$ be the Hilbert module obtained by completing $C_c(G; \mathcal{A})$ for the resulting norm.

For each $\phi \in C_c(G;\mathcal{A})$ define a <u>twisted convolution</u> *operator L_ϕ on $\mathcal{E}$ by*

$$L_\phi f(x) = \int_G \theta_{x^{-1}}(\phi(y)) \cdot f(y^{-1}x)\, dh(y).$$

The <u>reduced crossed product</u> *of $\mathcal{A}$ by G, denoted $\mathcal{A} \times_\theta G$, is the C*-subalgebra of $B(\mathcal{E})$ generated by the operators L_ϕ for $\phi \in C_c(G;\mathcal{A})$.*
(b) Now let $(\mathcal{M}, G, \theta)$ be a W-dynamical system. Define $\mathcal{E}$ and $L_\phi \in B(\mathcal{E})$ as above. Let $\mathcal{E}'$ be the dual module and let $\tilde{L}_\phi \in B(\mathcal{E}')$ be as given in Proposition 9.4.9. Then the* <u>reduced crossed product</u> *of $\mathcal{M}$ by G, denoted $\mathcal{M} \times_\theta G$, is the W*-subalgebra of $B(\mathcal{E}')$ generated by the operators $\tilde{L}_\phi$ for $\phi \in C_c(G;\mathcal{M})$.*

One can prove using Minkowski's inequality for integrals that for $f \in C_c(G;\mathcal{A})$ we have $\|L_\phi f\| \leq a\|f\|$ where $a = \int \|\phi(x)\|\, dh(x)$. Thus L_ϕ does extend to a bounded module operator on $\mathcal{E}$.

The term "reduced" is used to distinguish this crossed product, which is represented on $L^2(G;\mathcal{A})$, from the "full" crossed product, which arises from a universal representation of $C_c(G;\mathcal{A})$. The reduced crossed product is generally more important.

In fact, $\mathcal{A} \times_\theta G$ ($\mathcal{M} \times_\theta G$) is really just the norm closure (weak* closure) of the set of operators L_ϕ ($\tilde{L}_\phi$). That is because this set is closed under sums, products, and adjoints. Indeed, simple calculations show that $L_\phi + L_\psi = L_{\phi+\psi}$, $L_\phi L_\psi = L_{\phi\times_\theta\psi}$ where

$$(\phi \times_\theta \psi)(x) = \int_G \phi(y)\theta_y(\psi(y^{-1}x))\, dh(y),$$

and $L_\phi^* = L_{\phi^*}$ where

$$\phi^*(x) = \theta_x(\phi(x^{-1}))^*\Delta(x)^{-1}.$$

The quantum plane and tori can be viewed as crossed product algebras.

Example 9.5.3

(a) Let $\hbar > 0$ and define an action θ of $\mathbf{Z}$ on $C(\mathbf{T})$ by letting θ_n be translation by $n\hbar$. That is, $\theta_n f(t) = f(t - n\hbar)$. Now define $\phi, \psi : \mathbf{Z} \to C(\mathbf{T})$ by

$$\phi = 1_{\mathbf{T}} \cdot \chi_{\{1\}} \qquad \text{and} \qquad \psi = e^{it} \cdot \chi_{\{0\}}.$$

Then we have

$$(\phi \times_\theta \psi)(n) = \sum \phi(m)\theta_m(\psi(n-m)) = e^{i(t-\hbar)} \cdot \chi_{\{1\}}(n)$$

and

$$(\psi \times_\theta \phi)(n) = \sum \psi(m)\theta_m(\phi(n-m)) = e^{it} \cdot \chi_{\{1\}}(n).$$

Thus $L_\phi L_\psi = e^{-i\hbar} L_\psi L_\phi$. Similar computations show that L_ϕ and L_ψ are unitary. Finally, it is not too hard to show that $C(\mathbf{T}) \times_\theta \mathbf{Z}$ is generated by L_ϕ and L_ψ, so if $\hbar$ is an irrational multiple of π then Corollary 5.5.9 implies that $C(\mathbf{T}) \times_\theta \mathbf{Z} \cong C^\hbar(\mathbf{T}^2)$. In fact, this is true for all $\hbar > 0$. Likewise, $L^\infty(\mathbf{T}) \times_\theta \mathbf{Z} \cong L^\infty_\hbar(\mathbf{T}^2)$.
(b) Similarly, for $\hbar > 0$ we define an action θ of $\mathbf{R}$ on $C_0(\mathbf{R})$ by letting θ_t be translation by $\hbar t$. For $f_1, f_2 \in C_c^\infty(\mathbf{R})$ let ϕ be the function from $\mathbf{R}$ to $C_0(\mathbf{R})$ defined by $\phi(x)(t) = f_1(x)f_2(t)$; then the operators L_ϕ generate $C_0(\mathbf{R}) \times_\theta \mathbf{R}$, and there is a $*$-isomorphism from $C_0(\mathbf{R}) \times_\theta \mathbf{R}$ onto $C_0^\hbar(\mathbf{R}^2)$ which takes L_ϕ in the sense of Definition 9.5.2 to $L_{f_1 \check{f}_2}$ in the sense of Definition 5.4.1. Likewise, $L^\infty(\mathbf{R}) \times_\theta \mathbf{R} \cong L^\infty_\hbar(\mathbf{R}^2)$.

9.6 Hilbert $*$-bimodules

By Theorems 9.1.6 and 9.2.4, it is reasonable to regard Hilbert C*- and W*-modules as the quantum analog of continuous and measurable Hilbert bundles, that is, bundles for which each fiber has a complex Hilbert space structure. In geometric applications, however, we are often more interested in real Hilbert bundles (cf. Example 9.1.2). Their quantum version involves modules with an involution *, which is analogous to conjugation on the complexification of a real bundle (see Example 9.6.4). The self-adjoint part of such a module corresponds to the continuous sections of a bundle of real Hilbert spaces.

The existence of an involution has a surprising consequence: given a left module with involution we can use the equation $(\alpha\xi)^* = \xi^*\alpha^*$ to define a right action, and vice versa. Thus it is natural to work with bimodules, that is, modules which possess commuting left and right actions. Interestingly, bimodule structure is also crucial in geometric applications; see Section 10.3.

The precise definition of the relevant class of modules goes as follows.

DEFINITION 9.6.1 *Let $\mathcal{A}$ be a C*-algebra. A pre-Hilbert $*$-bimodule over $\mathcal{A}$ is an $\mathcal{A}$-$\mathcal{A}$-bimodule $\mathcal{E}$ together with a complex linear map $\langle\cdot,\cdot\rangle : \mathcal{E} \times \mathcal{E} \to \mathcal{A}$ such that*

(a) $\langle\alpha\xi, \eta\rangle = \alpha\langle\xi, \eta\rangle$
(b) $\langle\xi\alpha, \eta\rangle = \langle\xi, \alpha\eta\rangle$
(c) $\langle\xi, \eta\alpha\rangle = \langle\xi, \eta\rangle\alpha$

and an antilinear involution $* : \mathcal{E} \to \mathcal{E}$ such that

(d) $\langle\xi, \eta\rangle^* = \langle\eta^*, \xi^*\rangle$

(e) $(\alpha\xi)^* = \xi^*\alpha^*$
(f) $\langle \xi, \xi^* \rangle \geq 0$

for all $\alpha \in \mathcal{A}$ *and* $\xi, \eta \in \mathcal{E}$. *We define* left *and* right $\mathcal{A}$-valued pseudo inner products $\langle \xi, \eta \rangle_l = \langle \xi, \eta^* \rangle$ *and* $\langle \xi, \eta \rangle_r = \langle \xi^*, \eta \rangle$ *and* left, right, *and* middle seminorms $\|\xi\|_l^2 = \|\langle \xi, \xi \rangle_l\|$, $\|\xi\|_r^2 = \|\langle \xi, \xi \rangle_r\|$, *and* $\|\xi\|_m = \max(\|\xi\|_l, \|\xi\|_r)$.

A Hilbert $*$-bimodule *over* $\mathcal{A}$ *is a pre-Hilbert* $*$*-bimodule over* $\mathcal{A}$ *for which* $\|\xi\|_m$ *is a complete norm.*

We call $\langle \cdot, \cdot \rangle$ a bilinear form. The left and right pseudo inner products $\langle \cdot, \cdot \rangle_l$ and $\langle \cdot, \cdot \rangle_r$ that it gives rise to satisfy the appropriate versions of Definition 9.3.1; thus $\|\cdot\|_l$ and $\|\cdot\|_r$ are indeed both pseudonorms. It is important to note that these pseudonorms not only typically disagree, they often are not even comparable (though we always have $\|\xi\|_l = \|\xi\|_r$ if ξ is self-adjoint). Thus, in general it will not be possible to complete $\|\cdot\|_l$ or $\|\cdot\|_r$ without destroying the $*$-bimodule structure: if one factors out null vectors and completes for $\|\cdot\|_l$, for example, the result will only be an ordinary left Hilbert module. One can factor out null vectors and complete for $\|\cdot\|_m$, though (Proposition 9.6.3).

The following lemma is basic.

LEMMA 9.6.2
Let $\mathcal{A}$ *be a* C^**-algebra and* $\mathcal{E}$ *a pre-Hilbert* $*$*-bimodule over* $\mathcal{A}$. *Then for any* $\alpha \in \mathcal{A}$ *and* $\xi \in \mathcal{E}$ *we have*

$$\|\alpha\xi\|_l, \|\xi\alpha\|_l \leq \|\alpha\|\|\xi\|_l \qquad \text{and} \qquad \|\alpha\xi\|_r, \|\xi\alpha\|_r \leq \|\alpha\|\|\xi\|_r.$$

PROOF We have

$$\|\alpha\xi\|_l^2 = \|\langle \alpha\xi, (\alpha\xi)^* \rangle\| = \|\alpha\langle \xi, \xi^* \rangle \alpha^*\| \leq \|\alpha\|^2 \|\xi\|_l^2,$$

so $\|\alpha\xi\|_l \leq \|\alpha\|\|\xi\|_l$. For the second inequality, observe that $\beta \leq \gamma$ implies $\langle \xi\beta, \xi \rangle_l \leq \langle \xi\gamma, \xi \rangle_l$ since

$$\langle \xi(\gamma - \beta), \xi \rangle_l = \langle \xi(\gamma - \beta)^{1/2}, \xi(\gamma - \beta)^{1/2} \rangle_l \geq 0.$$

Without loss of generality suppose $\|\alpha\| \leq 1$; then letting $\beta = \alpha\alpha^*$ we have $\beta^2 \leq \beta$ and

$$\begin{aligned} 0 &\leq \langle \xi - \xi\beta, \xi - \xi\beta \rangle_l \\ &= \langle \xi, \xi \rangle_l - 2\langle \xi\beta, \xi \rangle_l + \langle \xi\beta^2, \xi \rangle_l \\ &\leq \langle \xi, \xi \rangle_l - \langle \xi\beta, \xi \rangle_l, \end{aligned}$$

so that $\|\xi\alpha\|_l^2 = \|\langle \xi\beta, \xi \rangle_l\| \leq \|\xi\|_l^2$, as desired. Taking adjoints yields the same inequalities for the seminorm $\|\cdot\|_r$. ∎

Using this lemma, we can give a Hilbert $*$-bimodule version of the technique of factoring out null vectors and completing.

PROPOSITION 9.6.3
Let $\mathcal{E}$ be a pre-Hilbert $*$-bimodule over a C-algebra $\mathcal{A}$ and define $\mathcal{E}_0 = \{\xi \in \mathcal{E} : \|\xi\|_m = 0\}$. Then $\mathcal{E}_0$ is a sub-bimodule of $\mathcal{E}$ and the inner product and involution on $\mathcal{E}$ descend to $\mathcal{E}/\mathcal{E}_0$ and extend to the completion of $\mathcal{E}/\mathcal{E}_0$. The completion of $\mathcal{E}/\mathcal{E}_0$ is a Hilbert $*$-bimodule over $\mathcal{A}$.*

PROOF By Lemma 9.6.2, $\mathcal{E}_0$ is a sub-bimodule of $\mathcal{E}$ and the left and right actions of $\mathcal{A}$ on $\mathcal{E}/\mathcal{E}_0$ extend continuously to its completion for $\|\cdot\|_m$. The inner product descends to $\mathcal{E}/\mathcal{E}_0$ and then extends to its completion by the Cauchy-Schwarz inequality for ordinary Hilbert modules (Proposition 9.3.2), and the corresponding assertions for the involution are trivial. ∎

Next we present some simple examples of Hilbert $*$-bimodules. The first generalizes Example 9.1.2, and the second is the $*$-bimodule version of Example 9.3.3.

Example 9.6.4

Let X be a compact Hausdorff space and let $\mathcal{X}$ be a real continuous Hilbert bundle over X; this is defined as in Definition 9.1.1, but using real scalars. If $\mathcal{S}(\mathcal{X})$ is the space of continuous sections of $\mathcal{X}$, then $\mathcal{S}(\mathcal{X}) \oplus i\mathcal{S}(\mathcal{X})$, with involution $(f + ig)^* = f - ig$ and bilinear form $\langle f, g\rangle$ defined by pointwise complexification of the inner product on fibers, is a pre-Hilbert $*$-bimodule over $C(X)$. It can be completed to a Hilbert $*$-bimodule by Proposition 9.6.3.

Example 9.6.5

(a) Let $\mathcal{A}$ be a C*-algebra and X a set. Let $l^2_{sym}(X;\mathcal{A})$ be the set of functions $f : X \to \mathcal{A}$ such that $\sum_x f(x)f(x)^*$ and $\sum_x f(x)^* f(x)$ both converge in $\mathcal{A}$. Then the natural actions and the bilinear form

$$\langle f, g\rangle = \sum_{x\in X} f(x)g(x)$$

make $l^2_{sym}(X;\mathcal{A})$ a Hilbert $*$-bimodule.
(b) Now let X be a σ-finite measure space and define $L^2_{sym}(X;\mathcal{A})$ to be the completion for $\|\cdot\|_m$ of the set of all simple functions from X to $\mathcal{A}$ whose support has finite measure (cf. Example 9.3.3 (b)). This is also a Hilbert $*$-bimodule over $\mathcal{A}$.

The following is a sort of converse to Example 9.6.4, as well as a two-sided version of Theorem 9.1.6 (a).

THEOREM 9.6.6
Let X be a compact Hausdorff space and let $\mathcal{E}$ be a Hilbert $*$-bimodule over $C(X)$. Suppose that any inner product of self-adjoint elements of $\mathcal{E}$ is a real-valued function in $C(X)$. Then there is a real continuous Hilbert bundle $\mathcal{X}$ over X such that $\mathcal{E} \cong \mathcal{S}(\mathcal{X}) \oplus i\mathcal{S}(\mathcal{X})$.

PROOF We want to show that the left and right actions of $C(X)$ on $\mathcal{E}$ coincide. Fix $x \in X$. For any $f \in C(X)$ such that $f \geq 0$ and $f(x) = 0$, and any $\phi \in \mathcal{E}$ such that $\phi = \phi^*$, set $g = f^{1/2}$; then

$$\langle g\phi + \phi g, g\phi + \phi g\rangle(x) = \langle \phi g, g\phi\rangle(x),$$

so $\langle \phi f, \phi\rangle(x) \geq 0$. But also

$$\langle g\phi + \phi g, ig\phi - i\phi g\rangle(x) = i\langle \phi g, g\phi\rangle(x);$$

since both $g\phi+\phi g$ and $ig\phi-i\phi g$ are self-adjoint, so is their inner product, so this shows that $\langle \phi f, \phi\rangle(x)$ must be purely imaginary. But we already showed that it is real, so it follows that $\langle \phi f, \phi\rangle(x) = 0$. By linearity, we have $\langle \phi f, \phi\rangle(x) = 0$ for any $f \in C(X)$ such that $f(x) = 0$.

Now let $f \in C(X)$ and $\phi \in \mathcal{E}$ and suppose f is real and $\phi = \phi^*$. Since the quantity

$$\langle f\phi + \phi f, if\phi - i\phi f\rangle = i\langle \phi f^2, \phi\rangle - if^2\langle \phi, \phi\rangle$$

is real, it follows that $\mathrm{Re}\langle \phi f^2, \phi\rangle = \mathrm{Re}\, f^2\langle \phi, \phi\rangle$. Therefore

$$\begin{aligned}\langle f\phi - \phi f, f\phi - \phi f\rangle &= \mathrm{Re}\langle f\phi - \phi f, f\phi - \phi f\rangle \\ &= 2\mathrm{Re}(f\langle \phi f, \phi\rangle - f^2\langle \phi, \phi\rangle),\end{aligned}$$

and evaluating this expression at x yields $2\mathrm{Re}(f(x)\langle \phi(f - f(x)), \phi\rangle(x))$, which is zero by the last paragraph. Since this is true for all $x \in X$, we have $f\phi - \phi f = 0$ as desired. Taking linear combinations, we conclude that this is true for any $f \in C(X)$ and $\phi \in \mathcal{E}$.

Define $\mathcal{E}_{sa} = \{\phi \in \mathcal{E} : \phi = \phi^*\}$, so that $\mathcal{E} = \mathcal{E}_{sa} + i\mathcal{E}_{sa}$. Then the real version of Theorem 9.1.6 (a) implies that $\mathcal{E}_{sa} \cong \mathcal{S}(\mathcal{X})$ for some real Hilbert bundle $\mathcal{X}$ over X. Thus $\mathcal{E} \cong \mathcal{S}(\mathcal{X}) + i\mathcal{S}(\mathcal{X})$. ∎

The left and right actions do not coincide in general in the commutative case; a counterexample is given by taking $\mathcal{E}$ to be $M_2(\mathbf{C})$ and $\mathcal{A}$ to be the diagonal matrices in $M_2(\mathbf{C})$, and for $A, B \in \mathcal{E}$ defining $\langle A, B\rangle \in \mathcal{A}$ to be the diagonal part of AB.

Also, the reality condition on inner products in Theorem 9.6.6 can fail even if the left and right actions agree. This is shown by taking $\mathcal{A} = \mathbf{C}$, $\mathcal{E} = \mathbf{C}^2$, and defining $(a,b)^* = (\bar{b}, \bar{a})$ and $\langle (a,b),(c,d)\rangle = bc$.

We conclude this section with a discussion of Hilbert $*$-bimodules which satisfy a sort of locality condition. This turns out to be the key property that enables us to prove a version of the structural characterization of self-dual Hilbert modules given in Theorem 9.4.11. Although we formulate only the von Neumann algebra version of the condition, there is an obvious C* version as well. First we define the von Neumann algebra version of a Hilbert $*$-bimodule.

DEFINITION 9.6.7 *Let $\mathcal{E}$ be a Hilbert $*$-bimodule over a von Neumann algebra $\mathcal{M}$. We define the* <u>$*$-weak topology</u> *on $\mathcal{E}$ to be the weakest topology such that the maps $\xi \mapsto \langle \xi, \eta\rangle$ and $\xi \mapsto \langle \eta, \xi\rangle$ are continuous from $\mathcal{E}$ into $\mathcal{M}$ for all $\eta \in \mathcal{E}$, with respect to the weak* topology on $\mathcal{M}$.*

We say that $\mathcal{E}$ is a <u>dual bimodule</u> *if it is a dual Banach space in such a way that the weak* topology and the $*$-weak topology agree on the unit ball. If for any bounded, weak* convergent net $\alpha_\kappa \to \alpha$ in $\mathcal{M}$ and any $\xi \in \mathcal{E}$ we have $\alpha_\kappa \xi \to \alpha\xi$ $*$-weakly, then we say that $\mathcal{E}$ is* <u>normal</u>. *Finally, if $\mathcal{E}$ is both normal and dual we call it a* <u>W* Hilbert $*$-bimodule</u>.

In fact, $*$-weak compactness of the unit ball of $\mathcal{E}$ is sufficient to imply that $\mathcal{E}$ is a dual bimodule. This can be established by the argument used to prove Proposition 9.4.2.

The desired locality condition is formulated as follows.

DEFINITION 9.6.8 *Let $\mathcal{E}$ be a W* Hilbert $*$-bimodule over a von Neumann algebra $\mathcal{M}$. The* <u>center</u> *of $\mathcal{E}$ is the set*

$$Z(\mathcal{E}) = \{\xi \in \mathcal{E} : \alpha\xi = \xi\alpha \text{ for all } \alpha \in \mathcal{M}\}.$$

We say that $\mathcal{E}$ is <u>centered</u> *if $\mathcal{M}\cdot Z(\mathcal{E})$ is weak* dense in $\mathcal{E}$, and we say that $\mathcal{E}$ is* <u>local</u> *if it is centered and the inner product of any two self-adjoint elements of $Z(\mathcal{E})$ is self-adjoint in $\mathcal{M}$.*

The main consequence of locality is given in the following lemma, which is analogous to Lemma 9.4.10. Here we say that two subspaces $\mathcal{E}_1, \mathcal{E}_2 \subset \mathcal{E}$ are <u>orthogonal</u> if $\langle \xi, \eta\rangle = \langle \eta, \xi\rangle = 0$ for all $\xi \in \mathcal{E}_1$ and $\eta \in \mathcal{E}_2$. We also use the notation $Z(\mathcal{M})_{sa}$ and $Z(\mathcal{E})_{sa}$ for the set of self-adjoint elements in $Z(\mathcal{M})$ and $Z(\mathcal{E})$.

LEMMA 9.6.9

Let $\mathcal{E}$ be a local W Hilbert $*$-bimodule over a von Neumann algebra $\mathcal{M}$*

and let $\mathcal{E}_1$ be a centered, weak closed, self-adjoint sub-bimodule of $\mathcal{E}$. Then there is another centered, weak* closed, self-adjoint sub-bimodule $\mathcal{E}_2$ of $\mathcal{E}$ which is orthogonal to $\mathcal{E}_1$, such that $\mathcal{E} = \mathcal{E}_1 \oplus \mathcal{E}_2$.*

PROOF Let $\mathcal{V}$ be a sub $Z(\mathcal{M})_{sa}$-module of $Z(\mathcal{E}_1)_{sa}$ which is finitely generated as a module over $Z(\mathcal{M})_{sa}$. We claim that we can find a finite set $\{\xi_1, \ldots, \xi_n\}$ which spans $\mathcal{V}$ over $Z(\mathcal{M})_{sa}$ such that $\langle \xi_i, \xi_j \rangle = 0$ whenever $i \neq j$. To see this let $\{\xi_1, \ldots, \xi_n\}$ be any finite set which spans $\mathcal{V}$ and assume inductively that $\langle \xi_i, \xi_j \rangle = 0$ for $i, j \leq n-1$, $i \neq j$. Since

$$\alpha\langle \xi, \eta \rangle = \langle \alpha\xi, \eta \rangle = \langle \xi\alpha, \eta \rangle = \langle \xi, \alpha\eta \rangle = \langle \xi, \eta\alpha \rangle = \langle \xi, \eta \rangle\alpha$$

for any $\xi, \eta \in Z(\mathcal{E}_1)$ and $\alpha \in \mathcal{M}$, it follows that the inner product of any two elements of $Z(\mathcal{E}_1)$ is in $Z(\mathcal{M})$. Thus $Z(\mathcal{E}_1)$ satisfies the hypothesis of Theorem 9.6.6 as a Hilbert $*$-bimodule over $Z(\mathcal{M})$. Using the conclusion of Theorem 9.6.6 it is easy to verify that

$$\eta = \xi_n - \sum_{i=1}^{n-1} \frac{\langle \xi_n, \xi_i \rangle}{\langle \xi_i, \xi_i \rangle} \xi_i$$

is well-defined and orthogonal to ξ_i $(i \leq n-1)$ and that $\{\xi_1, \ldots, \xi_{n-1}, \eta\}$ spans $\mathcal{V}$. This proves the claim.

Now fix $\eta \in Z(\mathcal{E})_{sa}$. For any $\mathcal{V} \subset Z(\mathcal{E}_1)_{sa}$ as above, let $\{\xi_1, \ldots, \xi_n\}$ verify the claim and define

$$\eta_{\mathcal{V}} = \sum_{i=1}^{n} \frac{\langle \eta, \xi_i \rangle}{\langle \xi_i, \xi_i \rangle} \xi_i \in \mathcal{V}.$$

This expression is sensible and $\|\eta_{\mathcal{V}}\| \leq \|\eta\|$ by Theorem 9.6.6. Direct the subspaces $\mathcal{V}$ by inclusion and let $P\eta$ be a cluster point of the net $(\eta_{\mathcal{V}})$; then $P\eta \in Z(\mathcal{E}_1)_{sa}$ and $\langle \eta - P\eta, \xi \rangle = 0$ for all $\xi \in Z(\mathcal{E}_1)_{sa}$, and as there is at most one element of $Z(\mathcal{E}_1)_{sa}$ which can have this property P is a well-defined projection from $Z(\mathcal{E})_{sa}$ onto $Z(\mathcal{E}_1)_{sa}$.

Let $\mathcal{E}_2$ be the weak* closure of the set $\mathcal{M} \cdot \ker(P)$. It is clear that $\mathcal{E}_2$ is orthogonal to $\mathcal{E}_1$. Also $Z(\mathcal{E}_1)_{sa} \oplus Z(\mathcal{E}_2)_{sa} = Z(\mathcal{E})_{sa}$, so $\mathcal{E}_1 \oplus \mathcal{E}_2$ is weak* dense in $\mathcal{E}$. So for any $\xi \in \mathcal{E}$ we can find a bounded net $(\eta_\kappa + \eta'_\kappa)$ such that $\eta_\kappa \in \mathcal{E}_1$ and $\eta'_\kappa \in \mathcal{E}_2$ and $\eta_\kappa + \eta'_\kappa \to \xi$ weak*. By orthogonality, the nets (η_κ) and (η'_κ) are also bounded and so they have cluster points $\eta \in \mathcal{E}_1$ and $\eta' \in \mathcal{E}_2$, and $\eta + \eta' = \xi$ by continuity. So $\mathcal{E} = \mathcal{E}_1 \oplus \mathcal{E}_2$. ∎

Let $(\mathcal{I}_\kappa)$ be a family of W*-ideals of a von Neumann algebra $\mathcal{M}$. We denote by $\bigoplus^\infty \mathcal{I}_\kappa$ the set of sequences $\bigoplus \alpha_\kappa$ $(\alpha_\kappa \in \mathcal{I}_\kappa)$ with the property that both of the sums $\sum \alpha_\kappa \alpha_\kappa^*$ and $\sum \alpha_\kappa^* \alpha_\kappa$ converge weak* (or equivalently, that both sums are bounded). This is a W* Hilbert

$*$-bimodule, and its center is $\bigoplus^{\infty} Z(\mathcal{I}_\kappa)$, so it is centered and local. The following result shows that this construction is general.

THEOREM 9.6.10
Let $\mathcal{E}$ be a local W Hilbert $*$-bimodule over a von Neumann algebra $\mathcal{M}$. Then there is a family $(\mathcal{I}_\kappa)$ of W*-ideals of $\mathcal{M}$ such that $\mathcal{E} \cong \bigoplus^{\infty} \mathcal{I}_\kappa$.*

PROOF As in the proof of the lemma, regard $Z(\mathcal{E})$ as a local W* Hilbert $*$-bimodule over $Z(\mathcal{M})$. Observe that for any $\xi \in Z(\mathcal{E})_{sa}$ the sequence

$$(\langle \xi, \xi \rangle^{1/2} + n^{-1} I_{\mathcal{M}})^{-1} \xi$$

is bounded, and if η is a weak* cluster point of this sequence then $\langle \eta, \eta \rangle$ is a projection in $Z(\mathcal{M})$. Now let $\{\xi_\kappa\}$ be a maximal family of orthogonal elements of $Z(\mathcal{E})_{sa}$ with the property that $\langle \xi_\kappa, \xi_\kappa \rangle$ is a projection in $Z(\mathcal{M})$. It follows from the lemma and the preceding observation that $\mathcal{M} \cdot \mathrm{span}\{\xi_\kappa\}$ is weak* dense in $\mathcal{E}$. Also the sub-bimodules $\mathcal{M}\xi_\kappa$ are pairwise orthogonal. So it suffices to show that each $\mathcal{M}\xi_\kappa$ is isomorphic to $\mathcal{I}_\kappa = p_\kappa \mathcal{M}$ where $p_\kappa = \langle \xi_\kappa, \xi_\kappa \rangle$. But the kernel of the map $\alpha \mapsto \alpha\xi_\kappa$ is a weak* closed ideal of $\mathcal{M}$, and hence is of the form $q_\kappa \mathcal{M}$ for some projection $q_\kappa \in Z(\mathcal{M})$, and clearly $q_\kappa = I_{\mathcal{M}} - p_\kappa$. Also, for any $\alpha, \beta \in \mathcal{I}_\kappa$ we have $\alpha\beta = \langle \alpha\xi_\kappa, \beta\xi_\kappa \rangle$ and $(\alpha\xi_\kappa)^* = \alpha^*\xi_\kappa$. So indeed $\mathcal{M}\xi_\kappa \cong \mathcal{I}_\kappa$ as Hilbert $*$-bimodules. ∎

9.7 Notes

For more on Hilbert modules over C*-algebras see [43].

The correspondence between continuous Hilbert bundles and Hilbert modules is given a very thorough treatment in [26].

Hilbert bundles over C*-algebras were originally considered in [56] and [59], and these papers are still a good source on the topic. The material in Section 9.4 is all from [56].

The approach to crossed products presented in Section 9.5 is taken from [41].

For more on Hilbert $*$-bimodules see [74].

There is also a quantum analog of Banach bundles, viz., operator modules; see [57].

Chapter 10

Lipschitz Algebras

10.1 The algebras $\mathrm{Lip}_0(X)$

The quantum version of a metric space is analogous to the quantum versions of topological and measure spaces. In each case there is a Hilbert space generalization of the algebra of scalar-valued maps on such an object which preserve the relevant structure. The topological and measure theoretic categories involve the algebras $C(X)$ and $L^\infty(X)$; in the metric case the appropriate algebra is $\mathrm{Lip}_0(X)$.

DEFINITION 10.1.1
(a) Let X and Y be metric spaces. A map $f : X \to Y$ is Lipschitz *if its* Lipschitz number

$$L(f) = \sup\left\{\frac{\rho(f(x), f(y))}{\rho(x, y)} : x, y \in X,\, x \neq y\right\}$$

is finite. (We use the generic notation $\rho(\cdot, \cdot)$ for distance in any metric space.)
(b) Let X be a complete metric space with finite diameter (that is, $\Delta(X) = \sup\{\rho(x, y) : x, y \in X\}$ is finite) and let $e \in X$ be a distinguished element. Then $\mathrm{Lip}_0(X)$ is the set of Lipschitz functions $f : X \to \mathbf{C}$ which satisfy $f(e) = 0$.

We call a metric space with a distinguished element pointed. In part (b) there is no loss of generality in requiring that X be complete, as scalar-valued Lipschitz functions always extend to the completion of a space without increasing their Lipschitz number.

It is straightforward to verify that $\mathrm{Lip}_0(X)$ is a Banach space for the norm $L(\cdot)$ and is closed under products. (The latter uses the fact that the diameter of X is finite; specifically, we have the bound $L(fg) \leq 2\Delta(X)L(f)L(g)$.) After this the most important fact about $\mathrm{Lip}_0(X)$ is that it is a dual space.

PROPOSITION 10.1.2

Let X be a complete pointed metric space with finite diameter. Then $\mathrm{Lip}_0(X)$ *is a dual Banach space in such a way that on bounded sets its weak* topology agrees with the topology of pointwise convergence.*

PROOF The argument is similar to the proof of Proposition 9.4.2. For every $x \in X$ the map $\hat{x} : f \mapsto f(x)$ satisfies

$$|\hat{x}(f)| = |f(x) - f(e)| \leq L(f) \cdot \rho(x, e)$$

and hence belongs to $\mathrm{Lip}_0(X)^*$. Let $\mathcal{V} \subset \mathrm{Lip}_0(X)^*$ be the closed span of the set $\{\hat{x} : x \in X\}$ and let $T : \mathrm{Lip}_0(X) \to \mathcal{V}^*$ be the natural map, i.e., $Tf(\omega) = \omega(f)$.

For distinct $x, y \in X$ we have $\|\hat{x} - \hat{y}\| \leq \rho(x, y)$. Since $Tf(\hat{x} - \hat{y}) = f(x) - f(y)$, it follows that $\|Tf\| \geq |f(x) - f(y)|/\rho(x, y)$. Taking the supremum over $x, y \in X$, we deduce that $\|Tf\| \geq L(f)$, and the reverse inequality is easy. So T is an isometry.

To show that T is surjective, let ω belong to the unit ball of $\mathcal{V}^*$. Define a function $f : X \to \mathbf{C}$ by setting $f(x) = \omega(\hat{x})$; then it is straightforward to verify that $f \in \mathrm{Lip}_0(X)$ (with $L(f) = \|\omega\|$) and $Tf = \omega$. Thus $\mathrm{Lip}_0(X) \cong \mathcal{V}^*$.

By evaluating on $\hat{x}$, it is clear that $f_\kappa \to f$ weak* implies $f_\kappa(x) \to f(x)$. Conversely, since the unit ball of $\mathrm{Lip}_0(X)$ is weak* compact and the topology of pointwise convergence is Hausdorff, the two topologies must agree on the unit ball. ∎

We now want to prove results analogous to those given in Sections 5.1, 6.1, and 6.2, and establish correspondences between metric properties of X and algebraic properties of $\mathrm{Lip}_0(X)$. As in Chapter 6, use of the weak* topology will be key. The main tool is the following version of the Stone-Weierstrass theorem.

LEMMA 10.1.3

Let X be a complete pointed metric space with finite diameter and let $\mathcal{M}$ be a W-subalgebra of* $\mathrm{Lip}_0(X)$. *Then for any set of real-valued functions $(f_\kappa) \subset \mathcal{M}^+ = \mathcal{M} + \mathbf{C} \cdot 1_X$ which is bounded in both Lipschitz and supremum norm, we have* $\sup f_\kappa, \inf f_\kappa \in \mathcal{M}^+$.

PROOF Let $f \in \mathcal{M}^+$ be real-valued. Then $f(e) \in \mathbf{R}$, so $f - f(e) \in \mathcal{M}$ is also real-valued. Let $a = \|f\|_\infty$. For any function $g \in C^1[-a, a]$, we have $g \circ f \in \mathrm{Lip}_0(X) + \mathbf{C} \cdot 1_X$ since $L(g \circ f) \leq L(f)L(g) = L(f)\|g'\|_\infty$; we claim that $g \circ f \in \mathcal{M}^+$. To see this, find a sequence of polynomials h_n such that $\|h_n - g'\|_\infty \to 0$. Let g_n be the indefinite integral of h_n which satisfies $g_n(0) = g(0)$, and let $f_n = g_n \circ f$. Then f_n is a polynomial in f, so $f_n \in \mathcal{M}^+$, and

$$L(f_n - g \circ f) \leq L(f)L(g_n - g) = L(f)\|g_n' - g'\|_\infty \to 0.$$

Thus $(f_n - f_n(e))$ is a bounded sequence in $\mathcal{M}$ and converges to $g \circ f - g \circ f(e)$, so we conclude that the latter function also belongs to $\mathcal{M}$, and hence $g \circ f \in \mathcal{M}^+$.

Now let (g_n) be a sequence of C^1 functions which converges pointwise to the absolute value function $t \mapsto |t|$ on $[-a, a]$ and is uniformly bounded in Lipschitz norm. By the claim, $g_n \circ f \in \mathcal{M}^+$ for all n, and $g_n \circ f \to |f|$ pointwise, so $|f| \in \mathcal{M}^+$ by weak* closure of $\mathcal{M}$. It follows that for any real-valued $f_1, f_2 \in \mathcal{M}^+$ the functions $\max(f_1, f_2) = \frac{1}{2}(f_1+f_2+|f_1-f_2|)$ and $\min(f_1, f_2) = \frac{1}{2}(f_1 + f_2 - |f_1 - f_2|)$ both belong to $\mathcal{M}^+$. Finally, for any Lipschitz and sup norm bounded set of real-valued functions in $\mathcal{M}^+$ we can take a pointwise limit of the max (respectively, min) of every finite subset to get the sup (respectively, inf) of the set, and this shows that $\mathrm{Re}\,\mathcal{M}^+$ is closed under suprema and infima of bounded sets. ∎

We say that a subspace $E \subset \mathrm{Lip}_0(X)$ separates points uniformly if there exists $C \geq 1$ such that for any $x, y \in X$ there exists $f \in E$ with $L(f) \leq C$ and $|f(x) - f(y)| = \rho(x, y)$.

THEOREM 10.1.4

Let X be a complete pointed metric space with finite diameter and let $\mathcal{M}$ be a W-subalgebra of $\mathrm{Lip}_0(X)$ which separates points uniformly. Then $\mathcal{M} = \mathrm{Lip}_0(X)$.*

PROOF Let $f \in \mathrm{Lip}_0(X)$ be real-valued and suppose $L(f) \leq 1$. It will suffice to show that $f \in \mathcal{M}^+ = \mathcal{M} + \mathbf{C} \cdot 1_X$.

Fix $C \geq 1$ as in the definition of uniform separation of points. For each $x, y \in X$ find $g \in \mathcal{M}$ such that $L(g) \leq C$ and $|g(x) - g(y)| = \rho(x, y)$. Multiplying g by the constant $(f(x) - f(y))/(g(x) - g(y))$, we may assume that $g(x) - g(y) = f(x) - f(y)$. Then adding the real constant $f(x) - g(x) = f(y) - g(y)$ to g, we get a function $h_{xy} \in \mathcal{M}^+$ such that $h_{xy}(x) = f(x)$, $h_{xy}(y) = f(y)$, $L(h_{xy}) \leq C$, and

$$\|h_{xy}\|_\infty \leq \|g\|_\infty + |f(x)| + |g(x)| \leq (2C + 1)\Delta(X).$$

Finally, we can replace h_{xy} with its real part without affecting the previous facts, and the lemma then implies that

$$f = \inf_{x \in X} \sup_{y \in X} h_{xy} \in \mathcal{M}^+,$$

as desired. ∎

Now we can establish the expected relationships between subalgebras of $\mathrm{Lip}_0(X)$ and quotients of X and between ideals of $\mathrm{Lip}_0(X)$ and subsets of X. As usual, given $\phi : X \to Y$ we let $C_\phi : \mathrm{Lip}_0(Y) \to \mathrm{Lip}_0(X)$ be the composition map $C_\phi f = f \circ \phi$.

Example 10.1.5

Let X and Y be complete pointed metric spaces with finite diameter and let $\phi : X \to Y$ be a contraction such that $\phi(e_X) = e_Y$. Assume that $C_\phi : \mathrm{Lip}_0(Y) \to \mathrm{Lip}_0(X)$ is an isometry. Then $C_\phi(\mathrm{Lip}_0(Y))$ is a W*-subalgebra of $\mathrm{Lip}_0(X)$.

PROPOSITION 10.1.6

Let X be a complete pointed metric space with finite diameter and let $\mathcal{M}$ be a W-subalgebra of $\mathrm{Lip}_0(X)$. Then there is a complete pointed metric space Y with finite diameter and a contraction $\phi : X \to Y$ such that $\phi(e_X) = e_Y$, $C_\phi : \mathrm{Lip}_0(Y) \to \mathrm{Lip}_0(X)$ is isometric, and $\mathcal{M} = C_\phi(\mathrm{Lip}_0(Y))$.*

PROOF Set $x \sim y$ if $f(x) = f(y)$ for all $f \in \mathcal{M}$, and let Y be the completion of $X/\sim$ for the metric

$$\rho_{\mathcal{M}}([x],[y]) = \sup\{|f(x) - f(y)| : f \in \mathcal{M},\ L(f) \le 1\}.$$

Let $e_Y = [e_X]$ and let $\phi : X \to Y$ be the quotient map.

It is clear that C_ϕ takes $\mathrm{Lip}_0(Y)$ nonexpansively into $\mathrm{Lip}_0(X)$. Conversely, it is easy to see that any $f \in \mathcal{M}$ descends to a function $\tilde{f}$ on Y with $L(\tilde{f}) \le L(f)$. Thus, the map $f \mapsto \tilde{f}$ is an isometry from $\mathcal{M}$ into $\mathrm{Lip}_0(Y)$. Since $\mathcal{M}$ is weak* closed in $\mathrm{Lip}_0(X)$, it is closed under pointwise convergence of bounded nets, and this remains true of its image in $\mathrm{Lip}_0(Y)$, so $C_\phi^{-1}(\mathcal{M})$ is a W*-subalgebra of $\mathrm{Lip}_0(Y)$. But by the definition of the metric on Y, $C_\phi^{-1}(\mathcal{M})$ separates points uniformly. Thus $C_\phi^{-1}(\mathcal{M}) = \mathrm{Lip}_0(Y)$, which is enough. ∎

Example 10.1.7

Let X be a complete pointed metric space with finite diameter and let K be a closed subset of X which contains e. Then $\{f \in \mathrm{Lip}_0(X) : f|_K = 0\}$ is a W*-ideal of $\mathrm{Lip}_0(X)$.

The converse of this example requires two lemmas. We use the following notation: X is a complete pointed metric space with finite diameter,

$\mathcal{I}$ is a W*-ideal of $\mathrm{Lip}_0(X)$, and $K = \{x \in X : f(x) = 0 \text{ for all } f \in \mathcal{I}\}$. We also define a pseudometric

$$\rho_{\mathcal{I}}(x,y) = \sup\{|f(x) - f(y)| : f \in \mathcal{I}, L(f) \leq 1\}$$

as in the proof of Proposition 10.1.6.

LEMMA 10.1.8
Let $x \in X$. If $\rho_{\mathcal{I}}(x,K) < a$ then for any $\epsilon > 0$ there exists $y \in X$ such that $\rho(x,y) < a$ and $\rho_{\mathcal{I}}(y,K) < \epsilon$.

PROOF If $\rho(x,K) < a$ then the conclusion is trivial, so assume $\rho(x,K) \geq a$. Let $\epsilon > 0$ and define

$$f(y) = \min(1, \rho_{\mathcal{I}}(y,K)/\epsilon)$$

and

$$g(y) = \max(0, a - \rho(x,y)).$$

Since $e \in K$ and $\rho(x,K) \geq a$, we have $f(e) = g(e) = 0$, so $f, g \in \mathrm{Lip}_0(X)$. From the construction of $X/\sim$ in the proof of Proposition 10.1.6 with $\mathcal{M} = \mathcal{I}$, it is seen that the function f descends to a Lipschitz function on the quotient $X/\sim$, so that result implies $f \in \mathcal{I}$. Thus $fg \in \mathcal{I}$. Now if the lemma fails then f must take the value 1 everywhere g is nonzero, which implies $g = fg \in \mathcal{I}$; but this implies $g|_K = 0$ and

$$\rho_{\mathcal{I}}(x,y) \geq |g(x) - g(y)| = a$$

for all $y \in K$, a contradiction. Thus the lemma holds. ∎

LEMMA 10.1.9
We have $\rho(x,K) = \rho_{\mathcal{I}}(x,K)$ for all $x \in X$.

PROOF Suppose $\rho_{\mathcal{I}}(x,K) < \rho(x,K)$ for some $x \in X$. Let $\delta = \frac{1}{2}(\rho(x,K) - \rho_{\mathcal{I}}(x,K))$. Take $a = \rho(x,K) - \delta = \rho_{\mathcal{I}}(x,K) + \delta$ and $\epsilon = \delta/3$ in Lemma 10.1.8; we get an element $y_1 \in X$ such that $\rho(x,y_1) < a$ (and hence $\rho(y_1,K) > \delta$) and $\rho_{\mathcal{I}}(y_1,K) < \delta/3$. Inductively, we can construct a sequence (y_n) such that $\rho(y_n, y_{n+1}) < \delta/3^n$, $\rho(y_n,K) \geq (1 + 3^{1-n})\delta/2$, and $\rho_{\mathcal{I}}(y_n,K) < \delta/3^n$. Thus $y_n \to y$ for some y which satisfies $\rho(y,K) \geq \delta/2$ but $\rho_{\mathcal{I}}(y,K) = 0$, which is impossible. This contradiction establishes the lemma. ∎

PROPOSITION 10.1.10
Let X be a complete pointed metric space with finite diameter and let $\mathcal{I}$ be a W-ideal of* $\mathrm{Lip}_0(X)$*. Then there is a closed subset K of X which contains e such that* $\mathcal{I} = \{f \in \mathrm{Lip}_0(X) : f|_K = 0\}$.

PROOF Let $K = \{x \in X : f|_K = 0 \text{ for all } f \in \mathcal{I}\}$. It is clear that $f \in \mathcal{I}$ implies $f|_K = 0$. Conversely, let $f \in \mathrm{Lip}_0(X)$ and suppose $f|_K = 0$; we must show that $f \in \mathcal{I}$.

We have $|f(x) - f(y)| \leq L(f)\rho(x, y)$ and

$$|f(x) - f(y)| \leq |f(x)| + |f(y)| \leq L(f)(\rho(x, K) + \rho(y, K))$$

for all $x, y \in X$. We claim that

$$\rho_{\mathcal{I}}(x, y) \geq \min(\rho(x, y), \frac{1}{2}(\rho(x, K) + \rho(y, K)));$$

this will imply that $|f(x) - f(y)| \leq 2L(f)\rho_{\mathcal{I}}(x, y)$, and we may then conclude that $f \in \mathcal{I}$ by Proposition 10.1.6.

To prove the claim, fix $x, y \in X$. Without loss of generality suppose $\rho(x, K) \geq \rho(y, K)$, and fix $0 < \epsilon < \rho(y, K)$. Define

$$g(z) = \max(0, \rho(x, K) - \rho(x, z) - \epsilon).$$

This function vanishes in the ϵ-neighborhood of K, so by Lemma 10.1.9 we have $gh = g$ where

$$h(z) = \min(1, \rho_{\mathcal{I}}(z, K)/\epsilon) = \min(1, \rho(z, K)/\epsilon).$$

But $h \in \mathcal{I}$ by Proposition 10.1.6, so we therefore have $g \in \mathcal{I}$.

If $\rho(x, y) \geq \rho(x, K)$ then $g(x) = \rho(x, K) - \epsilon$ and $g(y) = 0$, so

$$|g(x) - g(y)| = \rho(x, K) - \epsilon \geq \frac{1}{2}(\rho(x, K) + \rho(y, K)) - \epsilon.$$

Otherwise $\rho(x, y) < \rho(x, K)$ and we have

$$|g(x) - g(y)| \geq \rho(x, y) - \epsilon.$$

In either case, since $g \in \mathcal{I}$ and $L(g) \leq 1$, we conclude

$$\rho_{\mathcal{I}}(x, y) \geq \min(\rho(x, y), \frac{1}{2}(\rho(x, K) + \rho(y, K))) - \epsilon.$$

Taking $\epsilon \to 0$ completes the proof. ∎

We can use the preceding result to characterize quotients of $\mathrm{Lip}_0(X)$. This also requires a version of the Tietze extension theorem, which we present first.

LEMMA 10.1.11
Let X be a complete metric space, let $K \subset X$, and let $f : K \to \mathbf{R}$ be a Lipschitz function. Then there is an extension $\tilde{f} : X \to \mathbf{R}$ such that $L(\tilde{f}) = L(f)$.

PROOF Define

$$\tilde{f}(x) = \inf_{y \in K} (f(y) + L(f)\rho(x, y)).$$

A short computation verifies that $\tilde{f}|_K = f$ and $L(\tilde{f}) = L(f)$. ∎

PROPOSITION 10.1.12
Let X be a complete pointed metric space with finite diameter and let $\mathcal{I}$ be a W-ideal of* $\mathrm{Lip}_0(X)$. *Then* $\mathrm{Lip}_0(X)/\mathcal{I}$ *is isomorphic to* $\mathrm{Lip}_0(K)$ *where $K = \{x \in X : f(x) = 0$ for all $f \in \mathcal{I}\}$, and the isomorphism is isometric on real-valued functions.*

PROOF Let $\pi : \mathrm{Lip}_0(X) \to \mathrm{Lip}_0(K)$ be the restriction map. It is clear that π is nonexpansive, and $\ker(\pi) = \mathcal{I}$ by Proposition 10.1.10. For any real-valued $f \in \mathrm{Lip}_0(K)$, the lemma implies that there exists $\tilde{f} \in \mathrm{Lip}_0(X)$ such that $L(\tilde{f}) = L(f)$ and $\pi\tilde{f} = f$, so π must be surjective and isometric on real-valued functions. This is enough. ∎

Next we turn to the spectrum, and show that the metric space X can be recovered from the algebra $\mathrm{Lip}_0(X)$. Let $\hat{x} \in \mathrm{Lip}_0(X)^*$ be the evaluation map at the point x.

Since $1_X \notin \mathrm{Lip}_0(X)$, the weak* spectrum defined in Definition 6.1.1 is not appropriate here. Instead, we let $\mathrm{sp}_0^*(\mathrm{Lip}_0(X))$ be the set of all weak* continuous $*$-homomorphisms, including the zero map, from $\mathrm{Lip}_0(X)$ to $\mathbf{C}$. We give $\mathrm{sp}_0^*(\mathrm{Lip}_0(X))$ the metric it inherits from $\mathrm{Lip}_0(X)^*$.

PROPOSITION 10.1.13
Every complete pointed metric space X with finite diameter is isometric to $\mathrm{sp}_0^*(\mathrm{Lip}_0(X))$ *via the correspondence $x \leftrightarrow \hat{x}$.*

PROOF Let X be a complete pointed metric space with finite diameter. The map $x \mapsto \hat{x}$ is clearly nonexpansive, and applying $\hat{x} - \hat{y}$ to the function $f(z) = \rho(x, z) - \rho(x, e)$ shows that $\|\hat{x} - \hat{y}\| = \rho(x, y)$. To prove surjectivity, let $\omega : \mathrm{Lip}_0(X) \to \mathbf{C}$ be a weak* continuous $*$-homomorphism. If $\omega = 0 = \hat{e}$ we are done. Otherwise its kernel is a codimension one W*-ideal, so Proposition 10.1.12 implies that there exists $x \in X$, $x \neq e$, such that $\ker(\omega) = \{f \in \mathrm{Lip}_0(X) : f(x) = 0\}$. It follows that $\omega = a\hat{x}$ for some nonzero $a \in \mathbf{C}$; choosing $f \in \mathrm{Lip}_0(X)$ such that $f(x) = 1$, we then have

$$a = \omega(f) = \omega(f^2) = \omega(f)^2 = a^2,$$

so that $a = 1$. Thus $\omega = \hat{x}$. ∎

The desired correspondence between Lipschitz functions and weak* continuous $*$-homomorphisms follows easily.

Example 10.1.14

Let X and Y be complete pointed metric spaces with finite diameters. Let $\phi : Y \to X$ be Lipschitz and suppose $\phi(e_Y) = e_X$. Then the map $C_\phi : f \mapsto f \circ \phi$ is a weak* continuous $*$-homomorphism from $\mathrm{Lip}_0(X)$ into $\mathrm{Lip}_0(Y)$. The norm of C_ϕ is $L(\phi)$.

PROPOSITION 10.1.15

Let X and Y be complete pointed metric spaces with finite diameters and let $\pi : \mathrm{Lip}_0(X) \to \mathrm{Lip}_0(Y)$ be a weak continuous $*$-homomorphism. Then there is a Lipschitz map $\phi : Y \to X$ such that $\phi(e_Y) = e_X$ and $\pi = C_\phi$.*

PROOF For any $y \in Y$ the map $\hat{y} \circ \pi$ belongs to $\mathrm{sp}_0^*(\mathrm{Lip}_0(X))$, so there is a unique $x \in X$ such that $\hat{y} \circ \pi = \hat{x}$. Define $\phi(y) = x$. For any $f \in \mathrm{Lip}_0(X)$ and any $y \in Y$ we then have

$$f(\phi(y)) = \phi(y)\hat{}\,(f) = \hat{y}(\pi f) = \pi f(y),$$

so $f \circ \phi = \pi f$. It is clear that $\phi(e_Y) = e_X$. To see that ϕ is Lipschitz, let $x, y \in Y$ and define $f \in \mathrm{Lip}_0(X)$ by

$$f(z) = \rho(\phi(x), z) - \rho(\phi(x), e_X).$$

Then

$$|\pi f(x) - \pi f(y)| = |f(\phi(x)) - f(\phi(y))| = \rho(\phi(x), \phi(y)).$$

Since $L(\pi f) \leq \|\pi\| L(f) = \|\pi\|$, this implies $\rho(\phi(x), \phi(y)) \leq \|\pi\| \rho(x, y)$. So ϕ is Lipschitz. ∎

10.2 *Measurable metrics*

There is a measurable version of the notion of a metric and a corresponding version of the space of Lipschitz functions. In the special case of counting measure such "measurable metrics" reduce to ordinary metrics. However, we need to examine general measurable metrics before passing to the Hilbert space setting, because they, not pointwise metrics, represent the general commutative case.

Measurable metrics are defined as follows.

DEFINITION 10.2.1 *Let X be a σ-finite measure space and let Ω_+ be the family of positive measure subsets of X modulo null sets. A measurable pseudometric is a map $\rho : \Omega_+^2 \to [0, \infty]$ which satisfies*

(a) $\rho(S, S) = 0$
(b) $\rho(S, T) = \rho(T, S)$
(c) $\rho(\bigcup S_i, T) = \inf \rho(S_i, T)$
(d) $\rho(R, T) \leq \sup\{\rho(R, S') + \rho(S', T) : S' \subset S\}$

for all $R, S, T, S_i \in \Omega_+$.

A measurable metric is a measurable pseudometric with the property that the measurable σ-algebra is generated up to null sets by the sets $S \in \Omega_+$ with the property that, for every $T \in \Omega_+$, $S \cap T = \emptyset$ implies $\rho(S, T') > 0$ for some $T' \subset T$.

Intuitively, the quantity $\rho(S, T)$ is supposed to represent the minimum distance between the sets S and T, modulo null sets. Indeed, if we are given a genuine pointwise metric on a σ-finite measure space X, a careful interpretation of the preceding sentence does produce a measurable metric (or at least a pseudometric). The converse is true in a sense: given any measurable metric on a σ-finite measure space, we can replace the original space with a measurably equivalent space such that the corresponding measurable metric on the new space arises from a pointwise metric in the manner suggested above. We will not prove this result here.

Infinite distances are permitted in measurable metric spaces. This is done to accomodate general constructions such as the one in Theorem 10.3.6 which sometimes produce such a result. We will also find it more convenient to work with measurable pseudometrics rather than measurable metrics.

The following is a fundamental lemma that we will need later.

LEMMA 10.2.2
Let X be a σ-finite measure space, let ρ be a measurable pseudometric on X, and let S and T be positive measure subsets of X. Let $\epsilon > 0$. Then there exist positive measure sets $S_0 \subset S$ and $T_0 \subset T$ such that

$$\rho(S, T) \leq \rho(S', T') < \rho(S, T) + \epsilon$$

for all $S' \subset S_0$ and $T' \subset T_0$.

PROOF For any $S' \subset S$ and $T' \subset T$ we have $\rho(S,T) \leq \rho(S',T')$ by axiom (c) of Definition 10.2.1 applied successively to $S = S \cup S'$ and $T = T \cup T'$. To prove the remainder, we will find $S_0 \subset S$ such that $S' \subset S_0$ implies $\rho(S',T) \leq \rho(S,T) + \epsilon$; applying the same argument to T then produces the desired pair S_0, T_0.

Without loss of generality assume the measure μ on X is finite, and let

$$a = \sup\{\mu(R) : R \subset S \text{ and } \rho(R,T) \geq \rho(S,T) + \epsilon\}.$$

Then find a sequence (R_i) such that $\rho(R_i,T) \geq \rho(S,T) + \epsilon$ and $\mu(R_i) \to a$. Let $R = \bigcup R_i$; then we have $\rho(R,T) \geq \rho(S,T) + \epsilon$, and evidently R contains, up to a null set, every positive measure subset of S with this property. Thus $S_0 = S - R$ is a positive measure set (since $\rho(S,T) \neq \rho(R,T)$) and we have $\rho(S',T) < \rho(S,T) + \epsilon$ for any $S' \subset S_0$, as desired. ∎

Next we define Lipschitz spaces for measurable pseudometrics. Recall the definition of the essential range of a map from Example 3.2.2 (b).

DEFINITION 10.2.3 *Let X be a σ-finite measure space, let ρ be a measurable pseudometric on X, and let $f : X \to \mathbf{C}$ be measurable. For positive measure sets $S, T \subset X$ we write $\rho_f(S,T)$ for the distance in $\mathbf{C}$ between the essential ranges of $f|_S$ and $f|_T$. The Lipschitz number of f is then*

$$L(f) = \sup \frac{\rho_f(S,T)}{\rho(S,T)},$$

taking the supremum over all positive measure $S, T \subset X$ such that $\rho(S,T) > 0$; we say f is Lipschitz if $L(f) < \infty$.

$\mathrm{Lip}(X,\mu)$ is the subspace of $L^\infty(X,\mu)$ consisting of those functions f with the property that $L(f) < \infty$. We define the Lipschitz norm on this space to be $\|f\|_L = \max(\|f\|_\infty, L(f))$.

In the atomic case, where μ is counting measure, $\mathrm{Lip}(X,\mu)$ is a variation of the space $\mathrm{Lip}_0(X)$ discussed in the last section. Our first nonatomic example is the following.

Example 10.2.4

Let X be a σ-finite measure space. Define a measurable metric ρ on X by setting

$$\rho(S,T) = \begin{cases} 0 & \text{if } \mu(S \cap T) > 0 \\ 2 & \text{if } \mu(S \cap T) = 0. \end{cases}$$

One can check that $L(f) \le \|f\|_\infty$ for any $f \in L^\infty(X)$, so we therefore have $\mathrm{Lip}(X,\mu) = L^\infty(X)$, both as sets and as Banach spaces.

Just as for Lip_0 spaces, every $\mathrm{Lip}(X,\mu)$ is a dual Banach space.

LEMMA 10.2.5
Let X be a σ-finite measure space and let ρ be a measurable pseudometric on X. Then the unit ball of $\mathrm{Lip}(X,\mu)$ *is compact for the weak* topology inherited from $L^\infty(X)$.*

PROOF Let (f_κ) be a net in $\mathrm{Lip}(X,\mu)$ which converges weak* to a function f in $L^\infty(X)$, and suppose $\|f_\kappa\|_L \le 1$ for all κ. We must show that $\|f\|_L \le 1$; it is enough to check that $L(f) \le 1$.

Let $S, T \subset X$ be positive measure sets and fix $\epsilon > 0$. Find S_0 and T_0 as in Lemma 10.2.2. By passing to subsets of S_0 and T_0, we may assume that $|f(x_1) - f(x_2)| \le \epsilon$ and $|f(y_1) - f(y_2)| \le \epsilon$ for almost every $x_1, x_2 \in S_0$ and $y_1, y_2 \in T_0$. Then

$$\begin{aligned}\left|\int_{S_0\times T_0} (f(x) - f(y))\right| &\ge (\rho_f(S_0,T_0) - 2\epsilon)\mu(S_0)\mu(T_0)\\ &\ge (\rho_f(S,T) - 2\epsilon)\mu(S_0)\mu(T_0).\end{aligned}$$

There also exists κ such that $|\int_{S_0}(f_\kappa - f)| \le \epsilon\cdot\mu(S_0)$ and $|\int_{T_0}(f_\kappa - f)| \le \epsilon\cdot\mu(T_0)$; thus

$$\begin{aligned}\left|\int_{S_0\times T_0} (f(x) - f(y))\right| &\le \left|\int (f(x) - f_\kappa(x))\right| + \left|\int (f_\kappa(x) - f_\kappa(y))\right|\\ &\quad + \left|\int (f_\kappa(y) - f(y))\right|\\ &\le (3\epsilon + \rho(S,T))\mu(S_0)\mu(T_0).\end{aligned}$$

Combining the two inequalities yields $\rho_f(S,T) \le \rho(S,T) + 5\epsilon$, which is enough. ∎

PROPOSITION 10.2.6
Let X be a σ-finite measure space and let ρ be a measurable pseudometric on X. Then $\mathrm{Lip}(X,\mu)$ *is a dual Banach space in such a way that on bounded sets its weak* topology agrees with the weak* topology inherited from $L^\infty(X)$.*

The proof of this result proceeds along familiar lines. We refer the reader to the proof of Proposition 9.4.2.

We also have a Stone-Weierstrass theorem for $\mathrm{Lip}(X,\mu)$. This involves a measurable version of uniform separation of points.

THEOREM 10.2.7

Let X be a σ-finite measure space, let ρ be a measurable pseudometric on X, and let $\mathcal{M}$ be a unital W-subalgebra of* $\mathrm{Lip}(X, \mu)$*. Suppose there exists $C \geq 1$ such that for all positive measure sets $S, T \subset X$ there exists $f \in \mathcal{M}$ such that $L(f) \leq C$ and $\rho_f(S,T) = \rho(S,T)$. Then* $\mathcal{M} = \mathrm{Lip}(X, \mu)$.

PROOF Let $f \in \mathrm{Lip}(X, \mu)$ be positive and satisfy $L(f) \leq 1$. It will suffice to show that $f \in \mathcal{M}$. To do this, fix a positive measure set $S \subset X$. For every positive measure set $T \subset X$, we can find $g \in \mathcal{M}$ such that $L(g) \leq C$ and $\rho_g(S,T) = \rho(S,T)$. Since the essential ranges of $g|_S$ and $g|_T$ are compact, we can find representative elements a and b such that $|a - b| = \rho(S,T)$. Subtracting a from g, we may assume $a = 0$, and multiplying g by $|b|/b$ and taking the real part, we may assume g is real valued. Then $g|_S \leq 0$ and $g|_T \geq \rho(S,T)$.

Next, by exactly the same argument as in Lemma 10.1.3, but here with the simplification that we can take $a = 0$, we can show that the real part of $\mathcal{M}$ is closed under suprema and infima of bounded sets. Thus, replacing g successively with $\max(g, 0)$ and $\min(g, \rho(S,T))$ we can assume that $g|_S = 0$ and $g|_T = \rho(S,T)$ constantly.

Define $g_T = \min(g, \|f\|_\infty)$ and

$$h_S = \sup\{g_T : \mu(T) > 0\}.$$

Then $h_S|_S = 0$ and $h_S|_T \geq \min(\rho(S,T), \|f\|_\infty)$ for all $T \subset X$. Also, $h_S \in \mathcal{M}$ and $L(h_S) \leq 1$. Now for each positive measure set $S \subset X$, let $a_S = \|f|_S\|_\infty$. We claim that

$$f = \inf\{h_S + a_S : \mu(S) > 0\};$$

this will imply that $f \in \mathcal{M}$, as desired.

First, suppose $f \geq h_S + a_S + \epsilon$ on some positive measure set T, for some $S \subset X$. Then

$$f|_T \geq h_S|_T + a_S + \epsilon \geq \rho(S,T) + a_S + \epsilon$$

while $f|_S \leq a_S$ by the definition of a_S, so we deduce that $\rho_f(S,T) \geq \rho(S,T) + \epsilon$, contradicting the assumption that $L(f) \leq 1$. Thus $f \leq h_S + a_S$ almost everywhere, for all S.

Conversely, on any positive measure set S we have

$$f|_S \not\leq h_S|_S + a_S - \epsilon = a_S - \epsilon$$

for any $\epsilon > 0$. This shows that for any $\epsilon > 0$ we cannot have $f \leq \inf\{h_S + a_S : \mu(S) > 0\} - \epsilon$ on any positive measure set. We conclude that $f = \inf\{h_S + a_S : \mu(S) > 0\} \in \mathcal{M}$ as claimed. ∎

As an immediate corollary we deduce that unital W*-subalgebras of $\mathrm{Lip}(X, \mu)$ are themselves of the form $\mathrm{Lip}(Y, \nu)$.

COROLLARY 10.2.8

Let X be a σ-finite measure space, let ρ be a measurable pseudometric on X, and let $\mathcal{M}$ be a unital W-subalgebra of* $\mathrm{Lip}(X, \mu)$. *Then $\mathcal{M} \cong \mathrm{Lip}(Y, \nu)$ for some measurable pseudometric on the measure space $(Y, \nu) = (X, \mu)$.*

PROOF Define a measurable pseudometric on $(Y, \nu) = (X, \mu)$ by setting $\rho(S, T) = \sup\{\rho_f(S, T) : f \in \mathcal{M},\ L(f) \leq 1\}$. It is clear that $\mathcal{M} \subset \mathrm{Lip}(Y, \nu)$ isometrically, and the theorem then implies $\mathcal{M} = \mathrm{Lip}(Y, \nu)$. ∎

The structure of W*-ideals is less transparent; we do not have a simple analog of Propositions 5.1.5, 6.1.7, and 6.2.6. For example, give $[0, 1]$ Lebesgue measure and let $\mathcal{I} = \{f \in \mathrm{Lip}[0, 1] : f(0) = 0\}$. This is a proper W*-ideal but it does not vanish on any positive measure set.

10.3 The derivation theorem

Generally speaking, a derivation is a linear map d which satisfies the Leibniz identity $d(\alpha\beta) = \alpha d(\beta) + d(\alpha)\beta$. There is a remarkable connection between the algebras $\mathrm{Lip}(X, \mu)$ and derivations which states that the $\mathrm{Lip}(X, \mu)$ are precisely those spaces which arise as the domains of a certain type of derivation. This characterization of commutative Lipschitz algebras has an immediate noncommutative (Hilbert space) generalization.

In order for the derivation condition to make sense, we must be able to multiply elements of the range of the derivation on either side by elements of its domain; thus, the range must be a bimodule over the domain. Our examples in this section will involve L^∞ spaces, and in order to make a space $L^\infty(Y)$ a bimodule over $L^\infty(X)$, we will require a pair of weak* continuous $*$-homomorphisms from the latter into the former.

Here, for the first time, it really is convenient for us to use spaces which are not σ-finite. The next best condition is that X be locally finite, which means that it can be written as a disjoint union $X = \bigcup X_\kappa$ such that each X_κ is a finite measure space, a subset of X is measurable if and only if its intersection with each X_κ is measurable, and $\mu(S) = \sum \mu(S \cap X_\kappa)$ for all measurable subsets $S \subset X$. This class of measure spaces contains all discrete spaces and all σ-finite spaces, and much of what one can prove for σ-finite spaces actually holds for locally finite spaces with only minor modifications in the proofs. For instance, we have $L^\infty(X) \cong L^1(X)^*$ for any locally finite X. (Indeed, the spaces $L^\infty(X)$ for X locally finite are precisely, up to $*$-isomorphism, the abelian von Neumann algebras.)

DEFINITION 10.3.1 *Let X be a σ-finite measure space. An L^∞-bimodule over $L^\infty(X)$ is a weak* closed self-adjoint subspace $\mathcal{E}$ of $L^\infty(Y)$ for some locally finite measure space Y, together with a pair of weak* continuous unital $*$-homomorphisms $\pi_l, \pi_r : L^\infty(X) \to L^\infty(Y)$ such that $\pi_l(f)g, \pi_r(f)g \in \mathcal{E}$ whenever $f \in L^\infty(X)$ and $g \in \mathcal{E}$. We write $f_l = \pi_l(f)$ and $f_r = \pi_r(f)$.*

An (unbounded) L^∞-derivation from $L^\infty(X)$ into an L^∞-bimodule $\mathcal{E}$ is a linear map $d : \mathcal{L} \to \mathcal{E}$ such that

(a) $\mathcal{L}$ is a weak dense, unital $*$-subalgebra of $L^\infty(X)$;*
(b) d satisfies $d(\bar{f}) = \overline{d(f)}$ and $d(fg) = f_l d(g) + d(f) g_r$ for all $f, g \in \mathcal{L}$; and
(c) the graph of d, $\Gamma d = \{f \oplus df : f \in \mathcal{L}\}$, is a weak closed subspace of $L^\infty(X) \oplus \mathcal{E}$.*

We give $\mathcal{L} = \mathrm{dom}(d)$ the graph norm

$$\|f\|_D = \max(\|f\|_\infty, \|df\|_\infty).$$

When constructing L^∞-derivations it is always possible to take $\mathcal{E} = L^\infty(Y)$. However, the greater generality given in the definition will be convenient.

The basic example of an L^∞-derivation to keep in mind is differentiation on the unit interval (or on the unit circle). Here d is the map $d(f) = f'$. There are several natural domains for this map: it can be defined on $C^\infty[0,1]$ or on $C^1[0,1]$, for example. However, with these domains its graph is not weak* closed. In order to have a weak* closed graph, its domain must be expanded to $\mathrm{Lip}[0,1]$. Thus, the derivative map $d : L^\infty[0,1] \to L^\infty[0,1]$ with domain $\mathcal{L} = \mathrm{Lip}[0,1]$ is an L^∞-derivation. In higher dimensions, the range of the analogous map will no longer equal the domain, but will still be an L^∞-bimodule over it (see Example 10.4.1).

Next we show that every $\mathrm{Lip}(X, \mu)$ is the domain of an L^∞-derivation. This is where we need to use non σ-finite sets Y.

Example 10.3.2

Let X be a σ-finite measure space and let ρ be a measurable pseudometric on X. For $S, T \subset X$ positive measure sets define

$$\rho^+(S,T) = \sup\{\rho(S', T') : S' \subset S,\ T' \subset T\}.$$

Then for each $f \in \mathrm{Lip}(X,\mu)$ and each pair of positive measure subsets $S,T \subset X$ such that $\rho^+(S,T) > 0$, define $df \in L^\infty(S \times T)$ by

$$df(x,y) = \frac{f(x) - f(y)}{\rho^+(S,T)}.$$

(Note: this is not a pseudometric.) Also define $\pi_l f(x,y) = f(x)$ and $\pi_r f(x,y) = f(y)$. Letting Y be the disjoint union of the spaces $S \times T$, the maps $\pi_l, \pi_r : L^\infty(X) \to L^\infty(Y)$ make $L^\infty(Y)$ an L^∞-bimodule over $L^\infty(X)$, and the map $d : L^\infty(X) \to L^\infty(Y)$ with domain $\mathrm{Lip}(X,\mu)$ is an L^∞-derivation. Moreover, $\|df\|_\infty = L(f)$ for all $f \in \mathrm{Lip}(X,\mu)$.

Now we proceed to prove the converse of this example, namely, that the domain of any L^∞-derivation is of the form $\mathrm{Lip}(X,\mu)$. Fix the following notation. X is a σ-finite measure space, $L^\infty(Y)$ is an L^∞-bimodule over $L^\infty(X)$ (recall that we can assume $\mathcal{E} = L^\infty(Y)$), $d : L^\infty(X) \to L^\infty(Y)$ is an L^∞-derivation, and $\mathcal{L} = \mathrm{dom}(d)$. Observe that since $\mathcal{L}$ is isometric to the weak* closed subspace Γd of the dual space $L^\infty(X) \oplus L^\infty(Y)$, it is itself a dual space, with weak* topology defined by the condition that $f_\kappa \to f$ weak* in $\mathcal{L}$ if and only if both $f_\kappa \to f$ weak* in $L^\infty(X)$ and $df_\kappa \to df$ weak* in $L^\infty(Y)$.

LEMMA 10.3.3
Let $f \in \mathcal{L}$ be real-valued and let $g \in \mathrm{Lip}[-a,a]$ where $a = \|f\|_\infty$. Then $g \circ f \in \mathcal{L}$ and $\|d(g \circ f)\|_\infty \le L(g) \cdot \|df\|_\infty$.

PROOF If g is a polynomial then it is clear that $g \circ f \in \mathcal{L}$. The derivation identity then yields

$$d(f^n) = (f_l^{n-1} + f_l^{n-2} f_r + \cdots + f_r^{n-1})df.$$

Say $g(t) = \sum a_n t^n$ and define $h \in \mathrm{Lip}([-a,a]^2)$ by

$$h(s,t) = \sum a_n (s^{n-1} + s^{n-2}t + \cdots + t^{n-1}).$$

We have $h(s,t) = (g(s) - g(t))/(s-t)$ for $s \neq t$, so $\|h\|_\infty = L(g)$, but also $d(g \circ f) = h(f_l, f_r) \cdot df$, so $\|d(g \circ f)\|_\infty \le L(g) \cdot \|df\|_\infty$ as desired.

Next suppose $g \in C^1[-a,a]$ and let g_n be a sequence of polynomials such that $g_n' \to g'$ uniformly on $[-a,a]$ and $g_n(0) = g(0)$ for all n. Then $g_n \to g$ uniformly on $[-a,a]$, so $g_n \circ f \to g \circ f$ uniformly in $L^\infty(X)$, and the sequence $(d(g_n \circ f))$ is Cauchy in $L^\infty(Y)$ because

$$\begin{aligned} \|d(g_m \circ f) - d(g_n \circ f)\|_\infty &\le L(g_m - g_n) \cdot \|df\|_\infty \\ &= \|g_m' - g_n'\|_\infty \cdot \|df\|_\infty. \end{aligned}$$

Thus, norm closure of Γd implies that $g \circ f \in \mathcal{L}$ and $d(g_n \circ f) \to d(g \circ f)$. Since $L(g_n) = \|g_n'\|_\infty \to \|g'\|_\infty = L(g)$, the desired bound on $\|d(g \circ f)\|_\infty$ follows.

Finally, for any $g \in \mathrm{Lip}[-a, a]$ we have $g' \in L^\infty[-a, a]$, so we can find a sequence $g_n \in C^1[-a, a]$ such that $g_n(0) = g(0)$, $\|g_n' \to g'\|_1 \to 0$, and $\|g_n'\|_\infty \to \|g'\|_\infty$. This implies that $g_n \to g$ uniformly, and we have $\|d(g_n \circ f)\|_\infty \leq \sup \|g_n'\|_\infty \cdot \|df\|_\infty$, so we can find a subnet of $(g_n \circ f)$ which converges weak* in $\mathcal{L}$. Since $g_n \circ f \to g \circ f$ uniformly, we conclude that $g \circ f \in \mathcal{L}$ and $\|d(g \circ f)\|_\infty \leq L(g) \cdot \|df\|_\infty$. ∎

LEMMA 10.3.4

Let $f \in \mathcal{L}$ be real-valued, let S be a positive measure subset of X, and suppose $f \cdot \chi_S = 0$. Then $(\chi_S)_l df (\chi_S)_r = 0$.

PROOF Let $f_n = 1 - e^{-nf^2}$. Then $f_n \in \mathcal{L}$ by Lemma 10.3.3, and a short computation shows that $L(f \cdot f_n) \leq 2$. Thus $\|d(f \cdot f_n)\|_\infty \leq 2\|df\|_\infty$. But $f \cdot f_n \to f$ uniformly, and hence weak* in $\mathcal{L}$, and so $d(f \cdot f_n) \to d(f)$ weak* in $L^\infty(Y)$ by weak* closure of Γd. Thus

$$\begin{aligned} 0 &= (\chi_S)_l[f_l d(f_n) + d(f)(f_n)_r](\chi_S)_r \\ &= (\chi_S)_l d(f \cdot f_n)(\chi_S)_r \\ &\to (\chi_S)_l df (\chi_S)_r. \end{aligned}$$

So $(\chi_S)_l df (\chi_S)_r = 0$. ∎

LEMMA 10.3.5

Let $f, g \in \mathcal{L}$ be real-valued. Then $\max(f, g), \min(f, g) \in \mathcal{L}$ and

$$\|d(\max(f, g))\|_\infty, \|d(\min(f, g))\|_\infty \leq \max(\|df\|_\infty, \|dg\|_\infty).$$

PROOF The function $t \mapsto |t|$ is Lipschitz on $\mathbf{R}$, so $\max(f, g) = (f + g + |f - g|)/2$ and $\min(f, g) = (f + g - |f - g|)/2$ belong to $\mathcal{L}$ by Lemma 10.3.3. We claim that $|d(\max(f, g))| \leq \max(|df|, |dg|)$ almost everywhere on Y; this implies the desired inequality for $d(\max(f, g))$, and the inequality for $d(\min(f, g))$ can be proven similarly.

Let $S = \{x \in X : f(x) \geq g(x)\}$ and $T = \{x \in X : g(x) \geq f(x)\}$, and for $\epsilon > 0$ let $S_\epsilon = \{x \in X : f(x) \geq g(x) + \epsilon\}$ and $T_\epsilon = \{x \in X : g(x) \geq f(x) + \epsilon\}$. Then define sets $Y_S, Y_T, Y_\epsilon, Y_\epsilon' \subset Y$ by the equations $\chi_{Y_S} = (\chi_S)_l(\chi_S)_r$, $\chi_{Y_T} = (\chi_T)_l(\chi_T)_r$, $\chi_{Y_\epsilon} = (\chi_{S_\epsilon})_l(\chi_{T_\epsilon})_r$, and $\chi_{Y_\epsilon'} = (\chi_{T_\epsilon})_l(\chi_{S_\epsilon})_r$. We have

$$Y = Y_S \cup Y_T \cup \bigcup_\epsilon Y_\epsilon \cup \bigcup_\epsilon Y_\epsilon',$$

so it will suffice to prove the desired inequality separately on each of these sets.

First, observe that

$$(\chi_S)_l d(\max(f,g))(\chi_S)_r = (\chi_S)_l df(\chi_S)_r$$

by Lemma 10.3.4. This proves that $|d(\max(f,g))| \le \max(|df|,|dg|)$ on Y_S, and the same is true in Y_T by similar reasoning. We will conclude the proof by verifying this on Y_ϵ; the argument for Y'_ϵ is the same.

Define

$$h^S = \min\left(\max\left(\frac{f-g}{\epsilon},0\right),1\right)$$

and

$$h^T = \min\left(\max\left(\frac{g-f}{\epsilon},0\right),1\right).$$

Then $h^S, h^T \in \mathcal{L}$ and $h^S + h^T$ is identically 1 on $S_\epsilon \cup T_\epsilon$. Thus, using Lemma 10.3.4, we have

$$\begin{aligned} dk \cdot \chi_{Y_\epsilon} &= [d(k(h^S + h^T))] \cdot \chi_{Y_\epsilon} \\ &= [k_l(dh^S) + (dk)h_r^S + (h^T)_l(dk) + (dh^T)k_r] \cdot \chi_{Y_\epsilon} \\ &= [k_l(dh^S) + k_r(dh^T)] \cdot \chi_{Y_\epsilon} \\ &= h(k_l - k_r) \end{aligned}$$

for any $k \in \mathcal{L}$, where $h = (dh^S) \cdot \chi_{Y_\epsilon}$.

Examination of cases shows that $|f_l - g_r| \le \max(|f_l - f_r|, |g_l - g_r|)$ on Y_ϵ, and hence, with a double application of the result of the last paragraph, we get

$$\begin{aligned} |d(\max(f,g))| &= |h| \cdot |(\max(f,g))_l - (\max(f,g))_r| \\ &= |h| \cdot |f_l - g_r| \\ &\le |h| \cdot \max(|f_l - f_r|, |g_l - g_r|) \\ &\le \max(|df|,|dg|) \end{aligned}$$

on Y_ϵ. This completes the proof. ∎

THEOREM 10.3.6

Let (X,μ) be a σ-finite measure space, let $\mathcal{E}$ be an L^∞-bimodule over $L^\infty(X)$, let $d : L^\infty(X) \to \mathcal{E}$ be an L^∞-derivation, and let $\mathcal{L} = \mathrm{dom}(d)$. Then there exists a measurable pseudometric on X such that $\mathcal{L} = \mathrm{Lip}(X,\mu)$ isometrically and $L(f) = \|df\|_\infty$ for all $f \in \mathcal{L}$.

PROOF Define

$$\rho(S,T) = \sup\{\rho_f(S,T) : f \in \mathcal{L},\ \|df\|_\infty \le 1\}.$$

It follows from Lemma 10.3.5 and the fact that $\mathcal{L}$ is a dual space that suprema and infima of $\|\cdot\|_D$-bounded sets of real-valued functions in $\mathcal{L}$ also belong to $\mathcal{L}$. Using this fact, it is straightforward to verify that ρ is a measurable pseudometric, that $\mathcal{L} \subset \mathrm{Lip}(X)$, and that $L(f) = \|df\|_\infty$ for all $f \in \mathcal{L}$. The desired conclusion follows from Theorem 10.2.7. ∎

On the basis of the preceding result, we make the following definition.

DEFINITION 10.3.7 *Let $\mathcal{M}$ be a von Neumann algebra. A W*-bimodule over $\mathcal{M}$ is a weak* closed self-adjoint subspace $\mathcal{E}$ of a von Neumann algebra $\mathcal{N}$ together with a pair of weak* continuous unital *-homomorphisms $\pi_l, \pi_r : \mathcal{M} \to \mathcal{N}$ such that $\pi_l(\alpha)\xi, \xi\pi_r(\alpha) \in \mathcal{E}$ whenever $\alpha \in \mathcal{M}$ and $\xi \in \mathcal{E}$. We write $\alpha_l = \pi_l(\alpha)$ and $\alpha_r = \pi_r(\alpha)$.*

An (unbounded) W-derivation from $\mathcal{M}$ into a W*-bimodule $\mathcal{E}$ is a linear map $d : \mathcal{L} \to \mathcal{E}$ such that*

(a) $\mathcal{L}$ is a weak dense, unital *-subalgebra of $\mathcal{M}$;*
(b) d satisfies $d(\alpha^*) = d(\alpha)^*$ and $d(\alpha\beta) = \alpha_l d(\beta) + d(\alpha)\beta_r$ for all $\alpha, \beta \in \mathcal{L}$; and
(c) the graph of d, $\Gamma d = \{\alpha \oplus d\alpha : \alpha \in \mathcal{L}\}$, is a weak closed subspace of $\mathcal{M} \oplus \mathcal{E}$.*

We give $\mathcal{L} = \mathrm{dom}(d)$ the graph norm

$$\|f\|_D = \max(\|\alpha\|, \|d\alpha\|).$$

A (noncommutative) Lipschitz algebra is the domain of some W-derivation.*

10.4 Examples

In this section we describe various classes of examples of W*-derivations and their associated Lipschitz algebras. We begin with the most important class of commutative examples.

Example 10.4.1

Let X be a compact, connected Riemannian manifold. Thus X is a smooth manifold, and the tangent space at each point is equipped with an inner product. Let $\mathcal{X}$ be the complexified tangent bundle over X as in Example 9.1.2 and regard $\mathcal{X}$ as a measurable Hilbert bundle. This can be done by writing X minus a null set as a disjoint union of finitely many open sets X_i ($1 \le i \le k$) each of which is diffeomorphic to an open subset of $\mathbf{R}^n$. Then $\mathcal{X} \cong (X_1 \times \mathbf{C}^n) \cup \cdots \cup (X_k \times \mathbf{C}^n)$.

Let $L^\infty(X; \mathcal{X})$ be the Hilbert module of L^∞ sections of $\mathcal{X}$ as in Definition 9.2.2; this space can be identified with the complexification of the space of bounded measurable vector fields on X. Thus we may define $d : \mathrm{Lip}(X) \to L^\infty(X; \mathcal{X})$ by $d(f) = \nabla f$, the gradient vector field of f.

We regard $L^\infty(X, \mathcal{X})$ as a W* Hilbert $*$-bimodule. It can be realized as an L^∞-bimodule as follows. Let Y be the set of all pairs (x, v) such that $x \in X$ and v is a complex tangent vector at x of norm 1. That is, $Y = (X_1 \times S^{2n-1}) \cup \cdots \cup (X_k \times S^{2n-1})$ where S^{2n-1} is the unit sphere of $\mathbf{C}^n$. Then $L^\infty(Y)$ is an L^∞-bimodule over $L^\infty(X)$ via the actions $f_l g(x, v) = g f_r(x, v) = f(x)g(x, v)$. Finally, define $T : L^\infty(X; \mathcal{X}) \to L^\infty(Y)$ by $T\phi(x, v) = \langle \phi(x), v\rangle$; this is an isometric embedding of $L^\infty(X; \mathcal{X})$ in $L^\infty(Y)$ which respects the $*$-bimodule structure.

The fact that $\pi_l = \pi_r$ is an important feature of the preceding example; one says that $L^\infty(X; \mathcal{X})$ is a monomodule over $L^\infty(X)$. This is related to the differentiable character of the metric space X. The same relationship persists in the noncommutative case.

Monomodule derivations can also be constructed by differentiating one-parameter unitary groups, as in the next example.

Example 10.4.2

(a) Recall the von Neumann algebra $L^\infty_\hbar(\mathbf{R}^2)$ of the quantum plane from Definition 6.6.1 and the automorphism $\theta_{s,t}$ of $L^\infty_\hbar(\mathbf{R}^2)$. Define unbounded maps $d_i : L^\infty_\hbar(\mathbf{R}^2) \to L^\infty_\hbar(\mathbf{R}^2)$ $(i = 1, 2)$ by

$$d_1(A) = \lim_{s\to 0} \frac{1}{s}(\theta_{s,0}(A) - A)$$

and

$$d_2(A) = \lim_{t\to 0} \frac{1}{t}(\theta_{0,t}(A) - A),$$

with domains the set of $A \in L^\infty_\hbar(\mathbf{R}^2)$ for which the limits exist in the weak* sense. If $f \in \mathcal{S}(\mathbf{R}^2)$ we have $d_1(L_f) = L_{\partial f/\partial x}$ and $d_2(L_f) = L_{\partial f/\partial y}$; this follows from Proposition 5.4.5. Informally, we can write $d_1(A) = -\frac{i}{\hbar}[\mathcal{P}, A]$ and $d_2(A) = \frac{i}{\hbar}[\mathcal{Q}, A]$ (see Section 4.4).

Now define $d : L^\infty_\hbar(\mathbf{R}^2) \to L^\infty_\hbar(\mathbf{R}^2) \oplus L^\infty_\hbar(\mathbf{R}^2)$ by setting $d(A) = d_1(A) \oplus d_2(A)$, with domain $\mathrm{Lip}_\hbar(\mathbf{R}^2) = \mathrm{dom}(d_1) \cap \mathrm{dom}(d_2)$ and range $L^\infty_\hbar(\mathbf{R}^2) \oplus L^\infty_\hbar(\mathbf{R}^2)$, regarded as a W* Hilbert $*$-bimodule over $L^\infty_\hbar(\mathbf{R}^2)$ as in Example 9.6.5. Then $\mathrm{Lip}_\hbar(\mathbf{R}^2)$ is a Lipschitz algebra.

(b) The same construction applies to the quantum tori von Neumann algebras $L^\infty_\hbar(\mathbf{T}^2)$, using the automorphism $\hat{\theta}_{s,t}$ in place of $\theta_{s,t}$. We define $\mathrm{Lip}_\hbar(\mathbf{T}^2)$ to be the domain of the corresponding derivation.

In the preceding example, setting $\hbar = 0$ reduces to the map $df = (\partial f/\partial x, \partial f/\partial y)$, i.e., to Example 10.4.1.

Putting this example into the format of Definition 10.3.7 requires that we be able to realize W* Hilbert $*$-bimodules as W*-bimodules.

Here this can be done in a non-isometric fashion by treating $L^\infty_\hbar(\mathbf{R}^2) \oplus L^\infty_\hbar(\mathbf{R}^2)$ as a von Neumann algebra and letting $\pi_l = \pi_r$ be the map $A \mapsto (A, A)$. A completely general, isometric result will be given in Corollary 10.5.2.

We can relate the Lipschitz algebras $\mathrm{Lip}_\hbar(\mathbf{T}^2)$ to the harmonic analysis of quantum tori as developed in Theorem 5.5.7 and Proposition 6.6.4, for example. The relevant notation was given in Definition 5.5.4.

Let $\mathrm{Lip}_{\hat\hbar}(\mathbf{T}^2)$ be the corresponding Lipschitz algebra inside $\hat{L}^\infty_\hbar(\mathbf{T}^2)$. We start with a straightforward result whose verification we omit. (Its first assertion follows from Proposition 6.6.5, as continuity of the map $(s,t) \mapsto \hat\theta_{s,t}(A)$ is necessary for $d_1(A)$ and $d_2(A)$ to be defined.)

PROPOSITION 10.4.3
We have $\mathrm{Lip}_{\hat\hbar}(\mathbf{T}^2) \subset \hat{C}^\hbar(\mathbf{T}^2)$. For all $s, t \in \mathbf{R}$ the map $\hat\theta_{s,t}$ restricts to a $*$-isomorphism from $\mathrm{Lip}_{\hat\hbar}(\mathbf{T}^2)$ onto itself. This defines an action of $\mathbf{R}^2$ by automorphisms of $\mathrm{Lip}_{\hat\hbar}(\mathbf{T}^2)$.

THEOREM 10.4.4
Let $A \in \mathrm{Lip}_{\hat\hbar}(\mathbf{T}^2)$. Then $s_N(A) \to A$ in norm.

PROOF Let D_N be the Dirichlet kernel,

$$D_N(t) = \sum_{n=-N}^{N} e^{int} = \frac{\sin(n+1/2)t}{\sin(t/2)},$$

and let V_N be the de la Valée Poussin kernel,

$$V_N = 2K_{2N+1} - K_N,$$

where K_N is the Fejér kernel defined in the proof of Theorem 5.5.7. Let $G_N(s,t) = D_N(s)D_N(t)$ and $W_N(s,t) = V_N(s)V_N(t)$ and define

$$s'_N(A) = \frac{1}{(2\pi)^2} \int\int \hat\theta_{s,t}(A) W_N(s,t)\, dsdt.$$

Also, observe that

$$s_N(A) = \frac{1}{(2\pi)^2} \int\int \hat\theta_{s,t}(A) G_N(s,t)\, dsdt.$$

We have $\|A - s'_N(A)\| \to 0$ by the argument in the proof of Theorem 5.5.7 which showed $\|A - \sigma_N(A)\| \to 0$, so it will suffice to show that $\|s_N(A) - s'_N(A)\| \to 0$.

Let $B_N = s_N(A) - s'_N(A)$ and let $a_{k,l} = a_{k,l}(A)$ and $b_{k,l} = b_{k,l}(B_N)$ be the Fourier coefficients of A and B_N, respectively. The $b_{k,l}$ have the following properties:

(a) $|b_{k,l}| \leq |a_{k,l}|$ for all k, l;
(b) $b_{k,l} = 0$ if $|k|, |l| \leq N$; and
(c) $b_{k,l} = 0$ if $\max(|k|, |l|) > 2N + 1$.

Therefore

$$\sum_{k,l} |b_{k,l}| \leq \sum_{(k,l)\in X_N} |a_{k,l}|$$

where $X_N = \{(k,l) : N < \max(|k|,|l|) \leq 2N+1\}$. For $(k,l) \in X_N$ we have $k^2 + l^2 > N^2$, so

$$\sum_{(k,l)\in X_N} (k^2 + l^2)^{-1} \leq C$$

for some universal constant C, and thus

$$\begin{aligned} \|B_N\| \leq \sum |b_{k,l}| &\leq \sum_{(k,l)\in X_N} |a_{k,l}| \\ &\leq \left(C \cdot \sum_{(k,l)\in X_N} (k^2 + l^2)|a_{k,l}|^2 \right)^{1/2}. \end{aligned}$$

(The last line follows from the Cauchy-Schwarz inequality.)

Finally, the Fourier coefficients of $d_1(A)$ are $ka_{k,l}$, so

$$\sum k^2 |a_{k,l}|^2 = \tau(d_1(A)^* d_1(A)) < \infty$$

where τ is the state defined in Proposition 6.6.7, and the same holds for $\sum l^2 |a_{k,l}|^2$. Thus

$$\sum_{(k,l)\in X_N} (k^2 + l^2)^2 |a_{k,l}|^2 \to 0$$

as $N \to \infty$, so $\|B_N\| \to 0$, as desired. ∎

Our last example is a Lipschitz version of the matrix-valued function spaces $l^\infty_{mat}(K)$ and $C_{mat}(K)$ discussed in Section 8.4. It may be viewed as a quantum analog of the space $\text{Lip}(K)$ where K is the unit ball of a dual Banach space. This example is not differential geometric in nature, and correspondingly the bimodule involved is not a monomodule.

Let $\mathcal{V}$ be an operator space and let K be the corresponding dual matrix unit ball. For any $f \in l^\infty_{mat}(K)$, each $f_n : K_n \to M_n$ is a map between metric spaces (the first is the unit ball of a Banach space, while the second is a Banach space) and hence has a Lipschitz number, possibly infinite. Define the Lipschitz number of $f \in l^\infty_{mat}(K)$ to be $L(f) = \sup L(f_n)$.

Observe that $L(f)$ is finite for $f \in V_{mat}(K)$, since if $X, Y \in K_n$ then, choosing $v \in \mathcal{V}$ such that $f[x_{ij}] = [x_{ij}(v)]$, we have

$$\frac{\|f_n(X) - f_n(Y)\|}{\|X - Y\|} = \frac{\|(X-Y)(v)\|}{\|X - Y\|} = \frac{\|\hat{v}^{(n)}(X-Y)\|}{\|X-Y\|} \leq \|v\|$$

by the comment following Definition 8.3.1. Also observe that the adjoint, sum, and product of Lipschitz functions is again Lipschitz.

Example 10.4.5

Let K be a dual matrix unit ball and define $\mathcal{M}_K$ to be the l^∞ direct sum

$$\mathcal{M}_K = \bigoplus_{(m,X,Y)\in\Lambda} M_m$$

where $\Lambda = \{(m, X, Y) : m \in \mathbf{N},\ X, Y \in K_m,\ X \neq Y\}$. This is a von Neumann algebra and a W*-bimodule over $l^\infty_{mat}(K)$ relative to the embeddings $f \mapsto f_l$ and $f \mapsto f_r$ of $l^\infty_{mat}(K)$ in $\mathcal{M}_K$ defined by $f_l(X,Y) = f(X)$ and $f_r(X,Y) = f(Y)$. Define an unbounded derivation $d : l^\infty_{mat}(K) \to \mathcal{M}_K$ by

$$df(X,Y) = \frac{f_n(X) - f_n(Y)}{\|X - Y\|},$$

with domain the set of $f \in l^\infty_{mat}(K)$ with finite Lipschitz number.

The graph of d is a weak* closed subspace of $l^\infty_{mat}(K) \oplus \mathcal{M}_K$, so it is a dual operator space. The domain of d is a Lipschitz algebra and it has a weak* topology which is transferred from the graph of d via the bijection $f \leftrightarrow f \oplus df$. We now define $\mathrm{Lip}_{mat}(K)$ to be the unital W*-subalgebra of dom(d) generated by $V_{mat}(K)$. Thus $\mathrm{Lip}_{mat}(K)$ inherits its algebraic structure from dom(d), but it inherits a dual operator space structure from the graph of d.

Let $\hom^*(\mathrm{Lip}_{mat}(K), M_n)$ be the space of weak* continuous unital $*$-homomorphisms from $\mathrm{Lip}_{mat}(K)$ into M_n, with completely bounded norm.

THEOREM 10.4.6

Let $\mathcal{V}$ be an operator space and let K be the corresponding dual matrix unit ball. Then

(a) $\mathrm{Lip}_{mat}(K)$ is a Lipschitz algebra;
(b) $\mathrm{Lip}_{mat}(K)$ completely isometrically contains $\mathcal{V}$; and
(c) $\hom^*(\mathrm{Lip}_{mat}(K), M_n)$ is canonically isometric to K_n.

PROOF

(a) Let d' be the restriction of d to $\mathrm{Lip}_{mat}(K)$. Then d' is also an unbounded derivation, and its graph is a weak* closed subspace of the graph of d and hence a weak* closed subspace of $l^\infty_{mat}(K) \oplus \mathcal{M}_K$. Thus $\mathrm{Lip}_{mat}(K)$ is indeed the domain of a W*-derivation.

(b) Note that this is not implied by the fact that $\mathcal{V}$ is completely isometrically contained in $l^\infty_{mat}(K)$ (Theorem 8.4.4 (b)) since the operator space structure of $\mathrm{Lip}_{mat}(K)$ is not inherited from $l^\infty_{mat}(K)$, but rather from the graph of d.

Let $\Xi = [\xi_{ij}] \in M_n(\mathcal{V})$ and let $T : \mathcal{V} \to V_{mat}(K)$ be the natural map. Then the Lipschitz norm of $T^{(n)}\Xi$ in $M_n(\mathrm{Lip}_{mat}(K))$ equals $\max(\|T^{(n)}\Xi\|, \|d^{(n)}(T^{(n)}\Xi)\|)$ where

$$d^{(n)}(T^{(n)}\Xi) = [d(T\xi_{ij})] \in M_n(\mathcal{M}_K).$$

We already know from Theorem 8.4.4 (b) that $\|T^{(n)}\Xi\| = \|\Xi\|$, so we must show that $\|d^{(n)}(T^{(n)}\Xi)\| \leq \|\Xi\|$. To see this let $m \in \mathbf{N}$ and $X, Y \in K_m$, $X \neq Y$, and observe that

$$d^{(n)}(T^{(n)}\Xi)(X,Y) = [d(T\xi_{ij})(X,Y)] = \frac{[(T\xi_{ij})(X-Y)]}{\|X-Y\|}.$$

But $[(T\xi_{ij})(X-Y)] = (X-Y)^{(n)}(\Xi)$ has norm at most $\|X-Y\|\|\Xi\|$, by the definition of the norm in $M_m(\mathcal{V}^*)$. Thus $\|d^{(n)}(T^{(n)}\Xi)\| \leq \|\Xi\|$, as desired.

(c) For $X \in K_n$ let $\hat{X} : \mathrm{Lip}_{mat}(K) \to M_n$ be evaluation at X, and define $\phi : K_n \to \mathrm{hom}^*(\mathrm{Lip}_{mat}(K), M_n)$ by $\phi(X) = \hat{X}$. This map is nonexpansive since for $X, Y \in K_n$ and $F \in M_m(\mathrm{Lip}_{mat}(K))$,

$$\|(\phi(X) - \phi(Y))^{(m)}(F)\| = \|F_n(X) - F_n(Y)\| \leq \|d^{(m)}F\|\|X-Y\|.$$

Conversely, for any $X, Y \in K_n$ and $\epsilon > 0$ there exists $m \in \mathbf{N}$ and $\Xi \in M_m(\mathcal{V})$ such that $\|\Xi\| = 1$ and $\|(X-Y)^{(m)}(\Xi)\| \geq \|X-Y\| - \epsilon$. Then $T^{(m)}\Xi \in M_m(\mathrm{Lip}_{mat}(K))$ and we have

$$\|(\phi(X) - \phi(Y))^{(m)}(T^{(m)}\Xi)\| = \|(X-Y)^{(m)}(\Xi)\| \geq \|X-Y\| - \epsilon,$$

which shows that $\|\phi(X) - \phi(Y)\| = \|X - Y\|$, i.e., ϕ is an isometry.

Surjectivity of ϕ follows as in Theorem 8.4.6 (b) from the fact that $V_{mat}(K)$ generates $\mathrm{Lip}_{mat}(K)$. ∎

COROLLARY 10.4.7

Let $\mathcal{V}$ and $\mathcal{W}$ be operator spaces and let K and L be the corresponding dual matrix unit balls. Then any completely contractive linear map from $\mathcal{V}$ to $\mathcal{W}$ extends uniquely to a completely contractive weak continuous unital $*$-homomorphism from $\mathrm{Lip}_{mat}(K)$ to $\mathrm{Lip}_{mat}(L)$.*

PROOF Let $T : \mathcal{V} \to \mathcal{W}$ be a completely contractive linear map and form $\tilde{T} : \mathrm{Lip}_{mat}(K) \to \mathrm{Lip}_{mat}(L)$ in the same way as in Corollary 8.4.7. The main point is to estimate the completely bounded norm of this map. Thus let $F \in M_n(\mathrm{Lip}_{mat}(K))$. Then for any $m \in \mathbf{N}$ and $X, Y \in L_m$ we have $\|X - Y\| \geq \|T^\# X - T^\# Y\|$, and hence

$$\frac{\|\tilde{T}^{(n)}F(X) - \tilde{T}^{(n)}F(Y)\|}{\|X - Y\|} \leq \frac{\|F(T^\# X) - F(T^\# Y)\|}{\|T^\# X - T^\# Y\|}.$$

This shows that $L(\tilde{T}^{(n)}F) \leq \|d^{(n)}F\|$. Also $\|\tilde{T}^{(n)}F\| \leq \|F\|$ from the fact that $\tilde{T}$ extends to a $*$-homomorphism from $l^\infty_{mat}(K)$ to $l^\infty_{mat}(L)$ as in Corollary 8.4.7. Thus $\|\tilde{T}^{(n)}F\|_D \leq \|F\|_D$, and we conclude that $\tilde{T}$ is a complete contraction. ∎

10.5 Quantum Markov semigroups

In this section we will present a very general construction that produces W*-derivations of von Neumann algebras into W* Hilbert $*$-bimodules. By analogy with Example 10.4.1 such structures can be thought of as quantum Riemannian manifolds (or perhaps quantum sub-Riemannian manifolds, but we will not elaborate on this). They are the basis of "noncommutative (quantum) geometry."

We first address the Hilbert $*$-bimodule aspect of the construction. Although in the von Neumann algebra setting the appropriate topological conditions on Hilbert $*$-bimodules are that they be normal and dual (i.e., W* Hilbert $*$-bimodules), in practice one often first constructs a Hilbert $*$-bimodule without these conditions and then attempts to complete it. Doing this requires that the uncompleted bimodule satisfy a certain topological condition, but assuming this is the case one can prove that a completion of the desired type does exist (Theorem 10.5.4). We will establish this fact by using a "linking algebra" construction. This technique also shows, incidentally, that W* Hilbert $*$-bimodules can be embedded in von Neumann algebras in a manner that is compatible with their $*$-bimodule structure. Thus, they can be viewed as W*-bimodules in the sense of Definition 10.3.7.

Let $\mathcal{A}$ be a C*-algebra and $\mathcal{E}$ a Hilbert $*$-bimodule over $\mathcal{A}$. To produce a right Hilbert module, let $\mathcal{E}_0 = \{\xi \in \mathcal{E} : \|\xi\|_r = 0\}$ and define $\mathcal{E}_r$ to be the completion of $\mathcal{E}/\mathcal{E}_0$ for $\|\cdot\|_r$. Then $\langle \cdot, \cdot \rangle_r$ extends to an $\mathcal{A}$-valued inner product on $\mathcal{E}_r$, and we use the same notation for the extension.

Let $\mathcal{E}^+ = \mathcal{A} \oplus \mathcal{E}_r$ be the direct sum of right Hilbert $\mathcal{A}$-modules and let $B(\mathcal{E}^+)$ be the space of bounded adjointable right $\mathcal{A}$-linear maps from $\mathcal{E}^+$ to itself, as in Definition 9.3.4; this is a C*-algebra by Proposition 9.3.6. It is called the <u>linking algebra</u> of the Hilbert module $\mathcal{E}_r$.

PROPOSITION 10.5.1

Let $\mathcal{A}$ be a C^*-algebra and let $\mathcal{E}$ be a Hilbert $*$-bimodule over $\mathcal{A}$. Then

$$\Phi(\alpha)(\beta \oplus \eta) = \alpha\beta \oplus \alpha\eta$$

defines a $*$-isomorphism $\Phi : \mathcal{A} \to B(\mathcal{E}^+)$ and

$$\Psi(\xi)(\beta \oplus \eta) = \langle \xi^*, \eta\rangle_r \oplus \xi\beta$$

defines an isometric linear embedding $\Psi : \mathcal{E} \to B(\mathcal{E}^+)$. For every $\alpha \in \mathcal{A}$ and $\xi, \eta \in \mathcal{E}$ we have $\Phi(\alpha)\Psi(\xi) = \Psi(\alpha\xi)$, $\Psi(\xi)\Phi(\alpha) = \Psi(\xi\alpha)$, and $\Psi(\xi)^* = \Psi(\xi^*)$.

PROOF For any $\alpha \in \mathcal{A}$ the map $\Phi(\alpha)$ is bounded by Lemma 9.6.2, and a short computation shows that $\Phi(\alpha)^* = \Phi(\alpha^*)$, so Φ is a $*$-homomorphism from $\mathcal{A}$ into $B(\mathcal{E}^+)$. It is clearly injective. For any $\xi \in \mathcal{E}$ the map $\Psi(\xi)$ is clearly $\mathcal{A}$-linear, and it is bounded because

$$\begin{aligned}\langle \Psi(\xi)(\beta \oplus \eta), \Psi(\xi)(\beta \oplus \eta)\rangle_r &= \langle \eta, \xi^*\rangle_r \langle \xi^*, \eta\rangle_r + \langle \xi\beta, \xi\beta\rangle_r \\ &\leq \|\xi^*\|_r^2 \langle \eta, \eta\rangle_r + \|\xi\|_r^2 \beta^*\beta \\ &\leq \|\xi\|_m^2 (\langle \eta, \eta\rangle_r + \beta^*\beta).\end{aligned}$$

This actually shows that $\|\Psi(\xi)\| \leq \|\xi\|_m$, and the converse inequality follows from the computations

$$\langle \Psi(\xi)(\langle \xi, \xi\rangle_r \oplus 0), \Psi(\xi)(\langle \xi, \xi\rangle_r \oplus 0)\rangle_r = \langle \xi, \xi\rangle_r^3$$

(hence $\|\Psi(\xi)\| \geq \|\xi\|_r$) and

$$\langle \Psi(\xi)(0 \oplus \xi^*), \Psi(\xi)(0 \oplus \xi^*)\rangle_r = \langle \xi^*, \xi^*\rangle_r^2$$

(hence $\|\Psi(\xi)\| \geq \|\xi^*\|_r = \|\xi\|_l$). Also, $\Psi(\xi)$ is adjointable and in fact $\Psi(\xi)^* = \Psi(\xi^*)$ by the computations

$$\begin{aligned}\langle \Psi(\xi)(\beta \oplus \eta), (\gamma \oplus \zeta)\rangle_r &= \langle \langle \xi^*, \eta\rangle_r \oplus \xi\beta, \gamma \oplus \zeta\rangle_r \\ &= \langle \eta, \xi^*\rangle_r \gamma + \langle \xi\beta, \zeta\rangle_r\end{aligned}$$

and

$$\begin{aligned}\langle (\beta \oplus \eta), \Psi(\xi^*)(\gamma \oplus \zeta)\rangle_r &= \langle \beta \oplus \eta, \langle \xi, \zeta\rangle_r \oplus \xi^*\gamma\rangle_r \\ &= \beta^* \langle \xi, \zeta\rangle_r + \langle \eta, \xi^*\gamma\rangle_r.\end{aligned}$$

Finally, it is trivial to check that $\Phi(\alpha)\Psi(\xi) = \Psi(\alpha\xi)$ and $\Psi(\xi)\Phi(\alpha) = \Psi(\xi\alpha)$. ∎

Given a W* Hilbert $*$-bimodule $\mathcal{E}$ over a von Neumann algebra, the linking algebra construction can be modified by using the dual module

$\mathcal{E}_r'$ introduced in Section 9.4 in place of $\mathcal{E}_r$; this has the consequence that $B(\mathcal{E}^+)$ is then a von Neumann algebra (Corollary 9.4.3), so that the construction of Proposition 10.5.1 makes $\mathcal{E}$ a W*-bimodule over $\mathcal{M}$ in the sense of Definition 10.3.7. We record this fact:

COROLLARY 10.5.2
Let $\mathcal{M}$ be a von Neumann algebra and let $\mathcal{E}$ be a W Hilbert $*$-bimodule over $\mathcal{M}$. Then $\mathcal{E}$ is a W*-bimodule over $\mathcal{M}$.*

Now if $\mathcal{E}$ is any Hilbert $*$-bimodule over $\mathcal{M}$, we can replace it with the weak* closure of $\Psi(\mathcal{E})$ in $B(\mathcal{E}^+)$. In this way we can turn an ordinary Hilbert $*$-bimodule into a W* Hilbert $*$-bimodule, provided the original bimodule satisfies a version of normality. We formulate this result next.

LEMMA 10.5.3
Let $\mathcal{M}$ and $\mathcal{N}$ be von Neumann algebras and let $\mathcal{A}$ be a weak dense $*$-subalgebra of $\mathcal{M}$. Suppose $\pi : \mathcal{A} \to \mathcal{N}$ is a $*$-homomorphism and suppose $a_\kappa \to 0$ boundedly and weak* in $\mathcal{A} \subset \mathcal{M}$ implies $\pi(a_\kappa) \to 0$ weak* in $\mathcal{N}$. Then π extends to a weak* continuous $*$-homomorphism from $\mathcal{M}$ to $\mathcal{N}$.*

PROOF Let $\mathcal{M}'$ be the weak* closure of $\mathcal{A}' = \{a \oplus \pi(a) : a \in \mathcal{A}\}$ in $\mathcal{M} \oplus \mathcal{N}$. Then the natural projection $\pi_1 : \mathcal{M}' \to \mathcal{M}$ has zero kernel and hence is a $*$-isomorphism, the map $\pi_2 \circ \pi_1^{-1} : \mathcal{M} \to \mathcal{N}$ is weak* continuous, and the restriction of this map to $\mathcal{A}$ agrees with π. ∎

THEOREM 10.5.4
Let $\mathcal{M}$ be a von Neumann algebra, let $\mathcal{A}$ be a weak dense $*$-subalgebra of $\mathcal{M}$, and let $\mathcal{E}$ be a pre-Hilbert $*$-bimodule over $\mathcal{A}$. Suppose that for any bounded net (a_κ) in $\mathcal{A}$ and any $\xi, \eta \in \mathcal{E}$, $a_\kappa \to 0$ weak* implies $\langle \xi a_\kappa, \eta\rangle \to 0$ weak*. Then $\mathcal{E}$ densely embeds in a unique W* Hilbert $*$-bimodule over $\mathcal{M}$.*

In particular, if $\mathcal{E}$ is a normal Hilbert $*$-bimodule over $\mathcal{M}$ then it densely embeds in a unique W Hilbert $*$-bimodule over $\mathcal{M}$.*

PROOF Here we use the concept of a pre-Hilbert $*$-bimodule over an uncompleted C*-algebra; this causes no difficulties. Let $\mathcal{E}^+ = \mathcal{M} \oplus \mathcal{E}_r'$ be constructed as in Proposition 10.5.1, but using the dual module $\mathcal{E}_r'$ in place of $\mathcal{E}_r$, and let $\Phi' : \mathcal{M} \to B(\mathcal{E}^+)$ and $\Psi' : \mathcal{E} \to B(\mathcal{E}^+)$ be the corresponding maps. For a bounded net (A_κ) in $B(\mathcal{E}^+)$, weak* convergence is equivalent to weak* convergence of $\langle A_\kappa \xi, \eta\rangle_r$ in $\mathcal{M}$ for all $\xi, \eta \in \mathcal{E}^+$. Thus the lemma implies that there is a weak* continuous extension of $\Phi'|_{\mathcal{A}}$ to $\mathcal{M}$, and the continuity hypothesis implies that this extension must be Φ'. So Φ' is weak* continuous.

Define $\mathcal{E}_1$ to be the weak* closure of $\Psi'(\mathcal{E})$ in $B(\mathcal{E}^+)$. This is a bimodule over $\mathcal{M} \cong \Phi'(\mathcal{M})$ via operator multiplication, and normality and duality are trivial. It is also straightforward to check that the bimodule structure of $\mathcal{E}_1$ extends that of $\Psi'(\mathcal{E}) \cong \mathcal{E}$. The bilinear form and adjoint can either be extended from $\mathcal{E}$ by continuity or defined directly by $\langle \xi, \eta \rangle \oplus 0 = \Psi'\xi(\Psi'\eta(I_{\mathcal{M}} \oplus 0))$ and operator adjoints.

For uniqueness, let $\mathcal{E}_2$ be any other bimodule with the same properties and define a map $T : \mathcal{E}_1 \to \mathcal{E}_2$ by $T(\lim_{\mathcal{E}_1} \xi_\kappa) = \lim_{\mathcal{E}_2} \xi_\kappa$ for any bounded weakly convergent net (ξ_κ) in $\mathcal{E}$. This map is well-defined and compatible with inner products since

$$\langle \lim_{\mathcal{E}_2} \xi_\kappa, \eta \rangle = \lim_{\mathcal{M}} \langle \xi_\kappa, \eta \rangle = \langle \lim_{\mathcal{E}_1} \xi_\kappa, \eta \rangle$$

for any $\eta \in \mathcal{E}$. It follows that $\mathcal{E}_1$ is unique up to isometric isomorphism. ∎

We now present a general method for constructing W*-derivations. It involves Markov semigroups, which we define next.

In this definition we use the following terms. Let $\mathcal{M}$ be a von Neumann algebra and let τ be a linear map from a $*$-subalgebra of $\mathcal{M}$ into $\mathbf{C}$. We say that τ is a trace if $\tau(\alpha\beta) = \tau(\beta\alpha)$ for all α, β, and $\alpha \geq 0$ implies $\tau(\alpha) \geq 0$; it is faithful if $\alpha > 0$ implies $\tau(\alpha) > 0$; it is normal if $\tau(\alpha_\kappa) \to 0$ whenever (α_κ) is a decreasing net that weak* converges to 0; and it is semifinite if for every $\alpha \geq 0$ in $\mathcal{M}$ there exists $\beta \in \mathrm{dom}(\tau)$ such that $0 \leq \beta \leq \alpha$. A faithful, normal, semifinite trace is called an fns trace.

DEFINITION 10.5.5 *Let $\mathcal{M}$ be a von Neumann algebra. A C_0^*-semigroup of operators on $\mathcal{M}$ is a family of linear maps $\Theta_t : \mathcal{M} \to \mathcal{M}$ $(t \geq 0)$ such that $\Theta_s\Theta_t = \Theta_{s+t}$ for all $s, t \geq 0$ and the maps $t \mapsto \Theta_t(\alpha)$ and $\alpha \mapsto \Theta_t(\alpha)$ are weak* continuous (from $\mathbf{R}$ into $\mathcal{M}$ and from $\mathcal{M}$ into $\mathcal{M}$, respectively). The generator of a C_0^*-semgroup (Θ_t) is the map*

$$\Delta(a) = \lim_{t \to 0} \frac{\Theta_t(\alpha) - \alpha}{t},$$

with domain the set of $\alpha \in \mathcal{M}$ for which the limit exists in the weak topology.*

A quantum Markov semigroup is a C_0^*-semigroup for which each Θ_t is completely positive and which satisfies

(a) $\Theta_0 = \mathrm{id}_{\mathcal{M}}$,
(b) $\Theta_t(I_{\mathcal{M}}) = I_{\mathcal{M}}$ for all t, and

(c) the set $\mathcal{A}_\infty = \{\alpha \in \mathcal{M} : \alpha \in D(\Delta^n)$ *for all* $n\}$ *is a* $*$-*algebra.*

It is symmetric *if there is an fns trace* τ *on* $\mathcal{M}$ *such that* $\mathrm{dom}(\tau) = \mathcal{A}_\infty$ *and* $\tau(\alpha\Theta_t(\beta)) = \tau(\Theta_t(\alpha)\beta)$ *for all* $\alpha, \beta \in \mathcal{A}_\infty$.

The construction of the associated W*-derivation goes as follows. Let (Θ_t) be a symmetric quantum Markov semigroup of operators on a von Neumann algebra $\mathcal{M}$. For $\alpha, \beta, \gamma, \delta \in \mathcal{A}_\infty$ define

$$(\alpha \otimes \beta)^* = \beta^* \otimes \alpha^*$$

and

$$\langle \alpha \otimes \beta, \gamma \otimes \delta \rangle = \alpha\Delta(\beta\gamma)\delta,$$

and extend both linearly to $\mathcal{A}_\infty \otimes \mathcal{A}_\infty$. We do not have $\langle \xi, \xi^* \rangle \geq 0$ on $\mathcal{A}_\infty \otimes \mathcal{A}_\infty$, but this does hold on the sub-$\mathcal{A}_\infty$-$\mathcal{A}_\infty$-bimodule

$$\mathcal{E}_0 = \mathrm{span}\{\alpha \otimes \beta\gamma - \alpha\beta \otimes \gamma : \alpha, \beta, \gamma \in \mathcal{A}_\infty\}.$$

To see this, let $\xi \in \mathcal{E}_0$ and write $\xi = \sum_1^n \alpha_i \otimes \beta_i\gamma_i - \alpha_i\beta_i \otimes \gamma_i$. Let

$$B = \begin{bmatrix} \beta_1\gamma_1 \\ \vdots \\ \beta_n\gamma_n \\ \gamma_1 \\ \vdots \\ \gamma_n \end{bmatrix} \qquad \text{and} \qquad C = [\,\alpha_1 \;\; \cdots \;\; \alpha_n \;\; -\alpha_1\beta_1 \;\; \cdots \;\; -\alpha_n\beta_n\,].$$

Then $C\Theta_t(BB^*)C^* \geq 0$ for all $t \geq 0$ since Θ_t is completely positive, with equality when $t = 0$; differentiating at $t = 0$ therefore yields $C\Delta(BB^*)C^* \geq 0$, which, written out, is the desired inequality $\langle \xi, \xi^* \rangle \geq 0$.

Next we apply Theorem 10.5.4. To verify its hypothesis, let $\pi : \mathcal{M} \to B(\mathcal{H}_\tau)$ be the GNS representation deduced from τ. The condition that τ be fns implies that π is a weak* continuous $*$-isomorphism. Thus if (α_κ) is a bounded net in $\mathcal{A}_\infty$ and $\alpha_\kappa \to 0$ weak*, then

$$\tau(\gamma^*\alpha_\kappa\beta) = \langle \alpha_\kappa\bar{\beta}, \bar{\gamma} \rangle_\tau \to 0$$

for any $\beta, \gamma \in \mathcal{A}_\infty$. Now let $\xi = \alpha \otimes \beta\gamma - \alpha\beta \otimes \gamma$ and $\eta = \alpha' \otimes \beta'\gamma' - \alpha'\beta' \otimes \gamma'$ and let $\delta, \delta' \in \mathcal{A}_\infty$. We have

$$\langle \pi(\langle \xi\alpha_\kappa, \eta \rangle)\bar{\delta}', \bar{\delta} \rangle_\tau = \tau(\delta^*\langle \xi\alpha_\kappa, \eta \rangle\delta');$$

writing this out, we get

$$\tau(\delta^*\langle (\alpha \otimes \beta\gamma - \alpha\beta \otimes \gamma)\alpha_\kappa, (\alpha' \otimes \beta'\gamma' - \alpha'\beta' \otimes \gamma') \rangle\delta')$$

$$\begin{aligned}&= \tau(\delta^*\alpha\Delta(\beta\gamma\alpha_\kappa\alpha')\beta'\gamma'\delta' - \delta^*\alpha\Delta(\beta\gamma\alpha_\kappa\alpha'\beta')\gamma'\delta' \\ &\quad -\delta^*\alpha\beta\Delta(\gamma\alpha_\kappa\alpha')\beta'\gamma'\delta' + \delta^*\alpha\beta\Delta(\gamma\alpha_\kappa\alpha'\beta')\gamma'\delta') \\ &= \tau(\beta\gamma\alpha_\kappa\alpha'\Delta(\beta'\gamma'\delta'\delta^*\alpha) - \beta\gamma\alpha_\kappa\alpha'\beta'\Delta(\gamma'\delta'\delta^*\alpha) \\ &\quad -\gamma\alpha_\kappa\alpha'\Delta(\beta'\gamma'\delta'\delta^*\alpha\beta) + \gamma\alpha_\kappa\alpha'\beta'\Delta(\gamma'\delta'\delta^*\alpha\beta)),\end{aligned}$$

and the latter converges to 0 by the previous observation. Since this is true for all $\delta, \delta' \in \mathcal{A}_\infty$ and π is one-to-one, we conclude that $\langle \xi\alpha_\kappa, \eta\rangle \to 0$ weak*. Taking linear combinations, we find that this holds for all $\xi, \eta \in \mathcal{E}_0$, so the hypothesis of Theorem 10.5.4 is verified and we get that $\mathcal{E}_0$ densely embeds in a unique W* Hilbert $*$-bimodule $\mathcal{E}$ over $\mathcal{M}$. $\mathcal{E}$ plays the role of the module of bounded measurable 1-forms, and we have an exterior derivative $d_0 : \mathcal{A}_\infty \to \mathcal{E}$ defined by

$$d_0(\alpha) = i(1 \otimes \alpha - \alpha \otimes 1).$$

It is easy to check that d_0 is a derivation and $d_0(\alpha^*) = d_0(\alpha)^*$. Moreover, it is weak* to weak* closable because if (α_κ) and $(d_0(\alpha_\kappa))$ are bounded and $\alpha_\kappa \to 0$ weak* then

$$\begin{aligned}&\tau(\delta^*\langle i(1 \otimes \alpha_\kappa - \alpha_\kappa \otimes 1), (\alpha \otimes \beta\gamma - \alpha\beta \otimes \gamma)\rangle\delta') \\ &\qquad = i\tau(\alpha_\kappa\alpha\Delta(\beta\gamma\delta'\delta^*) - \alpha_\kappa\alpha\beta\Delta(\gamma\delta'\delta^*) \\ &\qquad\quad -\alpha_\kappa\Delta(\alpha)\beta\gamma\delta'\delta^* + \alpha_\kappa\Delta(\alpha\beta)\gamma\delta'\delta^*)\end{aligned}$$

converges to zero for any $\alpha, \beta, \gamma, \delta, \delta' \in \mathcal{A}_\infty$, and similarly for the inner product in reverse order, which implies that $i(1 \otimes \alpha_\kappa - \alpha_\kappa \otimes 1) \to 0$ weak* in $\mathcal{E}$ by the same reasoning as in the last paragraph. Thus the closure d of d_0 is a W*-derivation and its domain is a noncommutative Lipschitz algebra.

We summarize this result in the following theorem.

THEOREM 10.5.6

Let $\mathcal{M}$ be a von Neumann algebra and let (Θ_t) be a symmetric quantum Markov semigroup of operators on $\mathcal{M}$. Then the map $d : \mathcal{M} \to \mathcal{E}$ constructed above is a W-derivation into a W* Hilbert $*$-bimodule.*

Example 10.5.7

Let X be a compact, connected Riemannian manifold. Then there is a Laplace operator on X which can be written at each point as

$$\Delta f(x) = -\frac{\partial f}{\partial v_1}(x) - \cdots - \frac{\partial f}{\partial v_n}(x)$$

where $v_1, \ldots, v_n$ is an orthonormal basis of the tangent space at x. A symmetric Markov semigroup $\Theta_t = e^{t\Delta}$ can be obtained by exponentiating the Laplacian; this is known as the <u>diffusion semigroup</u> on X. Then $\mathcal{A}_\infty = C^\infty(X)$, and carrying out the above construction yields

$\langle df, dg\rangle = \nabla f \cdot \nabla g$ for any $f, g \in C^\infty(X)$. Thus we recover the usual first-order exterior derivative on X described in Example 10.4.1.

Example 10.5.8

Let τ be the weight (actually an fns trace) defined in Example 5.6.7 (b) on the noncommutative plane algebra $L^\infty_\hbar(\mathbf{R}^2)$. A quantum Markov semigroup (Θ_t) can be defined by exponentiating the Laplace operator $\Delta = -d_1^2 - d_2^2$, where d_1 and d_2 are as defined in Example 10.4.2 (a). Then the above construction recovers the derivation given in Example 10.4.2. The noncommutative tori work similarly; here we take τ to be the trace defined in Proposition 6.6.7.

10.6 Notes

A thorough treatment of Lipschitz algebras is given in [73]. Noncommutative geometry is discussed in [12]; for the relation between that approach and ours see Section V of [71]. Examples 10.4.1 and 10.4.2 are also treated in greater detail in [71]. The operator space example discussed in Section 10.4 is from [70].

The material of Section 10.5 follows [74]. Our construction of a noncommutative Lipschitz algebra from a symmetric quantum Markov semigroup is based on [64]. See [66] for the general theory of fns traces.

Chapter 11

Quantum Groups

11.1 Finite dimensional C*-algebras

Quantum groups are the Hilbert space analog of topological groups, and thus their description naturally involves C*-algebras. (There is a von Neumann algebra version of the theory as well; it is more or less equivalent to the C*-algebra version, just as in the classical case where there is an equivalence between topological and measurable groups.) As we are going to discuss quantum groups first in the finite dimensional setting, we will begin by giving an explicit description of all finite dimensional C*-algebras. Here there is some room for variety, as the following example shows.

Example 11.1.1

Let $n_1, \ldots, n_m$ be natural numbers and let M_n be the matrix algebra of linear operators on $\mathbf{C}^n$. Then $\bigoplus_{i=1}^m M_{n_i}$ is a finite dimensional C*-algebra which acts on $\bigoplus_{i=1}^m \mathbf{C}^{n_i}$ in an obvious way. Thus $\bigoplus M_{n_1}$ is naturally realized as a block diagonal C*-subalgebra of M_n where $n = n_1 + \cdots + n_m$.

The converse of this example is also true: every finite dimensional C*-algebra has the above form.

THEOREM 11.1.2
Every finite dimensional C-algebra is a direct sum of matrix algebras.*

PROOF Let $\mathcal{A}$ be a finite dimensional C*-algebra; then it is also a von Neumann algebra. By Proposition 6.5.5, if $\mathcal{A}$ has a proper C*-ideal $\mathcal{I}$ then it decomposes as $\mathcal{A} = P\mathcal{A} \oplus (I_{\mathcal{A}} - P)\mathcal{A}$. We may therefore inductively reduce to the case that $\mathcal{A}$ is simple, i.e., has no proper C*-ideals.

Say $\mathcal{A} \subset B(\mathcal{H})$. Let v be a nonzero vector in $\mathcal{H}$; then $\mathcal{K} = \mathcal{A}v = \{Av : A \in \mathcal{A}\}$ is a finite dimensional subspace of $\mathcal{H}$ and $\mathcal{A}(\mathcal{K}) \subset \mathcal{K}$, so the restriction map takes $\mathcal{A}$ $*$-homomorphically into $B(\mathcal{K})$. Since $\mathcal{A}$ is simple, this map is a $*$-isomorphism.

In fact, we may assume that the representation $\mathcal{A} \to B(\mathcal{K})$ is irreducible, i.e., $\mathcal{A}w = \mathcal{K}$ for all nonzero $w \in \mathcal{K}$. If not, let w be a counterexample and replace $\mathcal{K}$ with $\mathcal{A}w \subset \mathcal{K}$. Since $\mathcal{K}$ is finite dimensional, this process must eventually terminate with a Hilbert space $\mathcal{K}$ that has no such vectors w.

We may suppose $\mathcal{A} \subset B(\mathcal{K})$. Let B lie in its commutant $\mathcal{A}^c$ and suppose B is self-adjoint. Let v be an eigenvector for B, say $Bv = \lambda v$, and for any $w \in \mathcal{K}$ find $A \in \mathcal{A}$ so that $Av = w$. Then

$$Bw = BAv = ABv = \lambda Av = \lambda w$$

so that w is also an eigenvector of B, with the same eigenvalue. This shows that $B = \lambda I$, and we conclude that $\mathcal{A}^c = \mathbf{C} \cdot I$. Then $\mathcal{A} = \mathcal{A}^{cc} = B(\mathcal{K})$ by Theorem 6.5.7. ∎

We can also explicitly describe all *-homomorphisms between finite dimensional C*-algebras.

Example 11.1.3

Let $\mathcal{A} = \bigoplus_{i=1}^{p} M_{m_i}$ and $\mathcal{B} = \bigoplus_{j=1}^{q} M_{n_j}$ be two finite dimensional C*-algebras. For each j let $i_1^j, \ldots, i_k^j$ (here k depends on j) be a family of possibly repeating indices such that $n_j = m_{i_1^j} + \cdots + m_{i_k^j}$. Then $M_{m_{i_1^j}} \oplus \cdots \oplus M_{m_{i_k^j}}$ naturally embeds in $M_{n_j} \subset \mathcal{B}$, and composing with the natural map from $\mathcal{A}$ into $M_{m_{i_1^j}} \oplus \cdots \oplus M_{m_{i_k^j}}$ gives rise to a unital $*$-homomorphism from $\mathcal{A}$ into M_{n_j}. Taking the direct sum of these maps produces a unital $*$-homomorphism from $\mathcal{A}$ into $\mathcal{B}$.

PROPOSITION 11.1.4

Let $\mathcal{A}$ and $\mathcal{B}$ be finite dimensional C-algebras and let $\pi : \mathcal{A} \to \mathcal{B}$ be a unital $*$-homomorphism. Then there are realizations of $\mathcal{A}$ and $\mathcal{B}$ as direct sums of matrix algebras such that π is expressed as a map of the form given in Example 11.1.3.*

PROOF It suffices to consider the case $\mathcal{B} = M_n$. Say $\mathcal{A} = \bigoplus_1^p M_{m_i}$ and for each i let P_i be the identity matrix in M_{m_i}. Then the P_i are commuting projections and $P_1 + \cdots + P_p = I_{\mathcal{A}}$. Since π is a unital $*$-homomorphism, similar relations hold for $Q_i = \pi(P_i)$.

Thus, the Q_i are projections onto a spanning set of orthogonal subspaces of $\mathbf{C}^n$. Let $\mathcal{H}_i$ be the range of Q_i and consider the map π_i from

M_{m_i} into $B(\mathcal{H}_i)$ given by composing π with the restriction to $\mathcal{H}_i$. This map is also a unital $*$-homomorphism.

Let $\{v_1, \ldots, v_r\}$ be a maximal set of nonzero vectors in $\mathcal{H}_i$ such that $\mathcal{K}_1, \ldots, \mathcal{K}_r$ are orthogonal, where $\mathcal{K}_j = \pi_i(M_{m_i})v_j$. As in Theorem 11.1.2 we may assume that M_{m_i} is irreducibly represented on each $\mathcal{K}_j$, and hence that π_i takes M_{m_i} $*$-isomorphically onto each $B(\mathcal{K}_j)$. Thus $\mathcal{H}_i = \mathcal{K}_1 \oplus \cdots \oplus \mathcal{K}_r$ realizes $\mathcal{H}_i$ in such a way that π_i takes the form $A \mapsto A \oplus \cdots \oplus A$ (r summands) on M_{m_i}. Doing this for each i, we obtain a realization of $\mathcal{B}$ such that π is in the desired form. ∎

Finally, we define tensor products of finite dimensional C*-algebras.

DEFINITION 11.1.5 *Let $\mathcal{A} \subset M_m$ and $\mathcal{B} \subset M_n$ be finite dimensional C*-algebras. For any operators $A \in \mathcal{A}$ and $B \in \mathcal{B}$, define $A \otimes B \in M_{mn}$ by $(A \otimes B)(v \otimes w) = Av \otimes Bw$. Let $\mathcal{A} \otimes \mathcal{B}$ be the C*-algebra generated by all of the operators $A \otimes B$ for $A \in \mathcal{A}$ and $B \in \mathcal{B}$.*

It is easy to see that according to this definition $M_m \otimes M_n \cong M_{mn}$, and more generally

$$\left(\bigoplus_{i=1}^{p} M_{m_i}\right) \otimes \left(\bigoplus_{j=1}^{q} M_{n_j}\right) \cong \left(\bigoplus_{i,j} M_{m_i n_j}\right).$$

11.2 Finite quantum groups

In this section we introduce quantum groups in the finite dimensional setting. One might think that in this setting topology will be irrelevant, but this is not the case because of the variety of possible C*-algebras of a given finite dimension. Of course, there is (up to $*$-isomorphism) only one n-dimensional abelian C*-algebra, namely $\mathbf{C}^n$, reflecting the fact that there is only one Hausdorff topology on an n-element set. But as we saw in the last section, there are many more, though not unmanageably many more, finite dimensional nonabelian C*-algebras.

Now topology passes from the classical to the quantum setting by letting a nonabelian C*-algebra play the role of a "quantum $C(X)$." So in order to incorporate group structure into the picture, we must find a way of expressing the group axioms in terms of functions on a group rather than directly in terms of its elements. The natural way to do this is to consider the composition of continuous functions with the group operations. For the inverse operation $^{-1} : G \to G$ this yields a map from $C(G)$ to $C(G)$. Likewise, the group product $\bullet : G^2 \to G$ gives rise to a map from $C(G)$ to $C(G \times G) \cong C(G) \otimes C(G)$. Finally, the identity element may be regarded as an operation with no arguments, i.e., a map $e : G^0 \to G$ where $G^0 = \{\emptyset\}$. Composition with this map yields a linear functional on $C(G)$.

We now describe the quantum analog of the preceding. A linear map $T : \mathcal{A} \to \mathcal{B}$ between C*-algebras is an antihomomorphism if $T(\alpha\beta) = T(\beta)T(\alpha)$ for all $\alpha, \beta \in \mathcal{A}$. Also, let $\iota : \mathcal{A} \to \mathcal{A}$ be the identity map and (assuming $\mathcal{A}$ is finite dimensional) let $m : \mathcal{A} \otimes \mathcal{A} \to \mathcal{A}$ be the multiplication map $m(\alpha \otimes \beta) = \alpha\beta$.

DEFINITION 11.2.1 *Let $\mathcal{A}$ be a finite dimensional C*-algebra. A coproduct on $\mathcal{A}$ is a unital $*$-homomorphism $\Phi : \mathcal{A} \to \mathcal{A} \otimes \mathcal{A}$ which satisfies the coassociativity axiom*

$$(\Phi \otimes \iota) \circ \Phi = (\iota \otimes \Phi) \circ \Phi$$

as maps from $\mathcal{A}$ into $(\mathcal{A} \otimes \mathcal{A}) \otimes \mathcal{A} \cong \mathcal{A} \otimes (\mathcal{A} \otimes \mathcal{A})$; A counit is a unital $*$-homomorphism $\varepsilon : \mathcal{A} \to \mathbf{C}$ such that

$$(\varepsilon \otimes \iota) \circ \Phi = \iota = (\iota \otimes \varepsilon) \circ \Phi;$$

and an antipode is a unital antihomomorphism $\kappa : \mathcal{A} \to \mathcal{A}$ such that

$$m \circ (\kappa \otimes \iota) \circ \Phi = \varepsilon \cdot I_{\mathcal{A}} = m \circ (\iota \otimes \kappa) \circ \Phi.$$

A finite quantum group is a finite dimensional C-algebra equipped with a coproduct, a counit, and an antipode.*

Why is the antipode, which plays the role of a coinverse, assumed to be an antihomomorphism? Because if we took it to be a $*$-homomorphism, the only examples would be commutative! In fact, if we assume only that κ is a linear map, it follows from the remaining axioms that κ must be an antihomomorphism, and it is also automatically surjective. Thus, requiring κ to also be a homomorphism forces $\mathcal{A}$ to be abelian. When we pass to the infinite dimensional setting in Section 11.3, we will adopt an approach that may be more satisfying; there κ is left out of the axiomatization altogether, and its existence and antihomomorphic nature is deduced only later. However, in the finite dimensional setting it is easier to simply assert the existence of κ.

The antipode interacts with adjoints by the equation $\kappa(\kappa(\alpha)^*)^* = \alpha$. This can be proven from the above axioms (and it incidentally shows that κ must be surjective).

Our first examples of quantum groups arise from ordinary groups.

Example 11.2.2

Let G be a finite group with identity element e. Then $\mathcal{A} = C(G)$ is a finite quantum group, with $\Phi : C(G) \to C(G \otimes G)$, $\varepsilon : C(G) \to \mathbf{C}$, and $\kappa : C(G) \to C(G)$ defined by

$$\Phi f(x,y) = f(x \bullet y) \qquad \varepsilon(f) = f(e) \qquad \kappa f(x) = f(x^{-1}).$$

And these are the only commutative finite quantum groups.

THEOREM 11.2.3

Let $(\mathcal{A}, \Phi, \varepsilon, \kappa)$ be a finite quantum group and suppose the C-algebra $\mathcal{A}$ is abelian. Then $(\mathcal{A}, \Phi, \varepsilon, \kappa)$ is isomorphic to the quantum group $C(G)$ for some finite group G.*

PROOF Say $\mathcal{A} = C(G)$ where G is a finite set and identify $\mathcal{A} \otimes \mathcal{A}$ with $C(G \times G)$. The coproduct is then a unital $*$-homomorphism from $C(G)$ into $C(G \times G)$, and hence it is given by composition with a map $\phi : G \times G \to G$ by Proposition 5.1.9. The coassociativity axiom then yields

$$\phi \circ (\phi \times \mathrm{id}_G) = \phi \circ (\mathrm{id}_G \times \phi)$$

as maps from $G \times G \times G$ to G; that is, $\phi(\phi(x,y),z) = \phi(x, \phi(y,z))$ for all $x, y, z \in G$. Thus ϕ is an associative binary operation which makes G into a semigroup.

Likewise, since ε is a $*$-homomorphism it is given by evaluation at an element e of G, and a short computation shows that the counit axiom implies $ex = xe = x$ for all $x \in G$. So e is an identity element of G.

Finally, although κ is only assumed to be an antihomomorphism (and hence a homomorphism in the present case), the computation

$$\kappa(\chi_S) = \kappa(\chi_S^2) = \kappa(\chi_S)^2$$

shows that it takes characteristic functions to characteristic functions, and thus by linearity it is actually a $*$-homomorphism. Since we do not assume κ is onto, this means that there is a subset S of G (namely, $\chi_S = \kappa(1_G)$) and a map $\psi : S \to G$ such that $\kappa(f)|_S = f \circ \psi$ on S and $\kappa(f)$ is zero off of S, for all $f \in C(G)$. Applying the antipode axiom to χ_e then yields $\sum \kappa(\chi_x)\chi_y = 1_G$, where the sum is taken over all pairs $x, y \in G$ such that $x \bullet y = e$; this implies that for each $y \in G$ there exists $x \in G$ such that $x \bullet y = e$ and $\psi(y) = x$. From this it follows that $S = G$ and ψ is an inverse operation on G. ∎

Next, we consider the group algebra construction. If the underlying group is nonabelian, this will involve a nonabelian C*-algebra and hence no longer be a quantum group of the preceding type.

Example 11.2.4

Let G be a finite group and for $x \in G$ define a unitary translation operator T_x on $l^2(G)$ by $T_x f(y) = f(x^{-1}y)$. Then

$$\begin{aligned} T_x T_y f(z) &= T_y f(x^{-1}z) = f(y^{-1}x^{-1}z) \\ &= f((xy)^{-1}z) = T_{xy} f(z), \end{aligned}$$

so $T_x T_y = T_{xy}$.

Let $C^*(G)$ be the C*-algebra generated by the operators T_x for all $x \in G$. Define Φ, ε, and κ by $\Phi(\chi_x) = \chi_x \otimes \chi_x$, $\varepsilon(\chi_x) = \langle \chi_x, \chi_e \rangle$, and $\kappa(\chi_x) = \chi_{x^{-1}}$. This makes $C^*(G)$ into a finite quantum group.

The quantum groups of Example 11.2.4 are dual to the quantum groups of Example 11.2.2 in the following sense.

DEFINITION 11.2.5 *Let $(\mathcal{A}, \Phi, \kappa, \varepsilon)$ be a finite quantum group. Define $\hat{\mathcal{A}}$ to be the dual vector space $\mathcal{A}^*$ with algebra and coalgebra structure given by*

$$\begin{aligned} \omega\rho(\alpha) &= \overline{(\omega \otimes \rho)}(\Phi(\alpha)) \\ \omega^*(\alpha) &= \overline{\omega(\kappa(\alpha))} \\ I_{\hat{\mathcal{A}}} &= \varepsilon_{\mathcal{A}} \end{aligned}$$

and

$$\begin{aligned} \Phi\omega(\alpha \otimes \beta) &= \overline{\omega(\alpha\beta)} \\ \kappa\omega(\alpha) &= \omega(\alpha^*) \\ \varepsilon(\omega) &= \omega(I_{\mathcal{A}}) \end{aligned}$$

for $\omega, \rho \in \hat{\mathcal{A}}$ and $\alpha, \beta \in \mathcal{A}$.

More or less straightforward computations show that this defines a $*$-algebra structure on $\mathcal{A}$ and that Φ, κ, and ε satisfy the quantum group axioms. However, it is not obvious that $\hat{\mathcal{A}}$ is a C*-algebra. This can be proven using the Haar state h on $\mathcal{A}$, which we will construct in Section 11.4. Letting $\mathcal{H}_h$ be the associated GNS Hilbert space, one can show that the representation $\pi(\omega)(\bar{\alpha}) = \overline{(\iota \otimes \omega)(\Phi\alpha)}$ takes $\hat{\mathcal{A}}$ $*$-isomorphically into $B(\mathcal{H}_h)$. This exhibits $\hat{\mathcal{A}}$ as a C*-algebra.

In any case, by Proposition 3.2.4 the norm of a self-adjoint element of a C*-algebra can be computed algebraically, and the equality $\|\alpha\|^2 = \|\alpha^*\alpha\|$ then determines the norm of an arbitrary element of the C*-algebra. In this way we can determine the norm on the dual quantum group $\hat{\mathcal{A}}$ without actually carrying out the above construction. But the fact that this does produce a C*-algebra (in particular, that the norm is not really a seminorm) still depends on that argument.

Modulo the fact that $\hat{\mathcal{A}}$ is a C*-algebra, the verification of the following result is straightforward but tedious, so we omit it.

THEOREM 11.2.6
If $\mathcal{A}$ is a finite quantum group then so is $\hat{\mathcal{A}}$. There is a natural isomorphism of $\mathcal{A}$ with $\hat{\hat{\mathcal{A}}}$.

We also record the fact, mentioned above, that Examples 11.2.2 and 11.2.4 are dual to each other.

PROPOSITION 11.2.7
Let G be a finite group. Then $C^*(G)$ is naturally isomorphic to $C(G)\hat{}$.

The proof of this result is straightforward; the desired isomorphism $\pi : C^*(G) \to C(G)\hat{}$ is given by $\pi(T_x)(f) = f(x)$.

Notice that if G is abelian then $C^*(G) \cong C(G)\hat{}$ is also abelian, and therefore is isomorphic to $C(\hat{G})$ for some finite group $\hat{G}$ by Theorem 11.2.3. In fact, this group $\hat{G}$ is the <u>dual group</u>, the set of all homomorphisms $\phi : G \to \mathbf{T}$, with pointwise product (i.e., $\phi\psi(x) = \phi(x)\psi(x)$). One can easily verify that if G is a finite abelian group then there is a natural isomorphism between $C^*(G)$ and $C(\hat{G})$ that relates T_x with evaluation at x.

Thus, when G is abelian we have $C(G)\hat{} \cong C(\hat{G})$. But when G is nonabelian the dual of $C(G)$ is no longer of the form given in Example 11.2.4. So the quantum group setting allows us to generalize the notion of duality in a way that includes nonabelian groups.

There are even more finite quantum groups besides those discussed above. We include one example, omitting the tedious verification that it actually is a quantum group.

Example 11.2.8

Let $\mathcal{A} = \mathbf{C} \oplus \mathbf{C} \oplus \mathbf{C} \oplus \mathbf{C} \oplus M_2$ and write a typical element of $\mathcal{A}$ as $(\lambda_e, \lambda_a, \lambda_b, \lambda_c, A)$. Also let $\alpha_e = (1,0,0,0,0)$, $\alpha_a = (0,1,0,0,0)$, $\alpha_b = (0,0,1,0,0)$, $\alpha_c = (0,0,0,1,0)$, $U_e = I_2$ (the 2×2 identity matrix), and

$$U_a = \begin{pmatrix} 0 & i \\ 1 & 0 \end{pmatrix}, \quad U_b = \begin{pmatrix} 0 & 1 \\ i & 0 \end{pmatrix}, \quad U_c = \begin{pmatrix} -1 & 0 \\ 0 & 1 \end{pmatrix}.$$

Give $G = \{e, a, b, c\} \cong \mathbf{Z}/2 \times \mathbf{Z}/2$ a group structure by setting $a^2 = b^2 = c^2 = e$ and $ab = c$.

$\mathcal{A}$ is a C*-algebra, and we make it into a quantum group by defining the counit by

$$\varepsilon(\lambda_e, \lambda_a, \lambda_b, \lambda_c, A) = \lambda_e;$$

the coinverse by

$$\kappa(\lambda_e, \lambda_a, \lambda_b, \lambda_c, A) = (\lambda_e, \lambda_a, \lambda_b, \lambda_c, A^T);$$

and the coproduct by

$$\Phi(\alpha_y) = \sum_{x \in G} \alpha_x \otimes \alpha_{x^{-1}y} + \frac{1}{2}\left[(u_y)_{im}(\bar{u}_y)_{jn}\right]_{ijmn}$$

and

$$\Phi(A) = \sum_{x \in G} \alpha_{x^{-1}} \otimes U_x A U_x^* + \frac{1}{2}\sum_{x \in G} \bar{U}_x A U_x^T \otimes \alpha_x.$$

Here T denotes transpose, $U_y = [(u_y)_{ij}]$, and $\bar{U}_y = [(\bar{u}_y)_{ij}]$.

11.3 Compact quantum groups

General quantum groups present several technical difficulties that do not arise in the finite dimensional case. For instance, the multiplication map $m : \mathcal{A} \otimes \mathcal{A} \to \mathcal{A}$ which appears in the antipode axiom is unbounded if $\mathcal{A}$ is infinite dimensional, and in some important examples the counit and antipode are also unbounded.

Consequently, there is room for argument as to what is the best general definition of quantum groups; more restrictive conditions will exclude some examples, but will presumably give rise to a more satisfying theory. In the compact case, however, there is a definition which both encompases all of the most important examples and supports a rich and satisfying theory.

This approach to compact quantum groups takes the coproduct Φ as fundamental. Classically, this corresponds to starting with a semigroup. We then identify a classical condition which forces a semigroup to be a group, and we use the quantum version of this condition to complete the definition of compact quantum groups.

Recall that an abelian C*-algebra is the algebra of continuous functions on a *compact* space if and only if it has a unit. Thus, the definition of compact quantum groups will involve unital C*-algebras. Also, we will need to use tensor products of C*-algebras. As in Definition 11.1.5, if $\mathcal{A} \subset B(\mathcal{H})$ and $\mathcal{B} \subset B(\mathcal{K})$ are C*-algebras then we define $\mathcal{A} \otimes \mathcal{B} \subset B(\mathcal{H} \otimes \mathcal{K})$ to be the closure of the span of the elementary tensors $A \otimes B$ for $A \in \mathcal{A}$ and $B \in \mathcal{B}$. This is called the spatial tensor product C*-algebra and is actually independent of the representations of $\mathcal{A}$ and $\mathcal{B}$, although we will not need this fact. It follows from the Stone-Weierstrass theorem that $C(X) \otimes C(Y) \cong C(X \times Y)$ for any compact Hausdorff spaces X and Y.

The following is the relevant classical condition which guarantees that a semigroup (i.e., a set equipped with an associative binary operation) will be a group.

PROPOSITION 11.3.1

Let G be a compact semigroup and let $\mathcal{A} = C(G)$. Let $\Phi : \mathcal{A} \to \mathcal{A} \otimes \mathcal{A}$ be composition with the semigroup operation. Then G is a group if and only if the sets $\Phi(\mathcal{A}) \cdot (1_G \otimes \mathcal{A})$ and $\Phi(\mathcal{A}) \cdot (\mathcal{A} \otimes 1_G)$ are both dense in $\mathcal{A} \otimes \mathcal{A}$.

PROOF Suppose G is a group. Then $\Phi(\mathcal{A}) \cdot (1_G \otimes \mathcal{A})$ is spanned by all continuous functions of the form $(x, y) \mapsto f(xy)g(y)$. It follows that this set is a $*$-subalgebra of $C(G \times G)$. To see that it separates points, let $(x_1, y_1), (x_2, y_2) \in G \times G$. If $y_1 \neq y_2$ then take $f = 1_G$ and choose g so that $g(y_1) \neq g(y_2)$. If $y_1 = y_2$ and $x_1 \neq x_2$ then we must have $x_1 y_1 \neq x_2 y_2$ (this is where we use the fact that G is a group), so let $g = 1_G$ and choose f so that $f(x_1 y_1) \neq f(x_2 y_2)$. This shows that $\Phi(\mathcal{A}) \cdot (1_G \otimes \mathcal{A})$ separates points, so it is dense in $\mathcal{A} \otimes \mathcal{A} \cong C(G \times G)$ by the Stone-Weierstrass theorem. Density of $\Phi(\mathcal{A}) \cdot (\mathcal{A} \otimes 1_G)$ is proven similarly.

Conversely, suppose the conclusion holds and let $x_1, x_2, y \in G$. Suppose $x_1 y = x_2 y$; then the points (x_1, y) and (x_2, y) are not separated by $\Phi(\mathcal{A}) \cdot (1_G \otimes \mathcal{A})$. Since the latter is dense in $C(G \times G)$, we must have $x_1 = x_2$. Similarly, from density of $\Phi(\mathcal{A}) \cdot (\mathcal{A} \otimes 1_G)$ we can prove that $xy_1 = xy_2$ implies $y_1 = y_2$.

Now fix $x \in G$ and let H be the closed semigroup generated by x. Note that H is abelian. Let K be the intersection of all closed nonempty semigroups $J \subset H$ such that $HJ, JH \subset J$. If J and J' are two such semigroups then $JJ' \subset J \cap J'$, so compactness implies that K is nonempty. Let $y \in K$; then $yK \subset K$, and minimality of K implies that $yK = K$ (using the fact that H is abelian to verify that yK is a semigroup and $H(yK), (yK)H \subset yK$). Thus $ye = y$ for some $e \in K$. For any $z \in G$, we then have $yez = yz$, and therefore $ez = z$ by the last paragraph. Similarly $ze = z$ for all $z \in G$. So G has a unit. Also, $xe = x$ implies that $x \in K$ since $HK \subset K$. So $xK = K$, and therefore $xz = e$ for some $z \in K$. As x was arbitrary, we conclude that G has inverses. So G is a group. ∎

This motivates the following definition of compact quantum groups.

DEFINITION 11.3.2 *A compact quantum group is a unital C*-algebra $\mathcal{A}$ together with a unital $*$-homomorphism $\Phi : \mathcal{A} \to \mathcal{A} \otimes \mathcal{A}$ such that*

$$(\Phi \otimes \iota) \circ \Phi = (\iota \otimes \Phi) \circ \Phi$$

(i.e., Φ is coassociative) and the sets $\Phi(\mathcal{A}) \cdot (I_{\mathcal{A}} \otimes \mathcal{A})$ and $\Phi(\mathcal{A}) \cdot (\mathcal{A} \otimes I_{\mathcal{A}})$ are dense in $\mathcal{A} \otimes \mathcal{A}$.

Thus neither a counit nor an antipode comes into the definition of compact quantum groups. However, it is possible to deduce the existence of such operations.

Just as in the finite dimensional case (Example 11.2.2), classical compact groups give rise to quantum compact groups for which the underlying C*-algebra is abelian.

Example 11.3.3

Let G be a compact group and let $\mathcal{A} = C(G)$. Define $\Phi : \mathcal{A} \to \mathcal{A} \otimes \mathcal{A}$ by $\Phi f(x, y) = f(x \bullet y)$. Then $(\mathcal{A}, \Phi)$ is a compact quantum group. It satisfies coassociativity because $\bullet$ is associative, and it satisfies the density conditions by Proposition 11.3.1.

The converse also follows easily from Proposition 11.3.1:

THEOREM 11.3.4

Let $(\mathcal{A}, \Phi)$ be a compact quantum group and suppose $\mathcal{A}$ is abelian. Then $(\mathcal{A}, \Phi)$ is isomorphic to $C(G)$ for some compact group G.

PROOF Say $A = C(G)$ where G is a compact Hausdorff space. Then Φ is a unital $*$-homomorphism from $C(G)$ to $C(G \otimes G)$, and therefore $\phi = C_\phi$ for some continuous map $\phi : G \times G \to G$ by Proposition 5.1.9. As in the proof of Theorem 11.2.3 the coassociativity axiom implies that ϕ is associative, so it makes G a semigroup. It then follows from Proposition 11.3.1 that G is a group. ∎

We will now describe an interesting example of a compact quantum group, the quantum $SU(2)$. It depends on a parameter $q \in (0, 1)$ and is denoted $SU_q(2)$. (Actually, one can also define it for other values of q.)

The idea is this. The classical $SU(2)$ group consists of those 2×2 complex unitary matrices with unit determinant. By the Stone-Weierstrass theorem, the algebra of continuous functions on $SU(2)$ is generated by the "coordinate" functions which evaluate at the four matrix entries. These four functions satisfy a small number of identities arising from the fact that the matrices lie in $SU(2)$, and it is possible to show that the algebra $C(SU(2))$ is the universal abelian C*-algebra generated by four functions satisfying these identities. In fact, since two of the entries of any matrix in $SU(2)$ are the complex conjugates of the other two entries, $C(SU(2))$ is generated by only two functions.

To define the quantum version of $SU(2)$, we modify these identities in the following way. Let α_{ij} be operators for $i, j = 1, 2$. These correspond to the functions which evaluate at the four entries of matrices in $SU(2)$.

In line with the condition that matrices in $SU(2)$ are unitary, we require

$$\alpha_{1i}^*\alpha_{1j} + \alpha_{2i}^*\alpha_{2j} = \alpha_{i1}\alpha_{j1}^* + \alpha_{i2}\alpha_{j2}^* = \delta_{ij}I;$$

the unit determinant condition, however, is "twisted" to say

$$\sum_{\sigma}(-q)^{|\sigma|}\alpha_{\sigma(1)\tau(1)}\alpha_{\sigma(2)\tau(2)} = (-q)^{|\tau|}I$$

where τ is either permutation of $\{1,2\}$ and the sum is taken over both permutations σ of $\{1,2\}$. Here $|\sigma|$ is the parity of σ. If $q = 1$ this formally reduces to the condition that $\det[\alpha_{ij}] = 1$.

We have written these conditions in a way that suggests their generalization to n dimensions. In the case $n = 2$ they can be simplified. Write $\alpha = \alpha_{11}$, $\beta = \alpha_{12}$, $\gamma = \alpha_{21}$, and $\delta = \alpha_{22}$. Then it is possible to deduce that $\beta = -q\gamma^*$ and $\delta = \alpha^*$, and the preceding relations become

$$\alpha^*\alpha + \gamma^*\gamma = I \qquad \alpha\alpha^* + q^2\gamma\gamma^* = I$$
$$\alpha\gamma = q\gamma\alpha \qquad \alpha\gamma^* = q\gamma^*\alpha \qquad \gamma\gamma^* = \gamma^*\gamma.$$

The C*-algebra $SU_q(2)$ is then the universal C*-algebra generated by two elements α and γ satisfying the above relations.

Concretely, $SU_q(2)$ can be described in the following way. Let $\mathcal{H} = l^2(\mathbf{N} \times \mathbf{Z})$ and define operators A and C on $\mathcal{H}$ by

$$Av_{nk} = \sqrt{1 - q^{2n}}v_{n-1,k} \qquad Cv_{nk} = q^n v_{n,k+1}.$$

Then $SU_q(2)$ is the C*-algebra generated by A and C.

The quantum group structure of $SU_q(2)$ is given by setting

$$\Phi(\alpha) = \alpha\otimes\alpha - q\gamma^*\otimes\gamma \qquad \Phi(\gamma) = \gamma\otimes\alpha + \alpha^*\otimes\gamma,$$

which specializes the general prescription $\Phi(\alpha_{ij}) = \sum_k \alpha_{ik}\otimes\alpha_{kj}$. It is straightforward to verify that the operators $\Phi(\alpha)$ and $\Phi(\gamma)$ satisfy the same relations as α and γ; thus, the universal property of $SU_q(2)$ implies that there is a unital $*$-homomorphism $\Phi : SU_q(2) \to SU_q(2)\otimes SU_q(2)$ taking the prescribed values on α and γ. (This can also be checked directly in the $l^2(\mathbf{N}\times\mathbf{Z})$ model.) It is straightforward to check that Φ is coassociative.

The universal property of $SU_q(2)$ guarantees the existence of a $*$-homomorphism $\varepsilon : SU_q(2) \to \mathbf{C}$ taking α to 1 and γ to 0. This is the counit. The antipode κ satisfies

$$\kappa(\alpha) = \alpha^* \qquad \kappa(\alpha^*) = \alpha$$
$$\kappa(\gamma) = -q\gamma \qquad \kappa(\gamma^*) = -q^{-1}\gamma^*$$

and it is defined only on the $*$-algebra generated by α and γ, not on the whole C*-algebra $SU_q(2)$.

To verify the density condition that is required of quantum groups, observe that

$$\Phi(\alpha) \cdot (I \otimes \alpha^*) + q^2\Phi(\gamma^*) \cdot (I \otimes \gamma) = \alpha \otimes I$$

and

$$\Phi(\gamma) \cdot (I \otimes \alpha^*) - q\Phi(\alpha^*) \cdot (I \otimes \gamma) = \gamma \otimes I.$$

Thus α and γ both belong to $\Phi(SU_q(2)) \cdot (I \otimes SU_q(2))$. Taking adjoints in the preceding computations shows that α^* and γ^* belong to this set as well. Moreover, for any $\tilde{\alpha}, \tilde{\beta} \in SU_q(2)$ such that

$$\tilde{\alpha} \otimes I = \sum_i \Phi(\tilde{\alpha}_i)(I \otimes \tilde{\alpha}_i') \qquad \text{and} \qquad \tilde{\beta} \otimes I = \sum_j \Phi(\tilde{\beta}_j)(I \otimes \tilde{\beta}_j')$$

belong to $\Phi(SU_q(2)) \cdot (SU_q(2) \otimes I)$, we have

$$\begin{aligned}
\tilde{\alpha}\tilde{\beta} \otimes I &= (\tilde{\alpha} \otimes I)(\tilde{\beta} \otimes I) \\
&= \sum_i \Phi(\tilde{\alpha}_i)(I \otimes \tilde{\alpha}_i')(\tilde{\beta} \otimes I) \\
&= \sum_i \Phi(\tilde{\alpha}_i)(\tilde{\beta} \otimes I)(I \otimes \tilde{\alpha}_i') \\
&= \sum_{i,j} \Phi(\tilde{\alpha}_i)\Phi(\tilde{\beta}_j)(I \otimes \tilde{\beta}_j')(I \otimes \tilde{\alpha}_i') \\
&= \sum_{i,j} \Phi(\tilde{\alpha}_i\tilde{\beta}_j)(I \otimes \tilde{\beta}_j'\tilde{\alpha}_i'),
\end{aligned}$$

so that $\tilde{\alpha}\tilde{\beta} \otimes I \in \Phi(SU_q(2)) \cdot (I \otimes SU_q(2))$ as well. It follows that $SU_q(2) \otimes I$ is contained in the closure of $\Phi(SU_q(2)) \cdot (I \otimes SU_q(2))$, and thus the latter is dense in $SU_q(2) \otimes SU_q(2)$. This verifies one of the density conditions; the other is similar. We have shown the following.

THEOREM 11.3.5
$SU_q(2)$ is a compact quantum group.

11.4 Haar measure

We already mentioned Haar measure in Section 9.5. On a compact group it can always be normalized to be a probability measure. Thus integration against Haar measure can be described as a left invariant state on $C(G)$. This description still makes sense in the setting of quantum groups, where we call it a "Haar state." In this section we will prove that every compact quantum group has a unique Haar state.

The correct notion of invariance is given in the following definition. We include the definition of right invariance because we will prove that, as in the classical case, the Haar state on a compact quantum group is also right invariant.

DEFINITION 11.4.1 *Let $(\mathcal{A}, \Phi)$ be a compact quantum group and let $\omega \in \mathcal{A}^*$. We say ω is* <u>left invariant</u> *if $(\iota \otimes \omega) \circ \Phi(\alpha) = \omega(\alpha) I_{\mathcal{A}}$ for all $\alpha \in \mathcal{A}$, and it is* <u>right invariant</u> *if $(\omega \otimes \iota) \circ \Phi(\alpha) = \omega(\alpha) I_{\mathcal{A}}$ for all $\alpha \in \mathcal{A}$.*

A <u>Haar state</u> *is a left invariant state.*

To motivate this definition, consider the classical case where $\mathcal{A} = C(G)$. Here left invariance of a measure μ is equivalent to the condition that

$$\int_G f(xy)\, d\mu(y) = \int_G f(y)\, d\mu(y)$$

for all $f \in C(G)$ and $x \in G$. Now $\Phi f(x, y) = f(xy)$, so this condition can be rewritten as

$$\int_G \Phi f(x, y)\, d\mu(y) = \int_G f(y)\, d\mu(y).$$

Instead of regarding this as a separate condition for each x, we can regard both sides as functions of x. Then letting ω be integration against μ, the right side becomes $\omega(f) \cdot 1_G$ and the left side becomes $(\iota \otimes \omega)(\Phi f)$.

We proceed to prove the existence and uniqueness of a Haar state. Let $(\mathcal{A}, \Phi)$ be a compact quantum group and define a product on the dual space $\mathcal{A}^*$ by setting

$$\omega\rho(\alpha) = (\omega \otimes \rho)(\Phi(\alpha)).$$

LEMMA 11.4.2

Let $\omega \in \mathcal{A}^*$ be a state. Then there is a state h such that $h\omega = \omega h = h$.

PROOF Define

$$\omega_n = \frac{1}{n}(\omega + \omega^2 + \cdots + \omega^n)$$

and let h be a weak* cluster point of the sequence (ω_n). Then each ω_n is a state, so h is a state, and

$$\|\omega_n\omega - \omega_n\| = \|\omega\omega_n - \omega_n\| \leq \frac{2}{n}$$

implies $h\omega = \omega h = h$. ∎

LEMMA 11.4.3

Let $\omega, h \in \mathcal{A}^*$ be states satisfying $\omega h = h$. If $\rho \in \mathcal{A}^*$ and $0 \leq \rho \leq \omega$ then $\rho h = \rho(I_{\mathcal{A}})h$.

PROOF Let $\alpha \in \mathcal{A}$ and set $\alpha' = (\iota \otimes h)(\Phi(\alpha))$. Then

$$(\iota \otimes \omega)(\Phi(\alpha')) = (\iota \otimes \omega h)(\Phi(\alpha)) = (\iota \otimes h)(\Phi(\alpha)) = \alpha',$$

which by a short computation implies

$$(h \otimes \omega)((\Phi(\alpha') - \alpha' \otimes I)^*(\Phi(\alpha') - \alpha' \otimes I)) = 0.$$

Now $h \otimes \omega$ is a state on $\mathcal{A} \otimes \mathcal{A}$, so the Cauchy-Schwarz inequality in the associated GNS Hilbert space implies that

$$(h \otimes \omega)((\beta \otimes \gamma) \cdot (\Phi(\alpha') - \alpha' \otimes I)) = 0$$

for all $\beta, \gamma \in \mathcal{A}$. Replacing α' with $(\iota \otimes h)(\Phi(\alpha))$ yields

$$(h \otimes \omega \otimes h)((\beta \otimes \gamma \otimes I) \cdot (\Phi \otimes \iota)\Phi(\alpha)) = \omega(\gamma)(h \otimes h)((\beta \otimes I) \cdot \Phi(\alpha)).$$

But

$$\begin{aligned}(\beta \otimes \gamma \otimes I) \cdot (\Phi \otimes \iota)(\Phi(\alpha)) &= (\beta \otimes \gamma \otimes I) \cdot (\iota \otimes \Phi)(\Phi(\alpha)) \\ &= (I \otimes \gamma \otimes I) \cdot (\iota \otimes \Phi)((\beta \otimes I) \cdot \Phi(\alpha))\end{aligned}$$

by coassociativity. By the density condition which defines compact quantum groups, we can replace $(\beta \otimes I) \cdot \Phi(\alpha)$ with $I \otimes \delta$ for an arbitrary $\delta \in \mathcal{A}$ in the previous two results to get

$$(\omega \otimes h)((\gamma \otimes I) \cdot \Phi(\delta)) = \omega(\gamma)h(\delta).$$

Passing to the GNS representation π associated to ω and letting $\delta' = (\iota \otimes h)\Phi(\delta)$, the preceding becomes

$$\langle \pi(\delta')\bar{I}, \pi(\gamma)^*\bar{I} \rangle = h(\delta)\langle \bar{I}, \pi(\gamma)^*\bar{I} \rangle,$$

so we must have $\pi(\delta')\bar{I} = h(\delta)\bar{I}$. Since $0 \leq \rho \leq \omega$ the map $\pi(\alpha)\bar{I} \mapsto \rho(\alpha)$ is bounded, so there exists $v \in \mathcal{H}_\omega$ such that $\rho(\alpha) = \langle \pi(\alpha)\bar{I}, v \rangle$ for all $\alpha \in \mathcal{A}$. In particular,

$$\rho h(\delta) = \rho(\delta') = \langle \pi(\delta')\bar{I}, v \rangle = h(\delta)\langle \bar{I}, v \rangle = \rho(I)h(\delta),$$

and we conclude that $\rho h = \rho(I)h$. ∎

THEOREM 11.4.4

Let $(\mathcal{A}, \Phi)$ be a compact quantum group. Then there is a unique Haar state h on $\mathcal{A}$. Moreover, it is right invariant.

PROOF For any $\omega \in \mathcal{A}^*$, $\omega \geq 0$, let S_ω be the set of states h on $\mathcal{A}$ such that $\omega h = \omega(I_\mathcal{A})h$. This set is nonempty for any such ω by Lemma 11.4.2. It is straightforward to check that S_ω is weak* compact, and Lemma 11.4.3 implies that $S_\omega \subset S_\rho$ if $\rho \leq \omega$. This implies that the intersection of finitely many S_{ω_i} $(1 \leq i \leq n)$ contains $S_{\omega_1+\cdots+\omega_n}$, and hence is always nonempty. Thus there exists a state h which belongs to every S_ω. So

$$(\omega \otimes h)(\Phi(\alpha)) = \omega(I_\mathcal{A})h(\alpha)$$

for all $\alpha \in \mathcal{A}$ and all $\omega \in \mathcal{A}^*$, and this implies that $(\iota \otimes h) \circ \Phi = h \cdot I_\mathcal{A}$. So h is left invariant.

A similar argument shows that there exists a right invariant state h' such that $h'\omega = \omega(I_\mathcal{A})h'$ for all $\omega \in \mathcal{A}^*$. But then $h'h = h(I_\mathcal{A})h' = h'$ and (by the same property for h) $h'h = h'(I_\mathcal{A})h = h$, so $h' = h$. This shows that h is right invariant and also shows that h is unique. ∎

The Haar state on $SU_q(2)$ can be expressed in terms of its representation on $\mathcal{H} = l^2(\mathbf{N} \times \mathbf{Z})$ given in Section 11.3. We have

$$h(\tilde{A}) = (1 - q^2) \sum_{n=0}^{\infty} q^{2n} \langle \tilde{A} v_{n0}, v_{n0} \rangle$$

for all $\tilde{A} \in SU_q(2)$.

11.5 Notes

The material of Section 11.1 can be found in most standard references on C*-algebras. Section 11.2 is based on [39].

Our treatment of compact quantum groups in Sections 11.3 and 11.4 closely follows [46]. The quantum group $SU_q(2)$ was introduced in [78] and most of its properties were established there. The fact that the representation given on $l^2(\mathbf{N} \times \mathbf{Z})$ is faithful (i.e., that the operators α and γ defined there are universal) follows from an analysis of the irreducible representations of $SU_q(2)$; see [67]. The formula for the Haar state on $SU_q(2)$ is proven in [79].

References

[1] S. T. Ali, J.-P. Antoine, J.-P. Gazeau, and U. A. Mueller, Coherent states and their generalizations: a mathematical overview, *Rev. Math. Phys.* **7** (1995), 1013-1104.

[2] H. Araki, *Ryoshiba no Suri (Introduction to the Mathematical Theory of Local Observables)*, Iwavami Shoten Publishers, Tokyo, 1993; English translation *Mathematical Theory of Quantum Fields*, Oxford Univ. Press, New York, 1999.

[3] W. Arveson, *An Invitation to C*-Algebras*, Springer-Verlag GTM **39**, New York, 1976.

[4] A. Ashtekar and A. Magnon, Quantum fields in curved spacetimes, *Proc. R. Soc. Lond. A* **346** (1975), 375-394.

[5] A. J. Ayer, *Language, Truth and Logic*, Dover, 1946.

[6] J. C. Baez, I. E. Segal, and Z. Zhou, *Introduction to Algebraic and Constructive Quantum Field Theory*, Princeton Univ. Press, Princeton, 1992.

[7] G. Birkhoff and J. von Neumann, The logic of quantum mechanics, *Ann. of Math.* **37** (1936), 823-843.

[8] J. Blank, P. Exner, and M. Havliçek, *Hilbert Space Operators in Quantum Physics*, American Institute of Physics, New York, 1994.

[9] D. Blecher, The standard dual of an operator space, *Pacific J. Math.* **153** (1992), 15-30.

[10] O. Bratteli and D. W. Robinson, *Operator Algebras and Quantum Statistical Mechanics*, vol. 2 (second edition), Springer-Verlag, New York, 1997.

[11] M.-D. Choi and E. G. Effros, Injectivity and operator spaces, *J. Funct. Anal.* **24** (1977), 156-209.

[12] A. Connes, *Noncommutative Geometry*, Academic Press, New York, 1994.

[13] J. B. Conway, *A Course in Functional Analysis* (second edition), Springer-Verlag, 1990.

[14] C. Corduneanu, *Almost Periodic Functions* (second edition), Chelsea, New York, 1989.

[15] I. Daubechies, A. Grossman, and Y. Meyer, Painless nonorthogonal expansions, *J. Math. Phys.* **27** (1986), 1271-1283.

[16] K. R. Davidson, *C*-Algebras by Example*, American Mathematical Society, Providence, 1996.

[17] E. B. Davies, *Quantum Theory of Open Systems*, Academic Press, New York, 1976.

[18] P. A. M. Dirac, *The Principles of Quantum Mechanics* (fourth edition), Clarendon Press, Oxford, 1958.

[19] J. Dixmier, *Von Neumann Algebras* (English edition), North-Holland, New York, 1981.

[20] M. J. Dupré and R. M. Gillette, *Banach Bundles, Banach Modules and Automorphisms of C*-algebras*, Pitman, Boston, 1983.

[21] E. G. Effros, Structure in Simplexes, *Acta Math.* **117** (1967), 103–121.

[22] E. G. Effros and Z.-J. Ruan, *Operator Spaces*, Oxford Univ. Press, New York, 2000.

[23] E. G. Effros and C. Webster, Operator analogues of locally convex spaces, in *Operator Algebras and Applications* (A. Katavolos, ed.), Kluwer Acad. Publ., Dordrecht, 1997, 167-207.

[24] D. E. Evans and Y. Kawahigashi, *Quantum Symmetries on Operator Algebras*, Clarendon Press, Oxford, 1998.

[25] H. Everett III, The theory of the universal wavefunction, in *The Many-Worlds Interpretation of Quantum Mechanics*, Princeton Univ. Press, Princeton, 1973, 3-140.

[26] J. M. G. Fell and R. S. Doran, *Representations of *-Algebras, Locally Compact Groups, and Banach *-Algebraic Bundles*, vol. 1, Academic Press, New York, 1988.

[27] R. P. Feynman and A. R. Hibbs, *Quantum Mechanics and Path Integrals*, McGraw-Hill, New York, 1965.

[28] G. B. Folland, *Real Analysis: Modern Techniques and Their Applications*, John Wiley & Sons, New York, 1984.

[29] G. B. Folland, *Harmonic Analysis in Phase Space*, Princeton Univ. Press, Princeton, 1989.

[30] S. A. Fulling, *Aspects of Quantum Field Theory in Curved Space-Time*, Cambridge Univ. Press, Cambridge, 1989.

[31] L. K. Grover, Quantum mechanics helps in searching for a needle in a haystack, *Phys. Rev. Lett.* **79** (1997), 325-328.

[32] A. Guichardet, *Algèbres D'Observables Associ'ees aux Relations de Commutation*, Armand Colin, Paris, 1968.

[33] R. Haag, *Local Quantum Physics: Fields, Particles, Algebras*, Springer-Verlag, New York, 1992.

[34] P. R. Halmos, *A Hilbert Space Problem Book* (second edition), Springer-Verlag, New York, 1982.

[35] G. C. Hegerfeldt, Remark on causality and particle localization, *Phys. Rev. D* **10** (1974), 3320-3321.

[36] E. M. Henley and W. Thirring, *Elementary Quantum Field The-*

ory, McGraw-Hill, New York, 1962.

[37] J. M. Jauch, *Foundations of Quantum Mechanics*, Addison-Wesley, Reading, 1968.

[38] V. F. R. Jones, *Subfactors and Knots*, American Mathematical Society, Providence, 1991.

[39] G. I. Kac and V. G. Paljutkin, Finite ring groups, in *Transactions of the Moscow Mathematical Society for the Year 1966*, American Mathematical Society, 1967.

[40] R. V. Kadison and J. R. Ringrose, *Fundamentals of the Theory of Operator Algebras*, vol. I, Academic Press, New York, 1983.

[41] G. G. Kasparov, Equivariant KK-theory and the Novikov conjecture, *Invent. Math.* **91** (1988), 147-201.

[42] B. S. Kay and R. M. Wald, Theorems on the uniqueness and thermal properties of stationary, nonsingular, quasifree states on spacetimes with a bifurcate Killing horizon, *Phys. Rep.* **207** (1991), 49-136.

[43] E. C. Lance, *Hilbert C*-modules: A Toolkit for Operator Algebraists*, London Mathematical Society Lecture Note Series **210**, Cambridge University Press, Cambridge, 1995.

[44] C. Le Merdy, On the duality of operator spaces, *Canad. Math. Bull.* **38** (1995), 334-346.

[45] G. W. Mackey, *The Mathematical Foundations of Quantum Mechanics*, W. A. Benjamin, New York, 1963.

[46] A. Maes and A. van Daele, Notes on compact quantum groups, *Nieuw Arch. Wisk.* **16** (1998), 73-112.

[47] Yu. I. Manin, *Quantum groups and noncommutative geometry*, Univ. de Montréal, Montreal, 1988.

[48] R. Mills, Tutorial on infinities in QED, in *Renormalization: From Lorentz to Landau (and Beyond)* (L. M. Brown, ed.), Springer-Verlag, New York, 1993, 57-85.

[49] F. J. Murray and J. von Neumann, On rings of operators, *Ann. of Math.* **37** (1936), 116-229.

[50] F. J. Murray and J. von Neumann, On rings of operators II, *Trans. Amer. Math. Soc.* **41** (1937), 208-248.

[51] F. J. Murray and J. von Neumann, On rings of operators IV, *Ann. of Math.* **44** (1943), 716-808.

[52] J. von Neumann, *Mathematische Grundlagen der Quantenmechanik*, Springer, Berlin, 1932; English translation *Mathematical Foundations of Quantum Mechanics*, Princeton Univ. Press, Princeton, 1955.

[53] J. von Neumann, On rings of operators III, *Ann. of Math.* **41** (1940), 94-161.

[54] D. J. Newman and H. S. Shapiro, A Hilbert space of entire functions related to the operational calculus, manuscript.

[55] D. J. Newman and H. S. Shapiro, Fischer spaces of entire functions, in *Entire Functions and Related Parts of Analysis* (J. Korevaar, ed.), AMS Proc. Symp Pure Math. **XI**, Providence, 1968, 360-369.

[56] W. L. Paschke, Inner product modules over B*-algebras, *Trans. Amer. Math. Soc.* **182** (1973), 443-468.

[57] N. C. Phillips and N. Weaver, Modules with norms which take values in a C*-algebra, *Pacific J. Math.* **185** (1998), 163-181.

[58] M. Reed and B. Simon, *Functional Analysis* (revised edition), Academic Press, New York, 1980.

[59] M. A. Rieffel, Induced Representations of C*-algebras, *Advances in Math.* **13** (1974), 176-257.

[60] M. A. Rieffel, *Deformation Quantization for Actions of* $\mathbf{R}^d$, Mem. Amer. Math. Soc. **106**, 1993.

[61] R. Rochberg and N. Weaver, Noncommutative complex analysis and Bargmann-Segal multipliers, *Proc. Amer. Math. Soc.*, to appear.

[62] Z.-J. Ruan, Subspaces of C*-algebras, *J. Funct. Anal.* **76** (1988), 217-230.

[63] J. J. Sakurai, *Modern Quantum Mechanics*, Addison-Wesley, New York, 1985.

[64] J.-L. Sauvageot, Tangent bimodule and locality for dissipative operators on C*-algebras, *Quantum Probability and App. IV*, Springer LNM **1396** (1989), 322-338.

[65] P. W. Shor, Polynomial-time algorithms for prime factorization and discrete logarithms on a quantum computer, *SIAM J. Comput.* **26** (1997), 1484-1509.

[66] M. Takesaki, *Theory of Operator Algebras I*, Springer-Verlag, New York, 1979.

[67] L. L. Vaksman and Ya. S. Soĭbel'man, An algebra of functions on the quantum group $SU(2)$ (Russian), *Funct. Anal. Appl.* **22** (1988), 170-181.

[68] V. S. Varadarajan, *Geometry of Quantum Theory* (second edition), Springer-Verlag, New York, 1985.

[69] R. M. Wald, *Quantum Field Theory in Curved Spacetime and Black Hole Thermodynamics*, Univ. of Chicago Press, Chicago, 1994.

[70] N. Weaver, Operator spaces and noncommutative metrics, manuscript, 1995.

[71] N. Weaver, Lipschitz algebras and derivations of von Neumann algebras, *J. Funct. Anal.* **139** (1996), 261-300.

[72] N. Weaver, Deformation quantization for Hilbert space actions, *Comm. Math. Phys.* **188** (1997), 217-232.

[73] N. Weaver, *Lipschitz Algebras*, World Scientific Press, Singapore,

1999

[74] N. Weaver, Hilbert bimodules with involution, *Canad. Math. Bull.*, to appear.

[75] N. Weaver, Quantum predicate calculus, in preparation.

[76] A. S. Wightman, On the localizability of quantum mechanical systems, *Rev. Mod. Phys.* **34** (1962), 845-872.

[77] G. Wittstock, On matrix order and convexity, in *Functional Analysis: Surveys and Recent Results, III*, North-Holland, New York, 1984, 175-188.

[78] S. L. Woronowicz, Twisted $SU(2)$ group. An example of a non-commutative differential calculus, *Publ. Res. Inst. Math. Sci.* **23** (1987), 117-181.

[79] S. L. Woronowicz, Compact matrix pseudogroups, *Comm. Math. Phys.* **111** (1987), 613-665.

Index